# Lecture Notes in Mathematics

An informal series of special lectures, seminars and reports on mathematical topics

Edited by A. Dold, Heidelberg and B. Eckmann, Zürich

13


## Emery Thomas

University of California, Berkeley
Forschungsinstitut für Mathematik, ETH




# Seminar on Fiber Spaces

Lectures delivered in 1964 in Berkeley
and 1965 in Zürich
Berkeley notes by J. F. Mc Clendon

1966

Springer-Verlag · Berlin · Heidelberg · New York

Research supported in part by the U. S. Air Force Office of Scientific Research.


Library of Congress Catalog Card Number 66—19122. Printed in Germany. Title No. 7333.

# Foreword

These notes are divided into seven sections. The material in the first five was given as a series of informal lectures at Berkeley in the Summer of 1964. The material in the last two was discussed in Professor Eckmann's seminar in Zurich during the Spring of 1965.

Mr. J.F. Mc Clendon took my lecture notes from the Berkeley lectures and wrote them up in a presentable form. In particular the formulation of Theorem 5 in section five is due to him.

The first five lectures develop the theory of the Postnikov resolution of a map. In particular the main results of Mahowald's paper [3] are covered. Section six applies the theory to a specific example - the universal fibration with fiber the Stiefel manifold $V_{n,2}$. The last section deals with the problem of computing Postnikov invariants. A method for doing this is outlined and a theorem proved that implies that every (4k+3)-spin manifold has a tangent 2-field [12].

I would like to take this opportunity to thank Professor Eckmann for making possible my visit to the Mathematics Research Institute, E.T.H., Zurich.

14 September, 1965

E. Thomas

Table of Contents


| | | |
|---|---|---|
| | Foreword | III |
| I. | Introduction | 1 |
| II. | Principal Fiber Spaces | 7 |
| III. | The classical method of decomposing a fibration | 12 |
| IV. | Oriented sphere bundles | 19 |
| V. | Killing homotopy groups | 23 |
| VI. | An illustration | 29 |
| VII. | Computing Postnikov invariants | 35 |
| | References | 45 |

## I. Introduction

We consider three classical problems concerning the following diagram of spaces and maps:

(1)
$$\begin{array}{ccc} & & E \\ & & \downarrow p \\ X & \xrightarrow{f} & B \end{array}$$

i) (Existence problem) Does there exist a map $g: X \to E$ such that $pg = f$ ?

ii) (Enumeration problem) How many homotopy classes* of such g's are there ?

iii) (Classification problem) Given two such g's, can we distinguish between them by algebraic invariants ?

An example of (1) is:

$$\begin{array}{ccc} & & BO(n-k) \\ & & \downarrow i \\ M^n & \xrightarrow{f} & BO(n) \end{array}$$

where $M^n$ is a smooth n-dimensional manifold, $BO(n)$ is the classifying space for the orthogonal group, $i: BO(n-k) \to BO(n)$, $1<k<n$, is the natural inclusion, and $f$ is the map inducing the tangent bundle over $M$. In this example a lifting of $f$ corresponds to a field of tangent orthonormal k-frames and questions (i), (ii), and (iii), can be reformulated as familiar questions about such fields.

Other problems in differential topology can be stated in a similar fashion.

One method of getting a <u>negative</u> answer to the Existence

---

* There are two possible meanings for "homotopy" in question (ii): "free homotopy" or "homotopy relative to f."

question (i) is to derive an algebraic diagram from (1) and show that it cannot be commutative. Affirmative answers to question (i) have been obtained via obstruction theory. That is, if $E \to B$ is a fiber bundle and $q: Y \to X$ is the bundle induced by $f$, then a lifting of $f$ corresponds to a cross-section of $q$. Obstruction theory for bundles and the theory of characteristic classes are then applicable. For example, if $X(M)$ is the Euler class of the manifold $M$, then $X(M) = 0$ if and only if $M$ has an everywhere continuous non-zero vector field. (i.e. the Euler class is the only obstruction to lifting $f$ to $BO(n-1)$.)

These lectures will discuss the elaboration of this positive method due initially to Postnikov, with important contributions by Moore, Hermann, and Mahowald. Mahowald's paper [3] will be discussed in some detail.

<u>General description of the method</u>

We hope to factor the map $p$ of (1) into a diagram

(2)

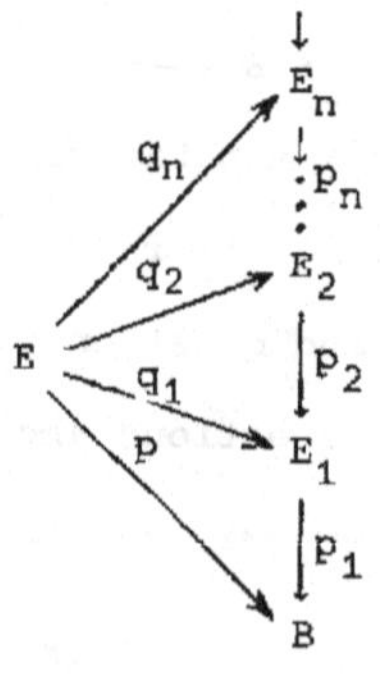


where $(p_1 p_2 \cdots p_n) q_n = p$ $(n \geq 1)$ and:

A) At each stage, the obstruction to lifting a map from $E_q$ to $E_{q+1}$ is given in terms of "computable" algebraic invariants (e.g., a finite set of cohomology classes.)

B) There exists a sequence $\{r_n\}$ of integers such that $1 < r_1 < r_2 < \ldots < r_n \ldots$ and such that the morphism $q_{r_n *}: \pi_i(E) \to \pi_i(E_{r_n})$ is bijective for $0 < i < n$ and surjective for $i = n$.

Let $X$ be a complex and consider the morphism[1] $q_{r_n *}: [X,E] \to [X,E_{r_n}]$. It follows from B) that if dim $X < n$, then $q_{r_n *}$ is bijective, while if dim $X = n$, then $q_{r_n *}$ is surjective. Thus if $X$ has dim $< n$, then a map $f: X \to B$ lifts to $E$ iff it lifts to $E_{r_n}$.

In order to achieve the factorization given in (2) we will use the following key construction:

(3)
$$\begin{array}{ccccc} & & \Omega C & = & \Omega C \\ & & \vdots & & \downarrow \\ & & E_w & \dashrightarrow & PC \\ & & \vdots & & \downarrow \\ X & \xrightarrow{f} & B & \xrightarrow{w} & C \end{array}$$

Here $PC$ is the space of paths of $C$ beginning at $*$ and $\Omega C$ is the space of loops of $C$ at $*$. Thus $PC \to C$, given by $\alpha \to \alpha(1)$, is a Hurewicz fibration with fiber $\Omega C$ [8] and $PC$ is contractible. $E_w \to B$ is the fibration induced by $w$, so $E_w = \{(b,\alpha) \in B \times PC: w(b) = \alpha(1)\}$. It is easily seen that $f$ lifts to $E_w$ iff $wf \simeq *$.

An important special case is when $C = K(\pi,n)$, an Eilenberg-MacLane space of type $(\pi,n)$. Since $H^n(B;\pi) = [B,C]$ and $w \in H^n(B,\pi)$,

---

1) We work in the category of spaces with base point (which is always denoted by $*$). $[X,Y]$ denotes the set of homotopy classes of base point preserving maps from $X$ to $Y$.

it follows that $wf \simeq *$ iff $f^*w = 0$ (provided that $B$ is a complex).

We now apply (3) to (1) taking $C_i = K(\pi_i, n_i)$, $i \geq 1$, and obtain the following diagram

(4)



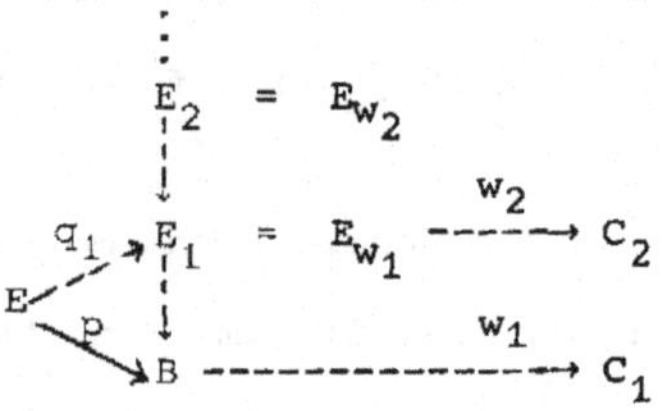

The $w_i$'s could be chosen as follows. First, consider ker $p^*$, where $p^*: H^{n_1}(B;\pi_1) \to H^{n_1}(E;\pi_1)$, choose a $w_1$ there and use it to construct the fibration $E_1 \to B$. Since $p^*w_1 = 0$ we can lift $p$ to $q_1$. Then we look in ker $q_1^*$ for a $w_2$, etc.

We have constructed a diagram (2) satisfying condition A. Our plan is to re-examine, under certain restrictions, construction (4) with condition B in mind. Assume that $p: E \to B$ is a fibration with an (n-1)-connected fiber $F$. It follows that $p_*: \pi_r(E) \to \pi_r(B)$ is bijective for $0 \leq r \leq n-1$ and surjective for $r = n$ and that $p^*: H^r(B,\pi) \to H^r(E,\pi)$ is bijective for $1 \leq r \leq n-1$ and injective for $r = n$. Now assume that $C = K(\pi, n+1)$ and choose $w \in H^{n+1}(B,\pi) \cap$ ker $p^*$. The situation after rearranging (4) slightly is:

(5)

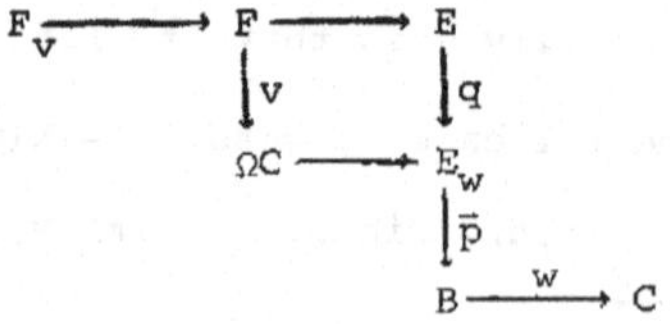

where $p = \bar{p}q$ and $p^*w = 0$. Note that $q(F) \subset \Omega C$; set $v = q|F$. Both the maps $v$ and $q$ are homotopically equivalent to fiber maps [7], and moreover the fibers themselves are homotopically equivalent [2]. Let $F_v$ denote the fiber of $v$. Then one obtains the commutative diagram given in (5), such that the triples

$$F_v \longrightarrow F \xrightarrow{v} \Omega C \ ,$$

$$F_v \longrightarrow E \xrightarrow{q} E_w \ ,$$

are homotopically equivalent to fibrations.

If $F_v$ were n-connected it would follow from the exact sequence for $q: E \to E_w$ that $q_*: \pi_i(E) \to \pi_i(E_w)$ is bijective for $0<i<n$, and surjective for $i = n+1$. Thus we would have gained a dimension in the passage to the second stage of the construction. This is a step in the direction of (B) - so we seek conditions under which $F_v$ is n-connected.

From the exact sequence for $v: F \to \Omega C$ we see that $\pi_i(F_v) \to \pi_i(F)$ is bijective for $i \neq n, n-1$ and that

$$\pi_n(F_v) = 0 \iff v_* \text{ is injective in dimension } n,$$

$$\pi_{n-1}(F_v) = 0 \iff v_* \text{ is surjective in dimension } n.$$

The question as to whether or not the construction (4) will satisfy B) now can be rephrased as:

1) Take $\pi = \pi_n(F)$. Can we choose $w \in H^{n+1}(B,\pi)$ such that the resulting $v$ gives $v_*: \pi_n(F) \approx \pi$ ?

2) If so, can we repeat the process - i.e., can we (a) compute $H^*(E_w)$ and $\ker\ q^*$; (b) determine which classes in $\ker q^*$

"kill" the higher homotopy groups of $F$ (in the sense that in going from $F$ to $F_v$ we kill the n'th homotopy group of F)?

In order to gain some insight into the problem let us consider part of (5) separately

$$(6)\qquad \begin{array}{ccc} F_v & \longrightarrow & F \\ & & \downarrow v \\ & & \Omega C = K(\pi,n) \end{array} \qquad F \text{ is } (n-1)\text{-connected}, \quad \pi_n(F) = \pi.$$

Since $[F,\Omega C] = H^n(F;\pi) = \mathrm{Hom}(\pi,\pi)$, taking $v = 1 \in \mathrm{Hom}(\pi,\pi)$ gives $v_*$ an isomorphism. The question is whether or not (6) can be fitted into (5) - that is, can we choose $w$ to produce such a $v$.

The answers to these questions are, more or less, yes, as we shall see later.

## II. Principal Fiber Spaces

If $w: (B,B_0) \to (C,C_0)$ is a (base point preserving) map then we have the diagram

$$\begin{array}{ccc}
(\Omega C, \Omega C_0) & = & (\Omega C, \Omega C_0) \\
\downarrow & & \downarrow \\
(E,E_0) & \xrightarrow{\bar{w}} & (PC,PC_0) \\
\downarrow & & \downarrow \\
(B,B_0) & \xrightarrow{w} & (C,C_0)
\end{array}$$

where $E = \{(b,\alpha) \in B \times PC: w(b) = \alpha(1)\}$

$E_0 = \{(b,\alpha) \in B_0 \times PC_0: w(b) = \alpha(1)\}$

<u>Definition</u>: $(\Omega C,\Omega C_0) \xrightarrow{i} (E,E_0) \xrightarrow{p} (B,B_0)$ is the <u>principal fibration induced by $w$</u>, with principal fibre space $(E,E_0)$.

<u>Lemma</u> 1. Let $g:(X,X_0) \to (B,B_0)$ be a map. Then $g$ lifts to $(E,E_0)$ iff $wg \simeq *$.

<u>Lemma</u> 2. Let $(X,X_0)$ be a pair of spaces. Then the following sequence of sets is exact:

$$(7)\qquad \begin{array}{l}[(X,X_0);(\Omega B,\Omega B_0)] \xrightarrow{(\Omega w)_*} [(X,X_0);(\Omega C,\Omega C_0)] \xrightarrow{i_*} [(X,X_0);(E,E_0)] \\
\xrightarrow{p_*} [(X,X_0);(B,B_0)] \xrightarrow{w_*} [(X,X_0);(C,C_0)].\end{array}$$

<u>Proof</u>: Exactness at $[(X,X_0);(B,B_0)]$ follows from the first lemma. The rest is not hard to verify (see [5]).

If $\alpha$ and $\beta$ are paths such that $\alpha(1) = \beta(0)$, denote their product by $\alpha \vee \beta$. There is a natural product

$$(\Omega C,\Omega C_0) \times (\Omega C,\Omega C_0) \longrightarrow (\Omega C,\Omega C_0)$$
$$(\alpha,\beta) \longrightarrow \alpha \vee \beta$$

and a natural action

$$(\Omega C,\Omega C_0) \times (E,E_0) \longrightarrow (E,E_0)$$
$$(\alpha,(b,\beta)) \longrightarrow (b,\alpha \vee \beta) .$$

They induce a product (denoted by $\vee$) and an action (denoted by $\cdot$)

$$[(X,X_0);(\Omega C,\Omega C_0)] \times [(X,X_0);(\Omega C,\Omega C_0)] \longrightarrow [(X,X_0);(\Omega C,\Omega C_0)]$$
$$[(X,X_0);(\Omega C,\Omega C_0)] \times [(X,X_0);(E,E_0)] \longrightarrow [(X,X_0);(E,E_0)].$$

For any pairs (with base point) $(X,X_0)$, $(Y,Y_0)$ let $0 \in [(X,X_0),(Y,Y_0)]$ denote the class of the constant map.

Lemma 3: (a) $0 \cdot q = q$ for $q \in [(X,X_0);(E,E_0)]$

(b) $u \cdot 0 = i_* u$ for $u \in [(X,X_0);(\Omega C,\Omega C_0)]$

(c) Let $q,q' \in [(X,X_0);(E,E_0)]$. Then $p_* q = p_* q'$ iff there is a $u \in [(X,X_0);(\Omega C,\Omega C_0)]$ such that $q' = u \cdot q$.

(d) The sequence (7) and the operations defined above are natural in the obvious ways.

Proof: For the proof in the absolute case see [6].

## Transgression in Fiber Spaces

Suppose that $F \xrightarrow{i} E \xrightarrow{p} B$ is a fibration and $B$ is path connected. Take cohomology with coefficients in a fixed group G. Denote reduced homology by $\bar{H}(\ )$. $\bar{H}^*(B)$ will sometimes be identified with $H^*(B,*)$ and sometimes with a subgroup of $H^*(B)$. Denote by $\bar{p}: (E,F) \to (B,*)$ the map defined by $p$.

Define $T^*(F) \subset H^*(F)$, $S^*(B) \subset \bar{H}^*(B)$ by

$$T^*(F) = \delta^{-1}\bar{p}^*\bar{H}^*(B)$$

$$S^*(B) = \bar{p}^{*-1}\delta H^*(F)$$

where $\delta: H^*(F) \to H^*(E,F)$. Note that $S^*(B) = \text{kernel } p^*$, $T^*(F) \supset \text{image } i^*$, $S^*(B) \supset \ker \bar{p}^*$.

Define $\tau: T^*(F) \to S^*(B)/\ker \bar{p}^*$

$\sigma: S^*(B) \to T^*(F)/\text{ im } i^*$

by $\tau(u) = [u']$, where $\delta u = \bar{p}^*u'$

$\sigma(v) = [v']$, where $\bar{p}^*v = \delta v'$ .

Clearly, kernel $\tau$ = image $i^*$, kernel $\sigma$ = kernel $\bar{p}^*$ so $\tau$, $\sigma$ induce inverse isomorphisms

$$T^*(F)/\text{im } i^* = S^*(B)/\ker \bar{p}^*.$$

$\tau$ is called the transgression in the fiber space, $\sigma$, the suspension. (We can define $\sigma: S^*(B) \to H^*(F)$, and then its values are cosets of $i^*H^*(E)$; similarly for $\tau$.)

<u>The Lifting Problem</u>

Suppose $F \xrightarrow{i} E \xrightarrow{p} B$ is a fibration, $B$ path connected, and $E$, $B$ have the homotopy type of complexes. We form the diagram

$$\begin{array}{ccccc} F & \xrightarrow{i} & E & & \\ \downarrow{\scriptstyle v_q} & & \downarrow{\scriptstyle q} & & \\ \Omega C & \longrightarrow & E_w & & \\ & & \downarrow{\scriptstyle p_1} & & \\ & & B & \xrightarrow{w} & C \end{array} \qquad (p_1 \circ q = p)$$

Suppose that $wp \simeq *$. By lemma 1, p lifts to $q: (E,F) \to (E_W, \Omega C)$. Let $v_q = q|F: F \to \Omega C$.

<u>Definition</u>: $v_q$ is (geometrically) <u>realized</u> by the pair $(w,q)$. A reasonable question is: which maps $v_q$ can be realized?

To answer this question we consider the sequence

$$(\Omega C, \Omega C) \xrightarrow{j} (E_W, \Omega C) \xrightarrow{p_1} (B,*) \xrightarrow{w} (C,C),$$

where $p_1$ is the principal fibration induced by $w$ (see page 7). Mapping the pair $(E,F)$ into the sequence we obtain the commutative diagram shown below:

$$\begin{array}{ccccccc}
[E,\Omega C] & \xrightarrow{j_*} & [(E,F),(E_W,\Omega C)] & \xrightarrow{p_{1*}} & [(E,F),(B,*)] & \xrightarrow{w_*} & [E,C] \\
\downarrow{\scriptstyle i^*} & & \downarrow{\scriptstyle i^*} & & & & \\
[F,\Omega C] & = & [F,\Omega C] \, . & & & &
\end{array}$$

Define $\Sigma w = i^* p_{1*}^{-1}[\bar{p}] \subset [F,\Omega C]$. Using Lemma 3(d), one easily shows that $\Sigma w$ is a coset of $i^*[E,\Omega C]$. Furthermore, $\Sigma w$ = all homotopy classes that can be geometrically realized by $(w,q)$ for some lifting $q$ of $p$.

If $C = K(G,n)$, $n>0$, this can be translated into cohomology. We have $w \in H^n(B;G)$ and $\Sigma w$ is a coset of $i^*H^{n-1}(E,G)$ in $H^{n-1}(F;G)$. Now $p^*w = 0$, so for the fibration $F \xrightarrow{i} E \xrightarrow{p} B$, $\sigma(w)$ is defined ( = the suspension of $w$, see page 9) and is a coset of $i^*H^{n-1}(E;G)$ in $H^{n-1}(F;G)$

<u>Theorem</u> 1. $-\sigma w = \Sigma w$.

<u>Proof</u>: It suffices to prove that $-\sigma w$ and $\Sigma w$ have a common representative. Suppose that $v \in \Sigma w$, so that $v: F \to \Omega C$ is the restriction of some $q: (E,F) \to (E_w, \Omega C)$ and $p_1 q = \bar{p}$.

If $\sigma^1$ denotes suspension in $\Omega C \to E_w \to B$ then it is easily checked that $\sigma^1(-w)$ is represented by $\iota_{n-1}$ = the characteristic class of $H^{n-1}(\Omega C;G)$. By the naturality of the suspension for the commutative diagram

$$\begin{array}{ccccc} \Omega C & \longrightarrow & E_w & \longrightarrow & B \\ \uparrow v & & \uparrow q & & \| \\ F & \longrightarrow & E & \longrightarrow & B \end{array}$$

we have

$$v = v^* \iota_{n-1} \in v^* \sigma^1(-w) \subset \sigma(-w),$$

which completes the proof.

We now can carry out the program of the first lecture.

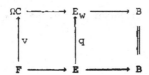

## III. The "classical" (Moore-Postnikov) method of decomposing a fibration

Suppose that $F \xrightarrow{i} E \xrightarrow{p} B$ is a fibration, that $\pi_1(B)$ acts trivially on $H_*(F;G)$[2], and that $F$ has non-zero homotopy groups $\pi_i = \pi_{n_i}(F)$ in dimension $n_1, n_2, \ldots$ with $0 < n_1 < n_2 < \ldots$ (if $n_1 = 1$ assume $\pi_1$ is abelian). Let $v_1 \in H^{n_1}(F,\pi_1)$ be the fundamental class of $F$. That is, $v_1$ corresponds to the Hurewicz map under $H^{n_1}(F,\pi_1) = \mathrm{Hom}\,(H_{n_1}(F),\pi_1)$. It follows from the spectral sequence of the fibration that $v_1$ is transgressive. Let $w_1 = -\tau(v_1)$. We have, as before, the diagram

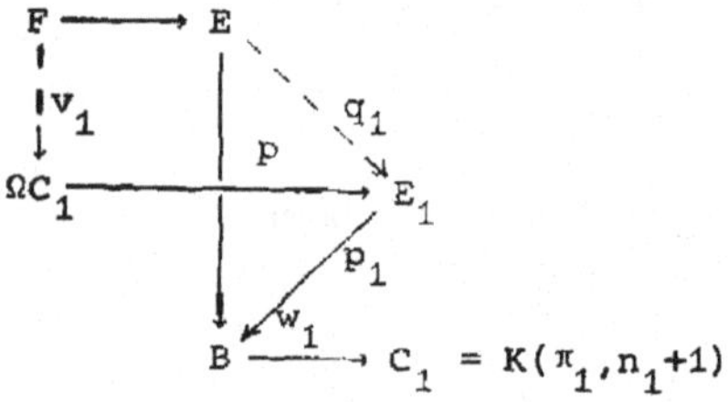


By Theorem 1 we can choose $q_1 : (E,F) \to (E_1, \Omega C_1)$ such that $v_{q_1} = v_1$.

Let $F_1$ = the fiber of $v_1$. Then in the homotopy sequence for the fibration $F_1 \to F \to \Omega C_1$, $v_{1*} : \pi_n(F) \to \pi_n(\Omega C_1) = \pi_1$ is an isomorphism, implying $\pi_r(F_1) = 0$, $r < n$, $\pi_r(F_1) \approx \pi_r(F)$, $r > n$.
Now let $v_2 \in H^{n_2}(F_1;\pi_2)$ be the fundamental class of $F_1$. Again it is transgressive (in the fibration $F_1 \to E \to E_1$).

Let $w_2 = -\tau(v_2) \in H^{n_2+1}(E_1, \pi_2)$ induce $E_2$ over $E_1$, etc.

---

2) Henceforth, "$F \to E \to B$ is a fibration" will include this condition.

Some questions about this construction are:

1) How can we compute the transgression?

2) Can we relate the classes $w_2, w_3, \ldots$ to the cohomology of F, E, B ? In particular, how do the Steenrod operations behave on the w's ?

3) If $f\colon X \to B$, $f^*w_1 = 0$, and $f_1, f_2\colon X \to E_1$, are two liftings of $f$, what is $f_1^*w_2 - f_2^*w_2$ ?

The answers to these questions will depend on 1) knowing the cohomology of the $E_i$'s and 2) knowing the action of $\Omega C_i$ on $E_i$ vis-a-vis cohomology.

Relative Transgression

Suppose $F \xrightarrow{i} E \xrightarrow{p} B$ a fiber space, $* \in B_0 \subset B$, $E_0 = p^{-1}(B_0) \subset E$, $p_0\colon (E, E_0) \to (B, B_0)$. In short:

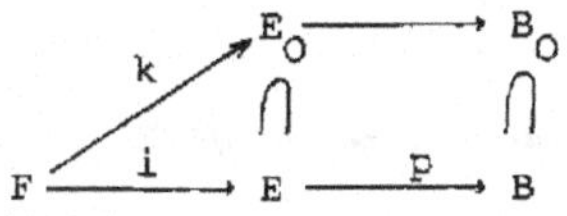


Define $U = U(B, B_0, E_0)$ = all pairs $(u, v) \in H^*(B, B_0) \oplus H^*(E_0)$ such that $\delta v = p_0^* u$.

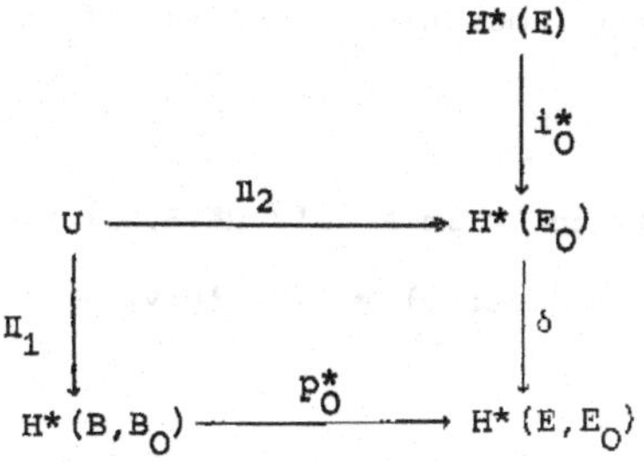


$\Pi_1, \Pi_2$ are the projections

Define $S^*(B,B_0) = \Pi_1 U$, $T^*(E_0) = \Pi_2 U$,

$$\sigma_0\colon S^*(B,B_0) \longrightarrow T^*(E_0)/\mathrm{Im}\, i_0^* \quad \text{induced by} \quad \Pi_2\Pi_1^{-1}$$

$$\tau_0\colon T^*(E_0) \longrightarrow S^*(B,B_0)/\ker p_0^* \quad \text{induced by} \quad \Pi_1\Pi_2^{-1}$$

$\sigma_0$ is the <u>relative suspension</u>, $\tau_0$ the <u>relative transgression</u> (We can think of $\tau_0$ as mapping $T^*(E_0)$ to $H^*(B,B_0)$; its values are then cosets of $\ker p_0^*$; similarly for $\sigma_0$.)

<u>Properties of the relative transgression $\tau_0$</u>

<u>Property</u> 1. Let $j\colon B \to (B,B_0)$, $k\colon F \to E_0$. Then $\tau k^* = j^*\tau_0$ where $\tau$ is the absolute transgression of the last lecture. (Proof obvious.)

Consider:

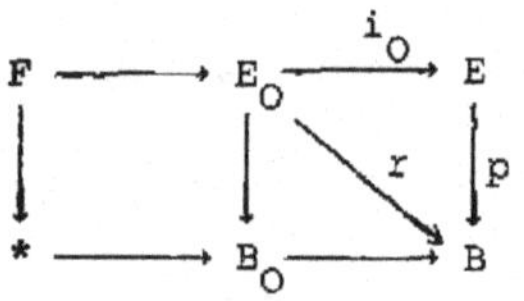


where $r = pi_0$. Using $r^*$ we can make $H^*(E_0)$ into an algebra over $H^*(B)$. That is, given $u \in H^*(E_0)$, $v \in H^*(B)$, define $u \cdot v = u \cup r^*v$. Also, $H^*(B,B_0)$ is an algebra over $H^*(B)$ and $\ker p_0^*$ is stable under $H^*(B)$.

<u>Property</u> 2. $\tau_0$ is an $H^*(B)$-morphism. That is, if $u \in T^*(E_0)$, $v \in H^*(B)$, then $u \cdot v \in T^*(E_0)$ and $\tau_0(u \cdot v) = \tau_0(u) \cdot v$.

<u>Proof</u>: $u \in T^*(E_0)$ implies there is a $u' \in H^*(B,B_0)$ such that $\delta u = p_0^* u'$. Then $\delta(u \cdot v) = \delta(u \cup r^*v) = \delta(u \cup i_0^* p^*v) = \delta u \cup p^*v = p_0^*u' \cup p^*v = p_0^*(u' \cup v)$.

Next, suppose that $\varphi$ is a primary cohomology operation, $\Psi$ its suspension, then:

Property 3: $\varphi\tau_0 = \tau_0\Psi$

Proof: Use the fact: $\varphi\delta = \delta\Psi$ .

For the next property we will need the following theorem of Serre [9]:

Theorem Let $p: E \to B$ be a fiber space with arcwise connected fiber $F$ and base $B$. Suppose $\pi_1(B)$ acts trivially on $H_*(F;Z)$. Let $* \in B_0 \subset B$ and set $E_0 = p^{-1}(B_0)$. Assume that $H_i(B,B_0;Z) = 0$ for $0<i<a$, $H_j(F;Z) = 0$, $0<j<b$. Then the homomorphism $p^*: H^k(B,B_0;G) \to H^k(E,E_0;G)$ is injective for $k < a+b$ and surjective for $k < a+b-1$, where $G$ is any abelian group.

Proof: Apply the universal coefficient theorem to the homology version [9, p.268].

Property 4. Under the hypotheses of the above theorem, the following sequence is exact:

$$\cdots \longrightarrow H^i(E_0) \xrightarrow{\tau_0} H^{i+1}(B,B_0) \xrightarrow{1^*} H^{i+1}(E) \xrightarrow{i_0^*} H^{i+1}(E_0)$$

$$\cdots \longrightarrow H^{a+b-1}(E_0), \quad \text{where } 1 = jp.$$

Proof: Start with the sequence for $(E,E_0)$ and insert $(B,B_0)$ by the theorem. Notice that $1^*$ can be obtained from the commutative diagram

$$\begin{array}{ccc} H^{i+1}(B,B_0) & \longrightarrow & H^{i+1}(B) \\ \Big\downarrow p_0^* \approx & \searrow l^* & \Big\downarrow p^* \\ H^{i+1}(E,E_0) & \longrightarrow & H^{i+1}(E) \end{array} .$$

Define $\tau_1 \colon T^*(E_0) \to H^*(B)/j^* \operatorname{Ker} p_0^*$ to be the composite homomorphism shown below:

$$T^*(E_0) \xrightarrow{\tau_0} H^*(B,B_0)/\operatorname{Ker} p_0^* \xrightarrow{j^*} H^*(B)/j^* \operatorname{Ker} p_0^* .$$

Notice that Property 1 can now be written $\tau k^* = \tau_1$ , and that $\tau_1$ continues to enjoy Properties 2 and 3.

<u>Property</u> 5. Let $k_0 \colon B_0 \subset B$ denote the inclusion. Let $t$ be an integer such that $0 < t < a+b-1$ , and suppose that

$$\begin{gathered} \text{Kernel } p^* \supset \text{Kernel } k_0^* \text{ in dim } t, \\ k_0^* \text{ is surjective in dim } t. \end{gathered}$$

Then the following sequence is exact:

$$0 \longrightarrow H^t(E) \xrightarrow{i_0^*} H^t(E_0) \xrightarrow{l^*} H^{t+1}(B) .$$

<u>Proof</u>: By exactness, Image $j^*$ = Kernel $k_0^*$ . Thus,

$$\text{Image } l^* = p^*(\text{Image } j^*) = p^*(\text{Kernel } k_0^*) = 0 ,$$

in dim t. Hence by the exactness of the sequence given in Property 4,

$$\text{Kernel } i_0^* = \text{Image } l^* = 0 \text{ in dim } t ,$$

as claimed.

If $k_0^*$ is surjective in dim t, then $j^*$ is injective in dim t+1 (using the cohomology sequence of the pair $(B,B_0)$), and so

$$\text{Kernel } \tau_0 = \text{Kernel } \tau_1 \text{ in dim } t,$$

which completes the proof.

Property 6: Given any fiber space commutative diagram:

$$\begin{array}{ccc} F & = & F \\ \downarrow & & \downarrow \\ E_0 & \xrightarrow{\bar{f}} & E \\ \downarrow & & \downarrow \\ B_0 & \xrightarrow{f} & B \end{array}$$

such that $H_j(F;Z) = 0$, $0<j<b$, and $f_*\colon H_r(B_0;Z) \to H_r(B;Z)$ is isomorphic for $0 < r < a-1$, and epic for $r = a-1$, then the sequence in (4) is still defined and exact with $i_0^* = \bar{f}^*$ and $(B,B_0)$ thought of as $(M_f,B_0)$ where $M_f$ is the mapping cylinder of $f$.

The proof follows from usual argument using the mapping cylinder.

Application

Consider the usual diagram with $w \in \ker p^*$.

$$\begin{array}{ccccc} F & \longrightarrow & E & & \\ \downarrow & & \downarrow{\scriptstyle q_1} & & \\ \Omega C & \longrightarrow & E_1 = E_w & & \\ & & \downarrow{\scriptstyle p_1} & & \\ & & B & \xrightarrow{w} & C = K(\pi,n+1) \end{array} \qquad p = p_1q_1$$

Lemma 4: There is a commutative diagram of fiber spaces

$$\begin{array}{ccccc} \Omega C & & = & & \Omega C \\ \downarrow & & & & \downarrow \\ \Omega C \times E & \xrightarrow{1\times q_1} & \Omega C \times E_1 & \xrightarrow{\mu} & E_1 \\ \downarrow{\scriptstyle \Pi} & & & & \downarrow{\scriptstyle p_1} \\ E & & \xrightarrow{p} & & B \end{array}$$

where $\Pi$ is the projection and $\mu$ is the action of $\Omega C$ in the principal fiber space $E_1$.

Proof: Recall that $q_1$ is defined by $q_1(e) = (p(e), \alpha_e)$ where $\alpha_e$ is a path from $*$ to $wp(e)$. If $\lambda \in \Omega C$ we have

$$p_1\mu(1 \times q_1)(\lambda, e) = p_1\mu(\lambda, (p(e), \alpha_e)) = p_1(p(e), \lambda \vee \alpha_e) = p(e)$$
$$= p\Pi(\lambda, e).$$

Corollary 1: Let $F \to E \xrightarrow{p} B$ be a fibration with $(n-1)$-connected fiber and let $C$ and $E_1$ be as above. Then there is an exact sequence

$$\cdots H^i(\Omega C \times E) \xrightarrow{\tau_0} H^{i+1}(B, E) \longrightarrow H^{i+1}(E_1) \xrightarrow{\nu^*} H^{i+1}(\Omega C \times E)$$

$\cdots \longrightarrow H^{2n}(\Omega C \times E)$ where $\nu = \mu \cdot (1 \times q_1)$. ($(B, E)$ should be thought of as $(M_p, E)$).

For future use define $s: E \to \Omega C \times E$ by $s(e) = (*, e)$ and note that

$$\nu \cdot s = \mu \cdot (1 \times q_1) \cdot s \simeq q_1 : E \to E_1$$

since $\mu(*, e_1) = e_1, e_1 \in E_1$.

IV. We will now illustrate the method by considering the classifying space for oriented (n-1)-sphere bundles $S^{n-1} \xrightarrow{i} BSO(n-1) \xrightarrow{p} BSO(n)$. Abbreviate $BSO(q) = B_q$. Note that $p$ is homotopically equivalent to the natural inclusion $B_{n-1} \subset B_n$; upon occasion it will be thought of as that inclusion.

We will use the scheme described in the last lecture in order to factor $p$. Let $0 < n_1 < n_2 < \cdots$ be the dimensions in which $S^{n-1}$ has non-zero homotopy groups and let $\Pi_i = \pi_{n_i}(S^{n-1})$. So $n_1 = n-1$, $n_2 = n$, $n_3 = n+1$, $\Pi_1 = Z$, $\Pi_2 = \Pi_3 = Z_2$. Let $s_{n-1}$ be a generator for the group $H^{n-1}(S^{n-1};Z)$. We know that $s_{n-1}$ is transgressive and we are interested in $\tau(s_{n-1})$. The Serre exact sequence [8;p.468] in this case is:

$$\cdots H^{n-1}(B_{n-1}) \longrightarrow H^{n-1}(S^{n-1}) \xrightarrow{\tau} H^n(B_n) \xrightarrow{p^*} H^n(B_{n-1}) \longrightarrow H^n(S^{n-1}),$$

so that in dimension $n$, $\operatorname{im} \tau = \ker p^*$. Now it follows from the Gysin sequence for the oriented sphere bundle $S^{n-1} \subset B_{n-1} \xrightarrow{p} B_n$ that $\ker p^* \cap H^n(B_n)$ is cyclic infinite generated by the Euler class $X_n$. Hence we can choose $s_{n-1}$ so that $\tau(s_{n-1}) = -X_n$. As in the general situation (see p.12) we can find $q_{n-1}$ such that the following diagram is commutative

$$\begin{array}{ccccc}
F_n \xrightarrow{i_n} & S_{n-1} & \longrightarrow & B_{n-1} & \\
 & \downarrow{\scriptstyle s_{n-1}} & & \downarrow{\scriptstyle q_{n-1}} & \\
 & K(Z,n-1) & \longrightarrow & E_{n-1} & \\
 & & & \downarrow{\scriptstyle p_{n-1}} & \\
 & & & B_n & \xrightarrow{X_n} K(Z,n)
\end{array}$$

where $F_n$ is the fiber of $q_{n-1}$, $F_n$ is (n-1)-connected, and the morphism $q_{n-1*}: \pi_r(B_{n-1}) \to \pi_r(E_{n-1})$ is bijective for $0<r<n-1$, and surjective for $r = n$. Moreover, $i_{n_*}: \pi_r(F_n) \approx \pi_r(S^{n-1})$, $r \geq n$.

Let $v_2$ = characteristic class of $F_n \in H^n(F_n, \Pi_2) = H^n(F_n, Z_2)$. We know that $v_2$ is transgressive and we are interested in $\tau(v_2) \in H^{n+1}(E_{n-1}, Z_2) \cap \ker q_{n-1}^*$.

We digress for a while to discuss $H^*(E_{n-1}, Z_2)$. Now for $q \geq 2$, $H^*(B_q; Z_2)$ is a polynomial algebra on the Stiefel-Whitney classes $w_2, \ldots, w_q$. Consequently,

i) The morphism $p^*: H^*(B_n; Z_2) \to H^*(B_{n-1}; Z_2)$ is surjective.

ii) The kernel of $p^*$ is the ideal in $H^*(B_n; Z_2)$ generated by $w_n$.

Since $w_n = X_n \bmod 2$, it follows that Kernel $p_{n-1}^*$ = Kernel $p^*$ in all dimensions. Therefore, by Property 5, we have the following exact sequence (with mod 2 coefficients) for $0<r<2n-2$.

$$0 \longrightarrow H^r(E_{n-1}) \xrightarrow{\nu^*} H^r\big((Z,n-1) \times B_{n-1}\big) \xrightarrow{\tau_1} H^{r+1}(B_n).$$

Let $\iota$ denote the mod 2 reduction of the fundamental class of $(Z,n-1)$; thus $\iota \in H^{n-1}(Z,n-1;Z_2)$ and $\tau(\iota) = w_n$. Therefore by Property 2 we have

iii) $\tau_1(\iota \otimes b) = w_n \cdot b$,

where $b \in H^i(B_n)$, $i<n-2$. Furthermore, by exactness of the above sequence we have

iv) $H^q(E_{n-1}) \approx \text{Kernel } \tau_1$, for $q<2n-2$.

Let $s: B_{n-1} \to (Z,n-1)\times B_{n-1}$ denote the inclusion. Since $q_{n-1} \simeq \nu\circ s$ (see page 18), our attention is shifted from

$$H^*(E_{n-1},Z_2)\cap \ker q_{n-1}^* \text{ to } \ker \tau_1 \cap \ker s^*.$$

Three facts of note are

<u>Fact</u> 1. Let $1 \in H^0(Z,(n-1))$. Then $\tau(1) = 0$ and thus in $H^*((Z,n-1)\times B_{n-1})$ any term of the form $1 \otimes b$, $b \in H^*(B_{n-1})$, is in $\ker \tau_1$.

<u>Fact</u> 2. By a formula of Wu [14], $Sq^i w_n = w_n \cdot w_i$, $0 < i < n$. Thus by Property 3, and iii) above,

$$\tau_1(Sq^i \iota \otimes 1) = Sq^i \tau_1(\iota \otimes 1) = Sq^i w_n = w_n \cdot w_i = \tau_1(\iota \otimes w_i),$$

and so

$$\tau_1(Sq^i \otimes 1 + \iota \otimes w_i) = 0 \qquad \text{for } i < n-2$$

<u>Fact</u> 3. Let $u \in H^*(E_{n-1})$, set $\nu^*(u) = 1 \otimes b + \sum_j a_j \otimes c_j$ where $a_j \in \bar{H}^*(Z,n-1)$, $b, c_j \in H^*(B_{n-1})$. Then $q^*_{n-1}u = s^*\nu^*u = s^*(1 \otimes b + \Sigma a_j \otimes c_j) = 1 \otimes b$ (since $s^*\bar{H}^*((Z,n-1)) = 0$ ). So $u \in \ker q^*_{n-1} \Longleftrightarrow \nu^*u \in \ker s^* \Longleftrightarrow \nu^*u = \Sigma a_j \otimes c_j$, $\deg a_j > 0$.

Using the above facts one can easily calculate $H^q(E_{n-1})$ for $q < 2n-2$. For example, $\ker \tau_1 \cap \ker s^*$ is 0 in dimension $< n$ while in dimension $n+1$ it is $Sq^2\iota \otimes 1 + 1 \otimes w_2$.

Let $k^{n+1} \in H^{n+1}(E_{n-1})$ be the unique element such that $\nu^*(k^{n+1}) = Sq^2\iota \otimes 1 + 1 \otimes w_2$.

We now can proceed with the factorization of p. Since $\tau(v_2) = k_{n+1}$ we have

$$\begin{array}{ccccc}
F_{n+1} & \longrightarrow & F_n & \longrightarrow & B_{n-1} & & \\
 & & \downarrow{\scriptstyle v_2} & & \downarrow{\scriptstyle q_2} & & \\
 & & K(Z_2,n) & \longrightarrow & E_n & & \\
 & & & & \downarrow{\scriptstyle p_2} & & \\
 & & & & E_{n-1} & \xrightarrow{k_{n+1}} & K(Z_2,n+1)
\end{array}$$

where $F_{n+1}$ is the fiber of $q_2$, $F_{n+1}$ is n-connected, the morphism $q_2^*: \pi_r(B_{n-1}) \to \pi_r(E_n)$ is bijective for $0<r<n$ and surjective for $r = n+1$, and $i_{{n+1}_*}: \pi_r(F_{n+1}) \approx \pi_r(S^{n-1})$ $r > n+1$. We have gained one more stage.

An unresolved question is: by altering the method somewhat, can we kill several homotopy groups of $F_n$ at once instead of only one?

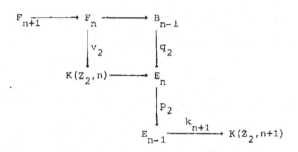

## V. Killing Homotopy Groups

Suppose that $v \in H^n(Y,J)$ where $J = Z$ or $Z_p$, $p$ prime, and $Y$ is a complex.

We have

$$\begin{array}{ccc} \Omega K & \longrightarrow & E_v \\ & & \downarrow \\ & & Y \xrightarrow{\ v\ } K(J,n) = K \end{array}$$

and we can ask: when does $v$ kill a factor $J$ of $\pi_n(Y)$ - i.e. when does $\pi_n(E_v)$ have one less factor $J$ than $\pi_n(Y)$ ?

We know [7] that any map is homotopically equivalent to a fibration and it is easy to check that if $v$ is changed to a fibration the fiber will be $E_v$. Hence we may consider $E_v \to Y \to K$ to be a fibration. The homotopy sequence of that fibration shows that if $v_*$ is surjective in dim n then $\pi_i(E_v) \simeq \pi_i(Y)$, $i \neq n$, and the sequence $0 \to \pi_n(E_v) \to \pi_n(Y) \to J \to 0$ is exact. In particular, if $\pi_n(Y)$ is finite and $J = Z_p$ we have: order $\pi_n(E_v)$ = order $\pi_n(Y)/p$.

Theorem 2. $v_*$ is surjective in dim n if and only if there exists a map $f: S^n \to Y$ such that $f^*(v)$ = a generator of $H^n(S^n;J)$.

Proof: We have $S^n \xrightarrow{f} Y \xrightarrow{v} K$ and $[Y,K] \xrightarrow{f^*} [S^n,K]$, $\pi_n(Y) = [S^n,Y] \xrightarrow{v_*} [S^n,K] = \pi_n(K)$, so that $f^*v = vf = v_*f$. Also

$$\begin{array}{ccccc} H^n(Y;J) & = & [Y,K] & & \\ \Big\downarrow{\scriptstyle f^*} & & & & \\ H^n(S^n;J) & = & [S^n,K] & = & \pi_n(K) \end{array}$$

so $f^*(v) = $ generator $H^n(S^n,J) = \Pi_n(K)$, iff $v_*(f) = $ generator $\Pi_n(K)$, iff $v_*$ is surjective.

Definition A class $v \in H^n(Y;J)$ is spherical if there exists $f: S^n \to Y$ such that $f^*(v) = $ generator of $H^n(S^n;J)$.[1)]

We are interested in finding spherical classes. In particular we are interested in spherical classes of $F$ where $F \to E \xrightarrow{p} B$ is a fibration and in finding spherical classes of the spaces $F_1, F_2, \cdots$ etc. which arise in the factorization of $p$.

## Irreducible Cohomology Operations

Suppose that $\varphi$ is a cohomology operation of type $(J,Z_p,n,n+q+1)$ with $q \geq 0$, i.e., for each space $X$, $\varphi: H^n(X;J) \to H^{n+q+1}(X;Z_p)$.

Definition [see 3.13 of 3] $\varphi$ is irreducible relative to $(X,w,\alpha)$ if $X$ is a space, $w \in H^n(X,J)$, $\alpha: S^{n+q} \to X$ and

(1) $\alpha^*(w) = 0, \quad \varphi(w) = 0, \quad \alpha^* H^{n+q}(X;Z_p) = 0$

(2) $\varphi_\alpha(w) = $ generator $H^{n+q}(S^{n+q};Z_p)$ (i.e. $\varphi_\alpha(w) \neq 0$) where $\varphi_\alpha$ is the functional cohomology operation (see [10]).

Theorem 3. [see 3.14 of 3]. Suppose $\varphi$ is irreducible relative to $(X,w,\alpha)$. Consider the fiber space $E_w \to X \xrightarrow{w} K(J,n) = K$ and think of $\varphi$ as an element of $H^{n+q+1}(K;Z_p)$. Then

(1) $w^*(\varphi) = 0$, i.e. $\varphi \in S^*(K)$ (see p.9, $w = p$)

(2) If $y \in H^{n+q}(E_w,Z_p) \cap \sigma(\varphi)$, then $y$ is spherical.

---

1) Notice that if $v$ is spherical, then so is $-v$ .

Proof:

$$\begin{array}{ccccc} & & E_w & & \\ & \overset{\beta}{\nearrow} & \big\downarrow i & & \\ S^{n+q} & \xrightarrow{\alpha} & X & \xrightarrow{w} & K(J,n) \end{array}$$

Since $\alpha^*(w) = 0$, $\alpha$ lifts to $\beta$. Since $wi \simeq *$ and $\varphi w \simeq *$, the functional operation $\varphi_i(w)$ is defined as a coset in $H^{n+q}(E_w, Z_p)$. Moreover, $\beta^* \varphi_i(w) = \varphi_\alpha(w)$ (naturality of the functional operation) = generator of $H^{n+q}(S^{n+q}; Z_p)$. Hence each element of $\varphi_i(w)$ is spherical and (since $\sigma(\varphi) \subset \varphi_i(w)$) the proof is complete.

This property of irreducibility is natural in the following sense. Suppose that $\varphi$ is irreducible relative to $(X,w,\alpha)$ and let $f: X \to Y$, $v \in H^n(Y,J)$, $f^*v = w$, and $\varphi(v) = 0$. Then $\varphi$ is irreducible relative to $(Y,v,f\alpha)$.

This naturality is particularly useful when, for a given $\varphi$, a simple $X$ (i.e., having few cells) can be found. For example, if $\varphi$ is irreducible relative to $(S^n, s, \alpha)$, where $s \in H^n(S^n, J)$ is the generator, and if $v \in H^n(Y,J)$ is spherical, then the map $f$ is available and we need only check the condition: $\varphi(v) = 0$, to insure that $\varphi$ is irreducible relative to $(Y,v,f\alpha)$. If $\varphi$ is irreducible relative to $(S^n \cup_\beta e^q, s, \alpha)$ and $v$ is spherical via $\gamma: S^n \to Y$ then if (1) $\gamma\beta \simeq *$, we can find an $f$, and if (2) $\varphi(v) = 0$ we then have that $\varphi$ is irreducible relative to $(Y,v,f\alpha)$.

Suppose a cohomology operation $\varphi$ is irreducible relative to $(X,w,\alpha)$ and $X$ is a CW-complex formed by attaching $i$-cells to a sphere. Call $\varphi$ irreducible of type $(i+1)$.

Theorem 4. The following operations are irreducible of type 1:

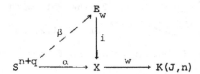

$Sq^{2^i}$, $i = 0,1,2,3$, $\beta_p, P_p^1$ $p$ = prime $> 2$. The following are irreducible of type 2: $Sq^{2^i}Sq^{2^{i+1}}, Sq^{2^{i+1}}Sq^{2^i}$, $i = 0,1,2$, $Sq^{16}$. The attaching maps for the type 2 operations are Hopf maps - except for $Sq^2Sq^1$ where it is a map of degree 2.

<u>Proof</u>: The type 1 results are well known. For example, if $\varphi = Sq^2$ use $X = S^n$, $w$ = generator of $H^n(S^n;J)$, $\alpha: S^{n+1} \to S^n$ the suspension of the Hopf map. For $Sq^2Sq^1$ see [3, p.323]. The others can be done similarly or by using a lemma of Toda [13; pp.84,190]. For $Sq^{16}$ see [13; p.86].

<u>Definition</u> [cf 2.1.1 of 3] Let $w_i \in H^{n_i}(X;J_i)$, $J_i$ cyclic, $i = 1, \cdot\cdot, l$. $\{w_i\}$ is a <u>spherical set</u> if there exist $\alpha_i: S^{n_i} \to X$, $i = 1, \cdot\cdot, l$, such that $\alpha_i^* w_i$ = generator of $H^{n_i}(S^{n_i};J_i)$ and $\alpha_j^* w_i = 0$ if $i \neq j$.

Let $w_i \in H^{n_i}(X;J_i)$, $J_i$ cyclic, $i = 1, \cdot\cdot, l$, and let $C = \mathop{X}\limits_{i=1}^{l} K(J_i,n_i)$, $w = (w_1, \cdot\cdot, w_l): X \to C$. It is easy to check that $w_*: \pi_m(X) \to \pi_m(C)$ is surjective for each $m$ iff $\{w_i\}$ is a spherical set.

Let each $J_i = Z$ or $Z_p$, and suppose that $\Psi \in H^m(C;Z_p)$, where $m < 2 \min \{n_i\}$. Then we may write

(*) $\Psi = \Sigma\, \varphi_i(\iota_i)$ where $\varphi_i$ is a primary operation of type $(n_i,J_i,m,Z_p)$. Suppose also that for some $t$, $0 < t < l$, $\varphi_t$ (in the representation (*) of $\Psi$) is either:

(a) irreducible of type 1

or

(b) irreducible of type 2 relative to $M = S^{n_t} \cup_\gamma e^n$ such that $\alpha_t \gamma \simeq *$. (Here $n$ is an integer with $n_t < n < m$).

<u>Theorem</u> 5. [see 3.1.4 and 3.1.7 of 3] Let $\{w_i\}$ be a spherical set and let $\Psi$ be given as above. If $w^*(\Psi) = 0$, then $\sigma(\Psi) \subset H^{m-1}(E_w;Z_p)$ consists of spherical classes, where $\sigma$ is calculated in the fibration $E_w \to X \overset{w}{\to} C$.

<u>Proof</u>: (a) The proof of (a) is similar to, and easier than that of (b)

(b) Since $\alpha_t \gamma \simeq *$ we can extend $\alpha_t$ to $\alpha: M \to X$ so that $\alpha^* w_t = s$ where $s \in H^{n_t}(M;Z_p)$ corresponds to the generator of $H^{n_t}(S^{n_t};Z_p)$. It can be checked that $\alpha$ can be chosen so that $\alpha^* w_i = 0$, $i \neq t$. (Using a Puppe sequence, alter $\alpha$ by means of $\alpha_1,\ldots,\alpha_{t-1},\alpha_{t+1},\ldots,\alpha_l$, if necessary). We have a commutative diagram

$$\begin{array}{ccccc} E_s & \longrightarrow & M & \xrightarrow{s} & K(J_t,n_t) \\ \downarrow{\scriptstyle r} & & \downarrow{\scriptstyle \alpha} & & \downarrow{\scriptstyle p} \\ E_w & \longrightarrow & X & \xrightarrow{w} & C \end{array}$$

where $p$ is the natural inclusion. Denote by $\sigma'$ suspension in $E_s \to M \to K(J_t,n_t)$. Then $r^*\sigma(\Psi) \subset \sigma'(p^*\Psi) = \sigma'(\varphi_t) \Rightarrow r^*\sigma(\Psi)$ consists of spherical classes $\Rightarrow \sigma(\Psi)$ consists of spherical classes.

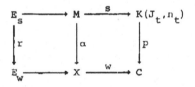

The above theorem gives a method of finding spherical classes. We are interested in spherical sets. Using the above notation, call $\Psi \in H^n(C;Z_p)$ <u>allowable</u> if $\sigma(\Psi)$ consists of spherical classes. Suppose $\{\Psi_1,\cdots,\Psi_a\} \subset H^n(C;Z_p)$ and $v_i \in \sigma(\Psi_i)$. Then we have the following criterion [see 3.4.2 of 3]: $\{v_i\}$ is a spherical set if every non-trivial linear combination of $\Psi_1,\cdots,\Psi_a$ is allowable. This is just a statement in terms of the $\Psi$'s of a fact about a collection of mod p cohomology classes of any space - namely, the collection forms a spherical set iff every non-trivial linear combination of its elements is a spherical class. Or, to put it somewhat differently, let $\Sigma_p \subset H_*(X;Z_p)$ denote the mod p reduction of the image of the Hurewicz homomorphism, and let $T \subset H^*(X;Z_p)$ denote any subspace such that

$$T \cap \text{Annihilator } \Sigma_p = 0.$$

Then any finite set of linearly independent elements in $T$ is a spherical set.

## VI. An illustration

We give an example that will illustrate the use of Theorem 5, page 27. Again let $B_q = BSO(q)$, $q \geq 1$, and consider the fibration

$$V_{n+2,2} \xrightarrow{i} B_n \xrightarrow{p} B_{n+2}, \qquad n \geq 2,$$

where $V_{n+2,2}$ denotes the Stiefel manifold of orthonormal 2 frames in $R^{n+2}$. Given a complex A, a map $\xi: A \to B_{n+2}$ can be regarded as an orientable (n+2)-plane bundle over A. Moreover, $\xi$ lifts to $B_n$ iff $\xi$ has two linearly independent (global) cross sections. In particular if A is an (n+2)-dimensional orientable manifold M and if $\xi$ = tangent bundle of M, then $\xi$ lifts to $B_n$ iff M has 2 linearly independent tangent vector fields.

We take the case $n = 4s+1$, $s \geq 1$, and construct the first three stages in a Postnikov resolution of the map p.

Set $V = V_{4s+3,2}$. By [4], V is 4s-connected and

$$\pi_{4s+1}(V) \approx \pi_{4s+2}(V) \approx Z_2, \quad \pi_{4s+3}(V) \approx Z_4 .$$

Take cohomology with mod 2 coefficients and consider the homomorphism

$$p^*: H^*(B_{4s+3}) \longrightarrow H^*(B_{4s+1}).$$

Now Kernel $p^*$ is the ideal in $H^*(B_{4s+3})$ generated by $w_{4s+2}$ and $w_{4s+3}$. Thus if $u \in H^{4s+1}(V)$ denotes the fundamental class, it follows by the Serre exact sequence [8] that $\tau u = w_{4s+2}$.

Following the method given in lecture III we construct the diagram shown below.

$$\begin{array}{ccccc}
V_1 \longrightarrow V & \xrightarrow{\ \ i\ \ } & B_{4s+1} & & \\
\big\downarrow u & & \big\downarrow q_1 & & \\
K(Z_2,4s+1) & \xrightarrow{\ \ j\ \ } & E_1 & & \\
 & & \big\downarrow p_1 & & \\
 & & B_{4s+3} & \xrightarrow[w_{4s+2}]{\qquad} & K(Z_2,4s+2)
\end{array}$$

Here $p_1q_1 = p$, $q_1$ is chosen so that $v_{q_1} = u$ (see pages 10,12), and $V_1$ is the fiber of the map $q_1$ ($\equiv$ fiber of the map $u$; see page 5). Because $u$ is spherical, the space $V_1$ is (4s+1)-connected. (See Theorem 2, page 23). Since $p_1^* w_{4s+2} = 0$ and $p_1^* w_{4s+3} = p_1^* Sq^1 w_{4s+2} = 0$, it follows that Kernel $p^*$ = Kernel $p_1^*$ in all dimensions. Also, $p^*$ is surjective in all dimensions. Thus by Property 5 and Corollary 1 we have an exact sequence

$$(*) \qquad 0 \to H^r(E_1) \xrightarrow{\nu^*} H^r(F\times B_{4s+1}) \xrightarrow{\tau_1} H^{r+1}(B_{4s+3}),$$

for all $0<r<8s-1$, where $F = K(Z_2,4s+1)$. Moreover, by Property 1,

$$\tau_1(\iota\otimes 1) = w_{4s+2},$$

where $\iota$ denotes the fundamental class of $F$. Recall (see page 18) that if $s\colon B_{4s+1} \to F\times B_{4s+1}$ denotes the canonical injection, then $\nu\circ s \simeq q_1$. Thus to compute Kernel $q_1^* \subset H^r(E_1)$ it suffices to compute Kernel $s^* \cap$ Kernel $\tau_1$ in $H^r(F\times B_{4s+1})$. Using Properties

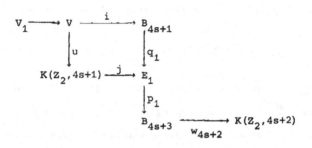

2 and 3 (as they apply to $\tau_1$) we obtain the following chart:

| dimension | Kernel $s^* \cap$ Kernel $\tau_1$ spanned by: |
|---|---|
| $\leq 4s+2$ | $0$ |
| $4s+3$ | $(Sq^2\iota\otimes 1 + \iota\otimes w_2) = A$ |
| $4s+4$ | $Sq^1A, \ (Sq^2Sq^1\iota\otimes 1 + Sq^1\iota\otimes w_2) = B$ . |

Moreover, one easily calculates that

$$Sq^1 B = Sq^2 A + A\cdot w_2 \ .$$

By sequence (*) there are classes $k_i \in H^{4s+i}(E_1)$, $i = 3,4$, such that

$$\nu^* k_3 = A, \quad \nu^* k_4 = B \ ,$$

$$Sq^1 k_4 + Sq^2 k_3 + k_3\cdot w_2 = 0.$$

Since $q_1^* = s^*\nu^*$, it follows that Kernel $q_1^*$ in dimension $< 4s+4$ looks as follows.

| dimension | Kernel $q_1^*$ spanned by: |
|---|---|
| $\leq 4s+2$ | $0$ |
| $4s+3$ | $k_3$ , |
| $4s+4$ | $Sq^1k_3, \ k_4$ . |

Let $\sigma$ denote th suspension in the fibration given by $q_1$. Since $k_i \in$ Kernel $q_1^*$ each $k_i$ is in domain $\sigma$. Choose any non-zero classes $\alpha_i \in \sigma(k_{i+1}) \subset H^{4s+i}(V_1)$, $i = 2,3$. By naturality, $\alpha_i \in \sigma_1(j^*k_{i+1})$, where $\sigma_1$ denotes the suspension in the fibration

$$(**) \qquad V_1 \to V \xrightarrow{u} K(Z_2, 4s+1).$$

By definition of $k_i$ we have

$$j^*k_3 = Sq^2\iota, \quad j^*k_4 = Sq^2Sq^1\iota$$

Now apply Theorem 5 to (**), taking $C = K(Z_2,4s+1)$, $X = V_1$, $w = u$, $E_u = V$. By the theorem it follows that $\{\alpha_2,\alpha_3\}$ is a spherical set. (Since $Sq^2Sq^1$ is an irreducible operation of type 2, we need to remark that $2\pi_{4s+1}(V) = 0$.) Therefore by Theorem 1 we can construct the following diagram:

$$\begin{array}{ccccc}
V_2 & \longrightarrow & V_1 & \longrightarrow & B_{4s+1} \\
 & & \downarrow{\scriptstyle(\alpha_2,\alpha_3)} & & \downarrow{\scriptstyle q_2} \\
 & & K(Z_2,4s+2)\times K(Z_2,4s+3) & \longrightarrow & E_2 \\
 & & & & \downarrow{\scriptstyle p_2} \\
 & & & & E_1 \xrightarrow{k_3\times k_4} K(Z_2,4s+3)\times K(Z_2,4s+4).
\end{array}$$

Here $q_1 = p_2q_2$, $q_2$ is chosen so that $v_{q_2} = (\alpha_2,\alpha_3)$ and $V_2$ denotes the fiber of $q_2$. Since $\alpha_2$ and $\alpha_3$ are spherical, and since $V_2$ can also be regarded as the fiber of the map $(\alpha_2, \alpha_3)$, it follows by Theorem 2 (page 23) that $V_2$ is $(4s+2)$-connected and that $\pi_{4s+3}(V_2) \cong Z_2$. We seek an invariant in $H^{4s+4}(E_2)$ to kill this group. Notice that Kernel $p_2^* =$ Kernel $q_1^*$ through dimension $4s+4$, and that $q_1^*$ is surjective in all dimensions since $p^*$ is. Thus by Property 5 and Corollary 1 we have the following exact sequence for $0<r<4s+4$.

$$(***) \qquad 0 \to H^r(E_2) \xrightarrow{\nu} H^r(F_1 \times B_{4s+1}) \xrightarrow{\tau_1} H^{r+1}(E_1).$$

where $F_1 = K(Z_2, 4s+2) \times K(Z_2, 4s+3)$. Let $\gamma_i$, $i = 2,3$, denote the fundamental class of the factor $K(Z_2, 4s+i)$ in the fiber $F_1$. Then by construction of the fibration $p_2$ it follows that

$$\tau\gamma_i = k_{i+1} \qquad\qquad (i = 2,3)$$

and so by Property 1,

$$\tau_1(\gamma_i \otimes 1) = k_{i+1} \quad , \quad i = 2,3.$$

Let $s_1: B_{4s+1} \to F_1 \times B_{4s+1}$ denote the injection. Using sequence (***) and the calculations given above for $k_3$, $k_4$ one finds that

Kernel $s_1^* \cap$ Kernel $\tau_1 = 0$ in dim. $<4s+3$, while in dim. $4s+4$, Kernel $s_1^* \cap$ Kernel $\tau_1$ is spanned by

$$(Sq^2\gamma_2 \otimes 1 + \gamma_2 \otimes w_2 + Sq^1\gamma_3 \otimes 1) = c$$

Let $\ell \in H^{4s+4}(E_2)$ be the class such that $\nu^*(\ell) = C$. Then in the fibration

$$V_2 \to B_{4s+1} \xrightarrow{q_2} E_2 ,$$

$q_2^* \ell = 0$ and $\sigma(\ell)$ consists of spherical classes, as is seen by applying Theorem 5 to the fibration

$$V_2 \to V_1 \xrightarrow{\alpha_2 \times \alpha_3} K(Z_2, 4s+2) \times K(Z_2, 4s+3) .$$

Thus $\ell$ can be used as the invariant for constructing the next fibration $p_3 \colon E_3 \to E_2$, and since $\sigma(\ell)$ consists of spherical classes there will be a map $q_3 \colon B_{4s+1} \to E_3$ such that $p_3 q_3 = q_2$ and such that the fiber of $q_3$ will be $(4s+3)$-connected.

## VII. Computing Postnikov invariants

Theorem 5 provides a satisfactory criterion for deciding at each stage which classes in $H^*(E_i)$ can be used as invariants for constructing the next fibration $p_i: E_{i+1} \to E_i$. However, we are still left with the problem of "computing" these invariants. For example, take the case $i = 1$ and let $k \in H^*(E_1)$ be such an invariant. Let $A$ be a complex and let $\xi: A \to B$ be a map that lifts to $E_1$. Define

$$k(\xi) = \bigcup_{\eta} \eta^* k ,$$

where the union is taken over all maps $\eta: A \to E_1$ such that $p_1\eta = \xi$, where $p_1: E_1 \to B$. Thus if deg $k = t$, then $k(\xi) \subset H^t(A)$. We consider the problem: compute the set $k(\xi)$. Only if a satisfactory method for solving this problem can be found is the theory of Postnikov invariants of more than limited use. In this section we consider a method that works in some situations.

We modify slightly the example given in lecture VI. Let Spin $(n)$, $n>2$, denote the universal covering group for $SO(n)$. Since $Z_2$ is the fiber of the covering homomorphism Spin $(n) \to SO(n)$, there is a principal fibration

$$K(Z_2,1) \to B\,\mathrm{Spin}\,(n) \xrightarrow{\Pi_n} BSO(n) , \qquad n>3 ,$$

induced by $w_2 \in H^2(BSO(n))$, where we regard $w_2$ as a map $BSO(n) \to K(Z_2,2)$. Thus given a complex $A$ and an orientable bundle $\xi: A \to BSO(n)$, $\xi$ lifts to $B\,\mathrm{Spin}(n)$ iff $w_2(\xi) = 0$.

Since the inclusion $SO(n) \subset SO(n+1)$ induces an isomorphism on the fundamental group, it follows that

$$\frac{SO(n+k)}{SO(n)} = \frac{Spin(n+k)}{Spin(n)} = V_{n+k,k}$$

for $n \geq 3$ , $k \geq 1$. In particular taking $k = 2$, we obtain the following commutative diagram of fiber spaces:

$$\begin{array}{ccc} V_{n+2,2} & = & V_{n+2,2} \\ \downarrow & & \downarrow \\ B\,Spin(n) & \xrightarrow{\Pi_n} & BSO(n) \\ \tilde{p}\downarrow & & \downarrow p \\ B\,Spin(n+2) & \xrightarrow{\Pi_{n+2}} & BSO(n+2) \end{array} .$$

(Here $\tilde{p}$ and $p$ are induced by inclusions). Thus if we have a $Spin(n+2)$-bundle on $A$ (i.e., a map $\xi: A \to B\,Spin(n+2)$), then $\xi$ has 2 linearly independent cross-sections iff there exists a map $\eta: A \to B\,Spin(n)$ such that $\tilde{p}\eta = \xi$.

We take the case $n = 4s+1$, $s > 0$, dim $A \leq 4s+3$. Set $\tilde{B}_q$ = $B\,Spin(q)$. Since the fiber map $\tilde{p}$ can be regarded as the fibration induced from $p$ by $\Pi_{n+2}$, so can the Postnikov resolution for $\tilde{p}$ be induced by $\Pi_{n+2}$ from the resolution for $p$, constructed in lecture VI. In particular we obtain the following commutative diagram:

$$\begin{array}{ccccc}
\tilde{B}_{4s+1} & \xrightarrow{\Pi_{4s+1}} & B_{4s+1} & & \\
\Big\downarrow{\scriptstyle \tilde{q}_1} & & \Big\downarrow{\scriptstyle q_1} & & \\
\tilde{E}_1 & \xrightarrow{\zeta} & E_1 & & \\
\Big\downarrow{\scriptstyle \tilde{p}_1} & & \Big\downarrow{\scriptstyle p_1} & & \\
\tilde{B}_{4s+3} & \xrightarrow[\Pi_{4s+3}]{} & B_{4s+3} & \xrightarrow[w_{4s+2}]{} & K(Z_2, 4s+2)
\end{array}$$

Here $p_1, q_1$ have the same meaning as in VI (so $p = p_1 q_1$), and $\tilde{p}_1$ is induced from $p_1$ by $\Pi_{4s+3}$. Since points in $\tilde{E}_1$ are pairs $(b,e)$, with $b \in \tilde{B}_{4s+3}$, $e \in E_1$, such that $p_1(e) = \Pi_{4s+3}(b)$, we define

$$\zeta \colon \tilde{E}_1 \to E_1 \, , \quad \tilde{q}_1 \colon \tilde{B}_{4s+1} \to \tilde{E}_1$$

by

$$\zeta(b,e) = e \, , \quad \tilde{q}_1(x) = (\tilde{p}x, \, q_1 \Pi_{4s+1} x) \, ,$$

for $x \in \tilde{B}_{4s+1}$. Then the diagram is commutative, $\tilde{p} = \tilde{p}_1 \tilde{q}_1$, and the fundamental class of $V_{4s+3,2}$ is geometrically realized by $\tilde{q}_1$. (See pages 10, 11).

Set $\tilde{k} = \zeta^* k_3 \in H^{4s+3}(\tilde{E}_1)$. Then $\tilde{k}$ is the only obstruction to lifting a map $\eta \colon A \to \tilde{E}_1$ into $\tilde{B}_{4s+1}$, provided that $\dim A < 4s+3$.

Recall that $k_3$ is characterized by the fact that

$$\nu^*(k_3) = Sq^2 \iota \otimes 1 + \iota \otimes w_2 \in H^{4s+3}(K(Z_2, 4s+1) \times B_{4s+1}) \, .$$

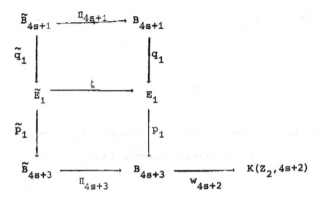

(See sequence (*) in VI.) Since $\Pi^*_{4s+1} w_2 = 0$, it follows by naturality that

$$\tilde{\nu}^*(\tilde{k}) = Sq^2\iota\otimes 1 \in H^{4s+3}(K(Z_2,4s+1)\times\tilde{B}_{4s+1}),$$

where $\tilde{\nu}^*$ denotes the operator $\nu^*$ in the fibration $\tilde{p}_1$. Consequently, the invariant $\tilde{k}$ is characterized uniquely by the properties

(1) $$\tilde{q}_1^*\,\tilde{k} = 0\ ,\quad \tilde{i}^*\,\tilde{k} = Sq^2\iota\ ,$$

where $\tilde{i}\colon K(Z_2,4s+1) \subset \tilde{E}_1$ denotes the inclusion of the fiber.

Now $\tilde{p}_1$ is a principal fibration with fiber $K(Z_2,4s+1)$. Let

$$m\colon K(Z_2,4s+1)\times\tilde{E}_1 \rightarrow \tilde{E}_1$$

denote the action in this fibration (see lecture II). Then by Lemma 3(a), (b), and (1) above, since $\tilde{E}_1$ is 3-connected, we have

(2) $$m^*\tilde{k} = Sq^2\iota\otimes 1 + 1\otimes\tilde{k}.$$

Given $\xi\colon A \rightarrow \tilde{B}_{4s+3}$, define as above,

$$\tilde{k}(\xi) = \bigcup_{\eta} \eta^*\tilde{k} \subset H^{4s+3}(A),$$

where the union is over all maps $\eta\colon A \rightarrow \tilde{E}_1$ such that $\tilde{p}_1\eta = \xi$. By Lemma 3(c) and (2) above we see that $\tilde{k}(\xi)$ is a coset of the subgroup

$$Sq^2 H^{4s+1}(A) \subset H^{4s+3}(A).$$

Summarizing, we have proved:

(3) Let $\xi: A \to \tilde{B}_{4s+3}$, where $A$ is a complex of dim $\leq$ 4s+3. Suppose that $w_{4s+2}(\xi) = 0$. Then $\xi$ lifts to $\tilde{B}_{4s+1}$, iff $0 \in \tilde{k}(\xi) \subset H^{4s+3}(A)$. Moreover, $\tilde{k}(\xi)$ is a coset of the subgroup $Sq^2 H^{4s+1}(A)$.

We seek a way to compute $\tilde{k}(\xi)$. Our method uses the secondary cohomology operations of Adams. (See Chapter 3 of [1]). Let $\Phi$ denote the operation associated with the Adem relation

(4) $$Sq^2\, Sq^2 + Sq^1(Sq^2\, Sq^1) = 0.$$

Thus $\Phi$ is defined on those classes $u \in H^*(A)$ such that

$$Sq^2(u) = Sq^2\, Sq^1(u) = 0.$$

And $\Phi(u)$ is then a coset of the subgroup

$$Sq^2\, H^{n+1}(A) + Sq^1\, H^{n+2}(A)$$

in $H^{n+3}(A)$, assuming dim $u = n$.

We prove

<u>Theorem</u> 6. Let $A$ and $\xi$ be as in (3), with $s > 0$. If $Sq^1\, H^{4s+2}(A) \subset Sq^2\, H^{4s+1}(A)$, then

$$\tilde{k}(\xi) = \Phi(w_{4s}(\xi)),$$

as cosets of $Sq^2H^{4s+1}(A)$ in $H^{4s+3}(A)$. Thus, by (3), $\xi$ has 2 linearly independent cross-sections, iff $0\in\Phi(w_{4s}(\xi))$.

Notice that $\Phi$ is indeed defined on $w_{4s}(\xi)$. For by the formulae of Wu,

$$Sq^2 w_{4s}(\xi) = w_2(\xi)\cdot w_{4s}(\xi) + w_{4s+2}(\xi) = 0,$$

since $w_{4s+2}(\xi) = 0$ by hypothesis and $w_2(\xi) = 0$ since $\xi$ is a Spin (4s+3)-bundle. Similarly,

$$Sq^2\, Sq^1\, w_{4s}(\xi) = w_2(\xi)\cdot w_{4s+1}(\xi) = 0.$$

The point of the theorem is that there are several good techniques for computing secondary operations, especially if A is a manifold. As an example we state (without proof) an important consequence of the theorem.

Theorem 7. Let M be a closed , connected, smooth manifold of dim 4s+3, $s\geq 0$. Suppose that $w_1(M) = w_2(M) = 0$. Then M has 2 linearly independent tangent vector fields.

Here $w_i(M)\in H^i(M)$, $i\geq 0$, denotes the $i^{th}$ Stiefel-Whitney class of the tangent bundle of M. Recall that by the classical theorem of H.Hopf, one knows that M has at least one non-zero vector field since dim M is odd. The proof of Theorem 7 consists in applying 6 to the tangent bundle of M and showing that $0\in\Phi(w_{4s}(M))$. For details see [12].

Proof of Theorem 6. The case $s = 1$ is somewhat anomalous, and so we assume that $s>1$. We begin by constructing the universal example for the operation $\Phi$. Consider the following (commutative) diagram of spaces and maps:

$$\begin{array}{ccccccc}
K(Z_2,4s+1) & \xrightarrow{i} & K(Z_2,4s+1)\times K(Z_2,4s+2) & \xrightarrow{j} & Z & & \\
 & & \Big\downarrow \rho & & \Big\downarrow g & & \\
 & & K(Z_2,4s+2) & \longrightarrow & Y & \xrightarrow{Sq^2\circ f} & K(Z_2,4s+2) \\
 & & & & \Big\downarrow f & & \\
 & & & & K(Z_2,4s) & \xrightarrow{Sq^2Sq^1} & K(Z_2,4s+3).
\end{array}$$

Here $f$ is the principal fibration induced by $Sq^2Sq^1$, and $g$ is the principal fibration induced by $Sq^2\circ f$. Thus the composite fibration $f\circ g\colon Z \to K(Z_2,4s)$ has $K(Z_2,4s+1)\times K(Z_2,4s+2)$ as fiber. The map $j$ denotes the inclusion of this fiber into the total space $Z$. The maps $\rho$ and $i$ denote, respectively, the projection and inclusion. By construction the fiber of $g$ is $K(Z_2,4s+1)$, imbedded by the composite map $j\circ i$. Because of the Adem relation (4) and the Serre exact sequence, there is a (unique) class $\emptyset\in H^{4s+3}(Z)$ such that

$$j^*\emptyset = Sq^2\iota_1\otimes 1 + 1\otimes Sq^1\iota_2,$$

where $\iota_i$ denotes the fundamental class of the factor $K(Z_2,4s+i)$, $i = 1,2$.

Let $A$ be a complex and $u\in H^{4s}(A)$ a class such that $Sq^2(u) = Sq^2Sq^1(u) = 0$. If we regard $u$ as a map $A \to K(Z_2,4s)$, then $u$ lifts to $Z$ and by definition

$$\Phi(u) = \bigcup_{v} v^* \emptyset,$$

where the union is taken over all maps $v: A \to Z$ such that $fgv = u$.

We consider again the Postnikov resolution for the Spin-fibration $\tilde{p}$. Regard $w_{4s} \in H^{4s}(\tilde{B}_{4s+3})$ as a map $\tilde{B}_{4s+3} \to K(Z_2, 4s)$. Since

$$Sq^2 Sq^1 w_{4s} = Sq^2 w_{4s+1} = w_2 \cdot w_{4s+1} = 0 ,$$

there is a map $h: \tilde{B}_{4s+3} \to Y$ such that $fh = w_{4s}$. (Here $Y$ and $f$ refer to the above diagram). But

$$Sq^2 \circ f \circ h = Sq^2 w_{4s} = w_{4s+2} .$$

Since $\tilde{p}_1: \tilde{E}_1 \to \tilde{B}_{4s+3}$ was defined as the fibration induced by $w_{4s+2}$, we can regard $\tilde{p}_1$ as the fibration induced by $h$ from the fibration $g: Z \to Y$, since $g$ is induced by the map $Sq^2 \circ f$ and $Sq^2 \circ f \circ h = w_{4s+2}$. Consequently, we obtain the commutative diagram below:

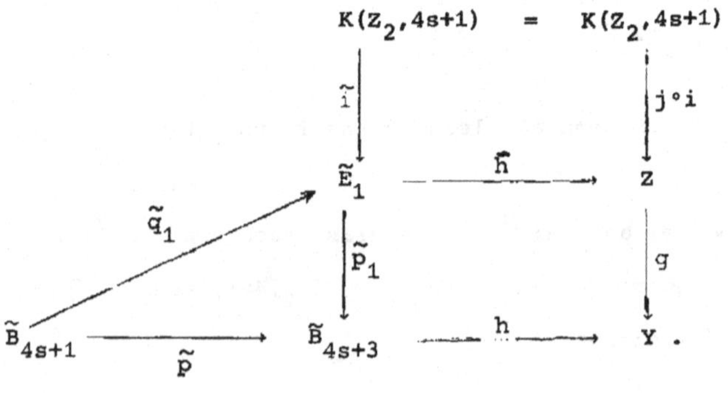

Set $k_1 = \bar{h}^*\emptyset \in H^{4s+3}(\tilde{E}_1)$. Then,

$$\tilde{i}^*k_1 = \tilde{i}^*\bar{h}^*\emptyset = j^*i^*\emptyset = Sq^2\iota_1 = \tilde{i}^*\tilde{k} .$$

Furthermore, by definition,

$$k_1 \in \Phi(\tilde{p}_1^* w_{4s}) ,$$

since $fh\tilde{p}_1 = w_{4s}\tilde{p}_1$, and so by naturality

(5) $$\tilde{q}_1^* k_1 \in \Phi(\tilde{p}^* w_{4s}) ,$$

since $\tilde{p} = \tilde{p}_1 \tilde{q}_1$ .

For convenience we define for any space $X$

$$I^q(X) = Sq^2 H^{q-2}(X) + Sq^1 H^{q-1}(X), \qquad q \geq 3.$$

Thus $I^q(X)$ is the indeterminacy subgroup of the operation $\Phi$, defined on classes of degree $q-3$.

We now need a fact whose proof is given in [ 12 ].

$$\Phi(\tilde{p}^*w_{4s}) = \tilde{p}^*I^{4s+3}(\tilde{B}_{4s+3}).$$

Assuming this we see by (5) that there is a class $\alpha \in I^{4s+3}(\tilde{B}_{4s+3})$ such that

$$\tilde{q}_1^* k_1 = \tilde{p}^* \alpha .$$

Set

$$k_2 = k_1 - \tilde{p}_1^*\alpha \in H^{4s+3}(\tilde{E}_1).$$

Then,

$$\tilde{i}^*k_2 = \tilde{i}^*k_1 - \tilde{i}^*\tilde{p}_1^*\alpha = \tilde{i}^*k_1 = \tilde{i}^*k ,$$
$$\tilde{q}_1^*k_2 = \tilde{q}_1^*k_1 - \tilde{q}_1^*\tilde{p}_1^*\alpha = \tilde{q}_1^*k_1 - \tilde{p}^*\alpha = 0 .$$

Therefore by (1), $k_2 = \tilde{k}$, and so we have shown that

$$\tilde{k} \in \Phi(\tilde{p}_1^* \, w_{4s}) ,$$

since $\tilde{k}$ differs from $k_1$ by an element of $I^{4s+3}(\tilde{E}_1)$.

Now let $\xi: A \to \tilde{B}_{4s+3}$ be a bundle as in (3) and let $\eta: A \to \tilde{E}_1$ be a map such that $\tilde{p}_1\eta = \xi$. Then, by definition,

$$\eta^*\tilde{k} \in \tilde{k}(\xi) .$$

But by naturality,

$$\eta^*\tilde{k} \in \Phi(\eta^*\tilde{p}_1^*w_{4s}) = \Phi(w_{4s}(\xi)) ,$$

which completes the proof, since by hypothesis $\tilde{k}(\xi)$ and $\Phi(w_{4s}(\xi))$ are cosets of the same subgroup.

The method of proof given for Theorem 6 is generalized in [12] to a method that handles other sorts of fibrations. In particular the method applies to give results on cross-sections of complex vector bundles, on immersions of certain manifolds in Euclidean space, and on the existence of almost-complex structures on certain 8-manifolds. Also, it is shown in [12] that by using "twisted" cohomology operations, one can remove the hypothesis $w_2(\xi) = 0$ in Theorem 6 (and hence remove the hypothesis $w_2(M) = 0$ in Theorem 7). For a resumé of these results, see [ 11 ].

References


1. J.F. Adams, On the non-existence of elements of Hopf invariant one, Ann. of Math., 72 (1960), pp. 20-104.

2. B. Eckmann and P.J. Hilton, Operators and co-operators in homotopy theory, Math. Ann., 141 (1960), 1-21.

3. M. Mahowald, On obstruction theory in orientable fiber bundles, Trans. Amer. Math. Soc., (1964), 315-349.

4. G. Paechter, The groups $\Pi_r(V_{n,m})$, I-IV, Quart. J. Math. (2) (7) (1956) pp. 249-268.

5. F.P. Peterson, Functional cohomology operations, Trans. Amer. Math. Soc., 86 (1957), 197-211.

6. F.P. Peterson and E. Thomas, A note on non-stable cohomology operations, Boletin de la Sociedad Matemática, 3 (1958), 13-18.

7. J.P. Serre, Seminaire Cartan: Algebres d'Eilenberg - MacLane et homotopie, 1954/55.

8. ——— Homologie singulière des espaces fibrés, Ann. of Math., 54 (1951), pp. 425-505.

9. ——— Groupes d'homotopie et classes de groupes abéliens, Ann. of Math. 58 (1953), pp. 258-244.

10. N.E. Steenrod, Cohomology invariants of mappings, Ann. of Math. 50 (1949), 954-988.

11. E. Thomas, On cross-sections to fiber spaces, Proc. Nat. Acad. Sci, U.S.A., 54 (1965), 40-41.

12. ——— Postnikov invariants and higher order cohomology operations, to appear.

13. H. Toda, Composition methods in homotopy groups of spheres, Princeton University Press, Princeton, New Jersey, (1962).

14. W.T. Wu, Les i-carrés dans une variété grassmannienne, C.R. Acad. Sci, Paris, 230 (1950), pp. 918-920.

# Lecture Notes in Mathematics

Edited by A. Dold, Heidelberg and B. Eckmann, Zürich

380



Mario Petrich

Pennsylvania State University, University Park, PA/USA

With an Appendix by Richard Wiegandt


# Rings and Semigroups


Springer-Verlag

Berlin · Heidelberg · New York 1974

AMS Subject Classifications (1970): 20-02, 20-M-20, 20-M-25, 20-M-30, 22-A-30, 16-02, 16A12, 16A20, 16A42, 16A56, 16A64, 16A80

ISBN 3-540-06730-2 Springer-Verlag Berlin · Heidelberg · New York
ISBN 0-387-06730-2 Springer-Verlag New York · Heidelberg · Berlin



Offsetdruck: Julius Beltz, Hemsbach/Bergstr.

# INTRODUCTION

Semigroup theory can be considered as one of the more successful offsprings of ring theory. The relationship of these two theories has been a subject of particular attention only within the last two decades and has generally taken the form of an investigation of the multiplicative semigroups of rings. The first and still the most fundamental work in this direction is due to L.M. Gluskin who studied certain dense rings of linear transformations from the multiplicative point of view. These Lectures represent an attempt to put selected topics concerning both rings of linear transformations and abstract rings, as well as their multiplicative semigroups, into a form suitable for presentation to students interested in algebra. The Lectures are divided into three parts according to the clusters of covered topics.

Part I consists of a study of certain semigroups and rings of linear transformations on an arbitrary vector space over a division ring. For dense rings of linear transformations containing a nonzero linear transformation of finite rank, two phenomena, from the present point of view, are of decisive importance: (a) its multiplicative semigroup is a dense extension of a completely 0-simple semigroup, and (b) it has unique addition. Because of (b), most information about these rings can be obtained by considering their multiplicative semigroups alone. This leads naturally to a study of semigroups of linear transformations and in particular of those satisfying (a) above. Hence in many instances we first establish the desired result for semigroups and then specialize it to rings of linear transformations. Even though the guiding idea adopted here is that first expounded by Gluskin [4], the principal references are the books of Baer [1] and Jacobson [7].

Part II contains an investigation of various abstract rings, their characterizations and representations. The classes of rings under study here are semiprime rings with minimal one-sided ideals subject to various other restrictions. For each of these classes of rings a multiplicative characterization is provided, this being possible in view of their unique addition. Their multiplicative semigroups are either dense extensions of completely 0-simple semigroups or of their orthogonal sums. Again some of the basic ideas here stem from Gluskin [4], but the main reference is Jacobson [7].

Part III represents a topological treatment of a left vector space, and its ring of linear transformations, in duality with a right vector space. Linear topology of modules and rings, linear compactness of vector spaces, various topologies on certain rings of linear transformations, as well as a completion of a topological vector space are covered here. Certain of the rings studied earlier appear here as topological rings. Many of the basic ideas discussed in this part are due to Dieudonné [2], but the chief references are the books of Jacobson [7] and Köthe [1].

The appendix contains a concise exposition of some principal achievements in the theory of linearly compact modules and semisimple rings. The two main results consist of several characterizations of linearly compact primitive and semisimple rings. The chief original reference here is Leptin [2]. Jacobson's Density Theorem is included with a short proof.

The method of maximum exploitation of the multiplicative structure of a ring often makes the use of various hypotheses more transparent and demonstrates the fact that, at least for the rings under study here, the addition is essentially extraneous.

The sources of these Lectures are numerous: in addition to the above mentioned references, they include a variety of papers which are generally referred to in the text. There is also a generous sprinkling of results published here for the first time. Among these are the results concerning the structure of simple rings with minimal one-sided ideals in terms of Rees matrix rings, isomorphisms of the latter, and some related results due to Dr. E. Hotzel; the remaining ones are due to the author. For the sake of clarity and uniformity of presentation, many known results have been rephrased and several new proofs have been provided. There are several exercises at the end of most sections; they are designed to test the understanding of the material and sometimes extend the subject covered in the text. However, the material in the main body of the text is independent of exercises.

Part I was the subject of a one semester course in linear algebra in the Summer of 1969; the entire Lectures formed the content of a two semester course in topics in ring theory in the school year 1971/72, both at the Pennsylvania State University. I am indebted to Dr. D.E. Zitarelli for taking the notes for Part I, to Mr. J.J. Streilein for taking the notes for Parts II and III, to Dr. R. Wiegandt for supplying the appendix, and to Professor B.M. Schein for reading the manuscript and suggesting several improvements. I am grateful to Dr. E. Hotzel for the permission to include his unpublished results, to students in the two classes for correcting many slips, as well as to all other persons who contributed to the existence of these Lectures.

Statements in the text are referred to only by number: if the statement is in the same Part, the Arabic numerals are used, say 5.6 which is statement 6 in

Section 5; if the statement is in a different Part, the number of the Part in Roman numeral in affixed, say I.5.6.

Since the work of Gluskin on the subject at hand has provoked considerable interest, it is hoped that a systematic and self-contained exposition will propagate the existing knowledge and stimulate new research in this highly promissing area of the ring-semigroup cooperation.

TABLE OF CONTENTS

## PART I

## SEMIGROUPS AND RINGS OF LINEAR TRANSFORMATIONS


| | | |
|---|---|---|
| I.1 | Definitions and notation | 1 |
| I.2 | Dense rings of linear transformations | 7 |
| I.3 | One-sided ideals of $\mathfrak{F}_U(V)$ | 16 |
| I.4 | Ideals of $\mathcal{S}(V)$ and principal factors of $\mathfrak{MF}_U(V)$ | 27 |
| I.5 | Semilinear isomorphisms | 34 |
| I.6 | Groups of semilinear automorphisms | 47 |
| I.7 | Extensions of semigroups and rings | 54 |


## PART II

## SEMIPRIME RINGS WITH MINIMAL ONE-SIDED IDEALS


| | | |
|---|---|---|
| II.1 | Prime rings | 65 |
| II.2 | Simple rings | 74 |
| II.3 | Maximal prime rings | 85 |
| II.4 | Semiprime rings | 91 |
| II.5 | Semiprime rings essential extensions of their socles | 95 |
| II.6 | Semiprime atomic rings | 102 |
| II.7 | Isomorphisms | 105 |


## PART III

## LINEARLY TOPOLOGIZED VECTOR SPACES AND RINGS


| | | |
|---|---|---|
| III.1 | A topology for a vector space | 112 |
| III.2 | Topological properties of subspaces | 120 |
| III.3 | Topological properties of semilinear transformations | 125 |
| III.4 | Completion of a vector space | 128 |
| III.5 | Linearly compact vector spaces | 131 |
| III.6 | A topology for $\mathfrak{L}_U(V)$ | 136 |
| III.7 | A topology for $\mathfrak{F}_U(V)$ | 139 |
| III.8 | Another topology for $\mathfrak{L}_U(V)$ | 143 |
| III.9 | Complete primitive rings | 146 |

APPENDIX

ON LINEARLY COMPACT PRIMITIVE AND SEMISIMPLE RINGS

0. Introduction 152
1. More about primitive rings 153
2. Inverse limits and linearly compact modules 156
3. Linearly compact primitive rings 159
4. Linearly compact semisimple rings 162

CURRENT ACTIVITY 167
BIBLIOGRAPHY 168
LIST OF SYMBOLS 175
INDEX 177

# PART I

## SEMIGROUPS AND RINGS OF LINEAR TRANSFORMATIONS

The subject of this part is a study of certain semigroups and rings of linear transformations on an arbitrary vector space over a division ring. In particular, dense rings of linear transformations containing a nonzero linear transformation of finite rank are characterized in several ways. For the ring of all linear transformations of finite rank having an adjoint in a vector space in duality with the given one, all one-sided ideals are found, its partially ordered set of idempotents is characterized, and a Rees matrix representation for its principal factors is constructed. For the semigroup and ring of all linear transformations all ideals are found. Isomorphisms between two semigroups of linear transformations each containing all linear transformations of rank 1 with an adjoint in a vector space in duality with the given one is expressed by means of semilinear transformations of the underlying vector spaces. A number of groups of semilinear automorphisms on a vector space are discussed. This part ends with a discussion of ideal extensions in semigroups and rings with application to certain semigroups and rings of linear transformations.

The emphasis here is one the multiplicative structure of the rings under study. Thus many statements which are usually established for these rings are first proved here for semigroups and are then specialized to rings.

### I.1 DEFINITIONS AND NOTATION

A division ring (or a skew field or a sfield) is an algebraic system satisfying all the axioms for a field except possibly commutativity of multiplication. (Thus a commutative division ring is a field and conversely.) It will be usually denoted by $\Delta$. The symbols $+$, $\cdot$, $1$, $0$ have their usual meaning.

Throughout this section, let $\Delta$ be a division ring and $V$ be an abelian group. Denote the elements of $\Delta$ by lower case Greek letters, and those of $V$ by lower case Roman letters $v, x, y, \ldots$ . We say that $\Delta$ acts on $V$ on the left if there is a function mapping $\Delta \times V$ into $V$, say $(\sigma, x) \to \sigma x$.

The ordered pair $(\Delta, V)$ is a left vector space if $\Delta$ acts on $V$ on the left in such a way that the following postulates are fulfilled:

$$\sigma(x+y) = \sigma x + \sigma y \qquad (\sigma \in \Delta,\ x,y \in V)$$
$$(\sigma+\tau)x = \sigma x + \tau x \qquad (\sigma,\tau \in \Delta,\ x \in V)$$
$$\sigma(\tau x) = (\sigma\tau)x \qquad (\sigma,\tau \in \Delta,\ x \in V)$$
$$1x = x \qquad (x \in V).$$

The action is called scalar multiplication, elements of $\Delta$ are called scalars and those of $V$ are called vectors.

If $\Delta$ acts on $V$ on the right, satisfying the right version of the above postulates, then the ordered pair $(V,\Delta)$ is called a right vector space. In this case, using the notation $(V,\Delta)$ makes it clear that the scalars act on the right. Hence the name vector space can be used in most cases without danger of confusion. We will write $V$ instead of $(\Delta,V)$ if it is clear that $V$ is to be regarded as a (left or right) vector space (rather than just an abelian group). In such a case the phrase "$V$ is a (left or right) vector space over $\Delta$" is customary.

Let $(\Delta,V)$ and $(\Delta,U)$ be two left vector spaces. A function $a:V \to U$ is called a linear transformation of $(\Delta,V)$ into $(\Delta,U)$ if

$$(x+y)a = xa + ya \qquad (x,y \in V)$$
$$(\sigma x)a = \sigma(xa) \qquad (\sigma \in \Delta,\ x \in V).$$

We write linear transformations of left vector spaces as operators on the right and denote them by lower case Roman letters $a,b,c,\dots$ .

We can consider $\Delta$ itself as a left vector space $(\Delta,\Delta^+)$ where $\Delta^+$ signifies the additive group of $\Delta$ and the scalar multiplication $\sigma x$ is the multiplication in $\Delta$ (here $\sigma \in \Delta$, $x \in \Delta^+$). A linear transformation of $(\Delta,V)$ into $(\Delta,\Delta^+)$ is called a linear form (or a linear function or functional) on $(\Delta,V)$.

The set of all linear transformations of $(\Delta,V)$ into $(\Delta,U)$ is easily seen to be an abelian group under addition defined by

$$x(f+g) = xf + xg \qquad (x \in V). \tag{1}$$

In fact, it is a subgroup of the abelian group of all homomorphisms of the abelian group $V$ into the abelian group $U$ under the usual addition of homomorphisms.

The set $V^*$ of linear forms on $(\Delta,V)$ can be made into a right vector space as follows. First $V^*$ is an abelian group under addition (1). We make $\Delta$ act on $V^*$ on the right by letting, for $f \in V^*$ and $\sigma \in \Delta$, $f\sigma$ be defined by

$$x(f\sigma) = (xf)\sigma \qquad (x \in V).$$

Then $(V^*,\Delta)$ is a right vector space, called the dual (or conjugate) space of $(\Delta,V)$. Elements of $V^*$ will be denoted by $f,g,h,\dots$ .

A semigroup is a nonempty set together with an associative binary operation (usually called multiplication). For any nonempty set $X$, the set of all transformations of $X$ into itself, written on the left, is a semigroup under the composition $(ab)x = a(bx)$ $(x \in X)$, to be denoted by $\mathfrak{J}(X)$. If the transformations are written on the right, the composition is given by $x(ab) = (xa)b$ $(x \in V)$ and the resulting semigroup will be denoted by $\mathfrak{J}'(X)$.

A linear transformation of $(\Delta,V)$ into itself is called an endomorphism of $(\Delta,V)$. The set of all endomorphisms of $(\Delta,V)$ is a subsemigroup of $\mathfrak{J}'(V)$; it is called the semigroup of endomorphisms of $(\Delta,V)$, and will be denoted by $\mathfrak{S}(\Delta,V)$ or simply $\mathfrak{S}(V)$. The same set endowed with the addition (1) forms a ring, called the ring of endomorphisms of $(\Delta,V)$ and will be denoted by $\mathfrak{L}(\Delta,V)$ or simply $\mathfrak{L}(V)$.

We have seen that the set of all linear transformations of $(\Delta,V)$ into $(\Delta,U)$ can be given the structure of an abelian group. In addition, for $U = \Delta^+$, this set can be made into a right vector space over $\Delta$, and for $U = V$ the same set can be made into a semigroup or a ring.

If $U$ is a subgroup of $V$ closed under multiplication by scalars in $\Delta$, then $(\Delta,U)$ is called a subspace of $(\Delta,V)$. It is clear that $(\Delta,U)$ is itself a vector space.

As in the case of a finite dimensional vector space over a field, one defines a linearly independent set of vectors (every finite subset is linearly independent), a generating set (every vector can be written as a finite linear combination ...), and a basis of a vector space $V$. A vector space always has a basis; it is characterized as a minimal generating set or a maximal linearly independent set. Any two bases have the same cardinality which is by definition the dimension of $V$, usually denoted by $\dim V$. Any linearly independent set of vectors can be completed to a basis of $V$, and from every generating set of vectors a basis can be extracted. In particular, every basis of a subspace can be completed to a basis of the whole space. To every subspace $A$ of $V$, there exists a subspace $B$ of $V$ such that every vector in $V$ can be uniquely written in the form $x+y$ with $x \in A$, $y \in B$; $B$ is called a complement of $A$ in $V$, and we write $V = A \oplus B$. We will use these properties of a vector space without express reference. Most of these statements require proof by transfinite methods (axiom of choice, Zorn's lemma). For the proofs and further discussion see the references at the end of this section.

If $V$ and $U$ are left vector spaces over $\Delta$, and $a$ is a linear transformation of $V$ into $U$, then $Va = \{xa \mid x \in V\}$ is a subspace of $U$, called the range of $a$; its dimenaion is the rank of $a$, denoted by rank $a$. The set $N_a = \{x \in V \mid xa = 0\}$ is a subspace of $V$ called the null space of $a$.

Linear transformations of right vector spaces are defined analogously and written as operators on the left; the dual of a right vector space is a left vector space; the range and the null space of a linear transformation on a right vector space are defined analogously.

For a given pair $(U,\Delta)$ and $(\Delta,V)$ of a right and a left vector space over $\Delta$, a function $\mathfrak{B}: V \times U \to \Delta$ is called a <u>bilinear form</u> if it satisfies:

i) $(v_1 + v_2, u) = (v_1, u) + (v_2, u)$,
ii) $(v, u_1 + u_2) = (v, u_1) + (v, u_2)$,
iii) $\sigma(v,u) = (\sigma v, u)$,
iv) $(v,u)\sigma = (v, u\sigma)$,

where $v, v_1, v_2 \in V$, $u, u_1, u_2 \in U$, $\sigma \in \Delta$, and $(v,u)$ stands for $\mathfrak{B}(v,u)$.

A bilinear form is <u>nondegenerate</u> if

v) $(v,u) = 0$ for all $u \in U$ implies that $v = 0$,
vi) $(v,u) = 0$ for all $v \in V$ implies that $u = 0$.

In such a case we say that $(U,V)$ is a <u>pair of dual vector spaces</u> (or a <u>dual pair</u>) over the division ring $\Delta$. More precisely, we should write $(U,\Delta,V;\mathfrak{B})$; we will variously use the notation $(U,\Delta,V)$ or $(U,V)$ according to the need for emphasizing the division ring ($\mathfrak{B}$ is tacitly assumed as given). Note that in Jacobson [7], the notation $(\mathfrak{M},\mathfrak{M}')$ is used where $\mathfrak{M}$ is a left and $\mathfrak{M}'$ is a right vector space. Property vi) is equivalent to:

vi′) $(v,u) = (v,u')$ for all $v \in V$ implies that $u = u'$,

which is more convenient for application; analogously for v).

Let $(\Delta,V)$ be a vector space. A subspace $U$ of $V^*$ is called a <u>total subspace</u> (or briefly a <u>t-subspace</u>) if for every $0 \neq x \in V$, there exists $f \in U$ such that $xf \neq 0$

I.1.1 <u>LEMMA</u>. <u>If</u> $U$ <u>is a t-subspace of</u> $(V^*,\Delta)$, <u>then</u> $(U,V)$ <u>is a dual pair over</u> $\Delta$ <u>with the bilinear form</u> $(v,f) = vf$ $(v \in V,\ f \in U)$. <u>Conversely, if</u> $(U,\Delta,V)$ <u>is a dual pair, then the function</u> $f: u \to f_u$ $(u \in U)$, <u>where</u> $f_u: V \to \Delta$ <u>is defined by</u> $vf_u = (v,u)$ $(v \in V)$, <u>is a linear isomorphism of</u> $(U,\Delta)$ <u>onto a t-subspace of</u> $(V^*,\Delta)$.

<u>PROOF</u>. The proof consists of a simple application of relevant definitions and is left as an exercise.

If convenient, we will identify $u$ and $f_u$, which amounts to considering $U$ as a t-subspace of $V^*$; we call $fU$ the <u>natural image</u> of $U$ in $V^*$ and write $fU = \text{nat } U$. A dual discussion is valid by interchanging the roles of $U$ and $V$, so we may consider $V$ as a t-subspace of $U^*$.

For a given $(U,\Delta,V)$ and $a \in \mathfrak{J}'(V)$, $b \in \mathfrak{J}(U)$, we say that $b$ is an adjoint of $a$ in $U$ (and $a$ is an adjoint of $b$ in $V$) if

$$(v,bu) = (va,u) \qquad (u \in U,\ v \in V). \tag{2}$$

I.1.2 LEMMA. *With the notation just introduced,* $a$ *has at most one adjoint in* $U$, *if so denote it by* $a^*$. *Then the function* $\xi: a \to a^*$ *is an isomorphism of a subring of* $\mathfrak{L}(\Delta,V)$ *onto a subring of* $\mathfrak{L}(U,\Delta)$.

PROOF. If both $b$ and $b'$ are adjoints of $a$ in $U$, then (2) yields

$$(va,u) = (v,bu) = (v,b'u) \qquad (v \in V,\ u \in U).$$

For a fixed $u$, we obtain $(v,bu) = (v,b'u)$ for all $v \in V$ which implies that $bu = b'u$. Since $u \in U$ is arbitrary, we have $b = b'$, which proves the uniqueness of the adjoint.

Suppose that $a$ has an adjoint. Let $\sigma,\tau \in \Delta$, $x,y \in V$; then for any $u \in U$, we get

$$\begin{aligned}((\sigma x + \tau y)a,u) &= (\sigma x + \tau y, a^*u) = \sigma(x,a^*u) + \tau(y,a^*u)\\ &= \sigma(xa,u) + \tau(ya,u) = (\sigma(xa) + \tau(ya),u)\end{aligned}$$

so that $(\sigma x + \tau y)a = \sigma(xa) + \tau(ya)$, and $a$ is linear.

Reversing the roles of the adjoints, we see that this also shows that $a$ is the unique adjoint of $a^*$ in $V$, and that $a^*$ is linear

Suppose that $a,c \in \mathfrak{J}'(V)$ have adjoints in $U$. Then for any $v \in V$, $u \in U$, we obtain

$$\begin{aligned}(v,(a^* + c^*)u) &= (v,a^*u + c^*u) = (v,a^*u) + (v,c^*u)\\ &= (va,u) + (vc,u) = (va + vc,u) = (v(a+c),u)\end{aligned}$$

which implies that $a^* + c^* = (a+c)^*$. Further,

$$(v,(a^*c^*)u) = (v,a^*(c^*u)) = (va,c^*u) = ((va)c,u) = (v(ac),u)$$

which implies that $a^*c^* = (ac)^*$. It follows that the set of all $a \in \mathfrak{J}'(V)$ which have an adjoint in $U$ forms a subring of $\mathfrak{L}(\Delta,V)$ and $\xi$ is a homomorphism of it onto a subring of $\mathfrak{L}(U,\Delta)$. By uniqueness of adjoints, if $a^*$ is the zero function on $U$, then $a$ must be the zero function on $V$. Thus $\xi$ is an isomorphism.

I.1.3 LEMMA. *For* $a,b \in \mathfrak{L}(V)$, *we have*

i) $\text{rank}\,(ab) \leq \min\{\text{rank}\,a,\ \text{rank}\,b\}$,

ii) $\text{rank}\,(a+b) \leq \text{rank}\,a + \text{rank}\,b$.

PROOF. i) Since $Vab \subseteq Vb$, we have $\text{rank}\,(ac) \leq \text{rank}\,b$. If $A$ is a basis of $Va$, then $Ab$ generates $Vab$ which implies that

$$\text{rank}\,(ab) = \dim Vab \leq \dim Va = \text{rank}\ a.$$

ii) Let $A$ and $B$ be bases of $Va$ and $Vb$, respectively. Then $A \cup B$ generates $V(a+b)$ and the desired inequality follows.

Let $\mathfrak{F}(\Delta,V) = \{a \in \mathfrak{L}(\Delta,V) \mid a \text{ has finite rank}\}$;

then $\mathfrak{F}(\Delta,V)$ is a subring of $\mathfrak{L}(\Delta,V)$.

For a given $(U,\Delta,V)$, let

$$\mathfrak{L}_U(\Delta,V) = \{a \in \mathfrak{L}(V) \mid a \text{ has an adjoint in } U\},$$
$$\mathfrak{F}_U(\Delta,V) = \{a \in \mathfrak{L}(V) \mid \text{rank } a \text{ is finite}\},$$

both with the ring structure inherited from $\mathfrak{L}(\Delta,V)$. Note that

$$\mathfrak{F}_U(\Delta,V) = \mathfrak{F}(\Delta,V) \cap \mathfrak{L}_U(\Delta,V).$$

Further let

$$\mathfrak{S}_U(\Delta,V) = \{a \in \mathfrak{S}(\Delta,V) \mid a \text{ has an adjoint in } U\},$$

and for $n = 1,2,3,\ldots$, let

$$\mathfrak{F}_{n,U}(\Delta,V) = \{a \in \mathfrak{L}_U(\Delta,V) \mid \text{rank } a < n\},$$

both with the semigroup structure inherited from $\mathfrak{S}(\Delta,V)$.

We will omit $\Delta$ from this notation if there is no need for stressing the division ring. Note that 1.3 implies that $\mathfrak{L}_U(V)$ is a ring, and $\mathfrak{S}_U(V)$ is a semigroup.

A nonempty subset $I$ of a semigroup $S$ is an <u>ideal</u> of $S$ if for any $a \in S$, $b \in I$, we have $ab, ba \in I$. It follows from 1.3 that $\mathfrak{F}_U(V)$ is an ideal of $\mathfrak{L}_U(V)$ and $\mathfrak{F}_{n,U}(V)$ is an ideal of $\mathfrak{S}_U(V)$.

I.1.4 <u>LEMMA</u>. $\mathfrak{L}_{V^*}(V) = \mathfrak{L}(V)$.

<u>PROOF</u>. It suffices to show that every element $a$ in $\mathfrak{L}(V)$ has an adjoint in $V^*$. For every $f \in V^*$, let $bf$ be the function given by $x(bf) = (xa)f$ $(x \in V)$. Then $bf: V \to \Delta$ and is linear since both $a$ and $f$ are. Hence $bf \in V^*$ and it follows easily that $b: f \to bf$ $(f \in V^*)$ is the adjoint of $a$ in $V^*$.

The adjoint of $a \in \mathfrak{L}(V)$ in $V^*$ is called the <u>conjugate</u> of $a$. We will denote by $a^*$ the adjoint of $a$ in any $U$, if it exists, and in particular the conjugate of $a$ (which should cause no confusion).

I.1.5 <u>LEMMA</u>. <u>Let</u> $U$ <u>be a</u> t-<u>subspace of</u> $V^*$; <u>then</u>

$$\mathfrak{L}_U(V) = \{a \in \mathfrak{L}(V) \mid a^*U \subseteq U \ \underline{\text{where}}\ a^* \ \underline{\text{is the conjugate of}}\ a\}.$$

<u>PROOF</u>. For $a \in \mathfrak{L}_U(V)$ let $b$ be its adjoint in $U$. Then for any $u \in U$,

$$(va,u) = (v,bu) = v(bu) = v(a^*u) \qquad (v \in V)$$

so that $a^*u = bu \in U$ and $a^*U \subseteq U$. For the opposite inclusion, it suffices to observe that $a^*|_U$ is an adjoint of $a$ in $U$.

Hence for $(U,\Delta,V)$, if $U$ is identified with its natural image in $V^*$, then 1.5 gives an easier way to identify $\mathcal{L}_U(V)$ as a subring of $\mathcal{L}(V)$.

For $A$ and $B$ any sets, we write $A\setminus B = \{a \in A \mid a \notin B\}$. The identity function on a set $X$ will be denoted by $\iota_X$. The Kronecker $\delta$-function is to take values $1$ or $0$ in any division ring. If $\Re$ is a ring, $\mathfrak{M}\Re$ will denote the multiplicative semigroup of $\Re$. If $S$ is a semigroup (or a ring or a vector space) with zero $0$, we let $S^- = \{s \in S \mid s \neq 0\}$. For $\Delta$ a division ring, $\mathfrak{M}\Delta^-$ means the multiplicative group of nonzero elements of $\Delta$.

The general references for this section are Baer ([1], Chapter II), Jacobson ([6], Chapters I,II,V,IX); ([7], Chapter IV) and Köthe ([1], §§7,8,9).

## I.2 DENSE RINGS OF LINEAR TRANSFORMATIONS

Throughout this section let there be given a pair of dual vector spaces $(U,\Delta,V)$. Let $I_U$ be an index set of all 1-dimensional subspaces $U_i$ of $U$, and for each $i \in I_U$, let $u_i$ be a fixed nonzero vector in $U_i$; let $I_V$ be an index set of all 1-dimensional subspaces $V_\lambda$ of $V$, and for each $\lambda \in I_V$, let $v_\lambda$ be a fixed nonzero vector in $V_\lambda$. For $i \in I_U$, $\gamma \in \Delta^-$, $\lambda \in I_V$, define the transformation $[i,\gamma,\lambda]$ on $V$ by

$$v[i,\gamma,\lambda] = (v,u_i)\gamma v_\lambda \qquad (v \in V), \tag{1}$$

and define $[i,0,\lambda]$ to be the zero function for any $i \in I_U$, $\lambda \in I_V$.

Note that any nonzero vector in $V$ can be uniquely written in the form $\sigma v_\lambda$ for some $\sigma \in \Delta^-$, $\lambda \in I_V$; similarly for $U$.

I.2.1 <u>LEMMA</u>. <u>We have</u>

$$[i,\gamma,\lambda]^*u = u_i\gamma(v_\lambda,u) \qquad (u \in U) \tag{2}$$

<u>where</u> $[i,\gamma,\lambda]^*$ <u>is the adjoint of</u> $[i,\gamma,\lambda]$ <u>in</u> $U$, <u>and</u>

$$[i,\gamma,\lambda]^*f = f_{u_i}\gamma(v_\lambda f) \qquad (f \in V^*) \tag{3}$$

<u>where</u> $[i,\gamma,\lambda]^*$ <u>is the conjugate of</u> $[i,\gamma,\lambda]$ <u>and</u> $f_{u_i}$ <u>is the natural image of</u> $u_i$ <u>in</u> $V^*$.

<u>PROOF</u>. For a fixed $u \in U$, we have for any $v \in V$,

$$(v[i,\gamma,\lambda],u) = ((v,u_i)\gamma v_\lambda,u) = (v,u_i)\gamma(v_\lambda,u) = (v,u_i\gamma(v_\lambda,u))$$

which implies (2). Further, for a fixed $f \in V^*$, we have for any $v \in V$,

$$(v[i,\gamma,\lambda])f = ((v,u_i)\gamma v_\lambda)f = (v,u_i)\gamma(v_\lambda f)$$
$$= (vf_{u_i})\gamma(v_\lambda f) = v(f_{u_i}\gamma(v_\lambda f))$$

which proves (3).

The next result is due to Gluskin [4].

I.2.2 THEOREM. *Every nonzero element of* $\mathfrak{F}_{2,U}(V)$ *can be uniquely written in the form* $[i,\gamma,\lambda]$ *for some* $i \in I_U$, $\gamma \in \Delta^-$, $\lambda \in I_V$, *and conversely, each such function is a nonzero element of* $\mathfrak{F}_{2,U}(V)$.

PROOF. Let $0 \neq a \in \mathfrak{F}_{2,U}(V)$. Since the range of $a$ is 1-dimensional, there exists $\lambda \in I_V$ such that $xa = (xf)v_\lambda$ $(x \in V)$, where $f$ is some function mapping $V$ into $\Delta$. Since $a$ is linear, so is $f$ and thus $f \in V^*$. For any $u \in U$, we get

$$(x,a^*u) = (xa,u) = ((xf)v_\lambda,u) = x(f(v_\lambda,u)) \qquad (x \in V).$$

Since $a \in \mathfrak{F}_{2,U}(V)$, we must have $a^*u \in U$ so that $f(v_\lambda,u) \in \text{nat}\, U$. Choosing $u \in U$ such that $(v,u) \neq 0$, we see that $f \in \text{nat}\, U$. Then there exists $i \in I_U$ such that $xf = (x,u_i\gamma)$ for some $\gamma \in \Delta^-$ and all $x \in V$. Consequently

$$xa = (x,u_i)\gamma v_\lambda = x[i,\gamma,\lambda] \qquad (x \in V),$$

so that $a = [i,\gamma,\lambda]$.

Suppose that $[i,\gamma,\lambda] = [j,\delta,\mu] \neq 0$. There exists $x \in V$ such that $(x,u_j) \neq 0$ so that $(x,u_i)\gamma v_\lambda = (x,u_j)\delta v_\mu \neq 0$ and thus $\lambda = \mu$. Similarly, there exists $u \in U$ such that $(v_\lambda,u) \neq 0$ so that $u_i\gamma(v_\lambda,u) = u_j\delta(v_\lambda,u) \neq 0$ and thus $i = j$. But then also $\gamma = \delta$.

The converse follows easily by (1), bilinearity and nondegeneracy of the bilinear form, and (2).

By direct calculation we see that multiplication in $\mathfrak{F}_{2,U}(V)$ is given by

$$[i,\gamma,\lambda][j,\delta,\mu] = [i,\gamma(v_\lambda,u_j)\delta,\mu] \tag{4}$$

for *all* elements of $\mathfrak{F}_{2,U}(V)$.

The next lemma is of fundamental importance. It will be used repeatedly and without express reference. This lemma can be found in Jacobson [5], and later in Gluskin [4].

I.2.3 SUPERLEMMA*. Let $(U,V)$ be a pair of dual vector spaces. For any linearly independent set $v_1,v_2,\ldots,v_n$ in $V$ there exists a linearly independent set $u_1,u_2,\ldots,u_n$ in $U$ such that $(v_i,u_j) = \delta_{ij}$ for $1 \leq i, j \leq n$. The corresponding statement holds if we interchange the roles of $U$ and $V$.

---

*Name given to it by some students in class in admiration of its many applications.

**PROOF.** The proof is by induction on $n$. If $n = 1$, then there exists $u \in U$ such that $(v_1, u) = \sigma \neq 0$, so $u_1 = u\sigma^{-1}$ has the property $(v_1, u_1) = 1$. Next let $v_1, v_2, \ldots, v_r, v_{r+1}$ be a set of linearly independent vectors in $V$ for which there exist $u_1, u_2, \ldots, u_r$ in $U$ such that $(v_i, u_j) = \delta_{ij}$ for $1 \leq i, j \leq r$. For every $u \in U$ define $\bar{u} = u - \sum_{i=1}^{r} u_i(v_i, u)$. Then for $1 \leq k \leq r$, we obtain

$$(v_k, \bar{u}) = (v_k, u) - \sum_{i=1}^{r} (v_k, u_i)(v_i, u) = (v_k, u) - (v_k, u) = 0 \tag{5}$$

and

$$\begin{aligned}(v_{r+1}, u) &= (v_{r+1}, \bar{u}) + \sum_{i=1}^{r} (v_{r+1}, u_i)(v_i, u) \\ &= (v_{r+1}, \bar{u}) + (\sum_{i=1}^{r} (v_{r+1}, u_i) v_i, u).\end{aligned} \tag{6}$$

If $(v_{r+1}, \bar{u}) = 0$ for every $u \in U$, then (6) implies $v_{r+1} = \sum_{i=1}^{r} (v_{r+1}, u_i) v_i$ contradicting linear independence of $v_1, v_2, \ldots, v_r, v_{r+1}$. Hence $(v_{r+1}, \bar{u}) = \tau \neq 0$ for some $u$; let $t_{r+1} = \bar{u}\tau^{-1}$. For $1 < k, j \leq r$, by (5) we have $(v_k, \bar{u}_j) = (v_k, t_{r+1}) = 0$, and by (6) we obtain $(v_{r+1}, u_k) = (v_{r+1}, \bar{u}_k) + (v_{r+1}, u_k)$ so that $(v_{r+1}, \bar{u}_k) = 0$. It follows that for $1 \leq i, j \leq r+1$, $(v_i, t_j) = \delta_{ij}$ where $t_j = \bar{u}_j$ if $1 \leq j \leq r$.

Now suppose that $v_1, v_2, \ldots, v_n$ is a linearly independent set of vectors in $V$ and $u_1, u_2, \ldots, u_n$ are vectors in $U$ for which $(v_i, u_j) = \delta_{ij}$. If $\sum_{j=1}^{n} u_j \sigma_j = 0$, then for $1 \leq i \leq n$, we have $0 = (v_i, \sum_{j=1}^{n} u_j \sigma_j) = \sum_{j=1}^{n} (v_i, u_j)\sigma_j = \sigma_i$ which establishes linear independence of $u_1, u_2, \ldots, u_n$.

The proof of the last statement of the lemma is symmetric.

I.2.4 **THEOREM.** For $n = 1, 2, \ldots$,

$$\mathfrak{F}_{n+1,U}(V) \setminus \mathfrak{F}_{n,U}(V) = \{\sum_{k=1}^{n} [i_k, \gamma_k, \lambda_k] \mid i_k \in I_U,\ \gamma \in \Delta^-,\ \lambda \in I_V,\ n \text{ minimal}\}.$$

**PROOF.** First let $a = \sum_{k=1}^{n} [i_k, \gamma_k, \lambda_k]$ with $i_k \in I_U$, $\gamma \in \Delta^-$, $\lambda_k \in I_V$, and $n$ minimal. By 2.2, each $[i_k, \gamma_k, \lambda_k]$ is an element of $\mathfrak{L}_U(V)$, so that $a \in \mathfrak{L}_U(V)$. We want to prove that rank $a = n$; we will show that $v_{\lambda_1}, \ldots, v_{\lambda_n}$ form a basis of $Va$.

Suppose that $v_{\lambda_1}, \ldots, v_{\lambda_n}$ are linearly dependent; we may assume that $\gamma_1 v_{\lambda_1} = \delta_2 v_{\lambda_2} + \ldots + \delta_m v_{\lambda_m}$ where all $\delta_i \neq 0$, $m \leq n$, $\gamma_1 \neq 0$ (by a suitable change of notation). Then for any $x \in V$,

$$(x, u_{i_1})\gamma_1 v_{\lambda_1} = (x, u_{i_1})\delta_2 v_{\lambda_2} + \ldots + (x, u_{i_1})\delta_m v_{\lambda_m}$$

so that

$$[i_1, \gamma_1, \lambda_1] = [i_1, \delta_2, \lambda_2] + \ldots + [i_1, \delta_m, \lambda_m].$$

Consequently

$$a = \sum_{k=2}^{n} [i_k, \gamma_k, \lambda_k] + \sum_{j=2}^{m} [i_1, \delta_j, \lambda_j].$$

For any $x \in V$ and $2 \leq j \leq m$,

$$\begin{aligned} x([i_j, \gamma_j, \lambda_j] + [i_1, \delta_j, \lambda_j]) &= (x, u_{i_j})\gamma_j v_{\lambda_j} + (x, u_{i_1})\delta_j v_{\lambda_j} \\ &= (x, u_{i_j}\gamma_j + u_{i_1}\delta_j) v_{\lambda_j} = (x, u_{\ell_j})\sigma_j v_{\lambda_j} \end{aligned}$$

for some $\ell_j \in I_U$ and $\sigma_j \in \Delta$, so that

$$a = \sum_{k=2}^{m} [\ell_k, \sigma_k, \lambda_k] + \sum_{k=m+1}^{n} [i_k, \gamma_k, \lambda_k]$$

contradicting the minimality of $n$. Thus $v_{\lambda_1}, \ldots, v_{\lambda_n}$ are linearly independent. A similar proof shows that $u_{i_1}, \ldots, u_{i_n}$ are linearly independent.

Further, for any $x \in V$, $xa = \sum_{k=1}^{n} (x, u_{i_k})\gamma_k v_{\lambda_k}$ and thus $a$ maps $V$ into the subspace generated by $v_{\lambda_1}, \ldots, v_{\lambda_n}$.

Let $1 \leq k \leq n$. There exists $x \in V$ such that $(x, u_{i_j}) = \delta_{jk}$ for $1 \leq j \leq n$. Let $y = \gamma_k^{-1} x$; then

$$ya = \sum_{j=1}^{n} (\gamma_k^{-1} x, u_{i_j})\gamma_j v_{\lambda_j} = \gamma_k^{-1}\gamma_k v_{\lambda_k} = v_{\lambda_k}$$

which proves that each $v_{\lambda_k}$ is in $Va$ and completes the proof that $v_{\lambda_1}, \ldots v_{\lambda_n}$ is a basis of $Va$. Therefore rank $a = n$ and thus $a \in \mathfrak{F}_{n+1,U}(V) \setminus \mathfrak{F}_{n,U}(V)$. A similar proof shows that each $u_{i_k}$ is in $a^*U$, so $u_{i_1}, \ldots, u_{i_n}$ is a basis of $a^*U$.

Conversely, let $a \in \mathfrak{F}_{n+1,U}(V) \setminus \mathfrak{F}_{n,U}(V)$ and let $\sigma_1 v_{\lambda_1}, \ldots, \sigma_n v_{\lambda_n}$ be a basis of $Va$. Then $v_{\lambda_1}, \ldots, v_{\lambda_n}$ is also a basis of $Va$. For any $x \in V$ we have $xa = \sum_{k=1}^{n} (xf_k) v_{\lambda_k}$ where $f_1, \ldots, f_n$ are some functions, $f_k : V \to \Delta$. By linearity of $a$ and linear independence of $v_{\lambda_1}, \ldots, v_{\lambda_n}$, it follows easily that each $f_k$ is linear, so that $f_k \in V^*$. For any $x \in V$, $u \in U$, we have

$$(x, a^*u) = (xa, u) = (\sum_{k=1}^{n} (xf_k) v_{\lambda_k}, u) = x(\sum_{k=1}^{n} f_k (v_{\lambda_k}, u))$$

where $a^*u \in U$ so that $\sum_{k=1}^{n} f_k (v_{\lambda_k}, u) \in \operatorname{nat} U$. There exists $u_k \in U$ such that $(v_{\lambda_j}, u_k) = \delta_{jk}$ for $1 \leq j,k \leq n$. Hence $f_k = \sum_{j=1}^{n} f_j (v_{\lambda_j}, u_k) \in \operatorname{nat} U$ and thus $f_k$ is the natural image of $u_{i_k}\gamma_k$ for some $i_k \in I_U$ and $\gamma_k \in \Delta^-$. It follows that

$xa = \sum_{k=1}^{n} (x, u_{i_k})\gamma_k v_{\lambda_k} = x(\sum_{k=1}^{n} [i_k, \gamma_k, \lambda_k])$ and thus $a = \sum_{k=1}^{n} [i_k, \gamma_k, \lambda_k]$. If $n$ is not minimal, then $a$ admits also a representation of the form $a = \sum_{j=1}^{m} [t_j, \delta_j, \mu_j]$

with $m < n$ and $m$ minimal. In the first part of the proof, we have seen that this implies that rank $a = m$, contradicting the hypothesis that rank $a = n$, $n > m$. Therefore $n$ is minimal.

I.2.5 COROLLARY. Let $a = \sum_{k=1}^{n} [i_k, \gamma_k, \lambda_k]$; then the following are equivalent.

i) $n$ is minimal.

ii) $\{v_{\lambda_1}, \ldots, v_{\lambda_n}\}$ and $\{u_{i_1}, \ldots, u_{i_n}\}$ are both linearly independent.

iii) rank $a = n$.

In such a case, $v_{\lambda_1}, \ldots, v_{\lambda_n}$ form a basis of $Va$ and $u_{i_1}, \ldots, u_{i_n}$ form a basis of $a^*U$, so rank $a$ = rank $a^* = n$.

PROOF. i) $\Rightarrow$ ii). This was established in the first part of the proof of 2.4.

ii) $\Rightarrow$ iii). There exist $x_k \in V$ such that $(x_k, u_{i_p}) = \delta_{kp}$ for $1 \leq k,p \leq n$. Hence $x_k a = \gamma_k v_{\lambda_k} \in Va$ and thus $v_{\lambda_k} \in Va$. Since also every element of $Va$ is a linear combination of $v_{\lambda_1}, \ldots, v_{\lambda_n}$ and these are linearly independent, it follows that $v_{\lambda_1}, \ldots, v_{\lambda_n}$ is a basis of $Va$. In particular, rank $a = n$.

iii) $\Rightarrow$ i). This follows immediately from 2.4.

The remaining statements follow from the proof of 2.4.

If $\mathfrak{R}$ is a ring and $A$ is a subsemigroup of $\mathfrak{M}\mathfrak{R}$, by $C(A)$ we denote the subring of $\mathfrak{R}$ generated by $A$. Since $A$ is a semigroup, the elements of $C(A)$ are of the form $\pm a_1 \pm \ldots \pm a_n$ with $a_i \in A$, so $C(A)$ is actually the additive group closure of $A$.

I.2.6 COROLLARY.

$$\mathfrak{F}_U(V) = \bigcup_{n=1}^{\infty} \mathfrak{F}_{n,U}(V) = C(\mathfrak{F}_{2,U}(V)) = \{a \in \mathfrak{F}(V) \mid a^*V^* \subseteq \text{nat } U\}$$

and $\mathfrak{F}_U(V)$ is a right ideal of $\mathfrak{L}(V)$.

PROOF. The first equality is obvious. The second equality follows from 2.4 and the fact that $-[i,\gamma,\lambda] = [i,-\gamma,\lambda]$ so that with each element, $\mathfrak{F}_{2,U}(V)$ contains its negative. If $a \in \mathfrak{F}_U(V)$, then $a$ is of the form $a = \sum_{k=1}^{n} [i_k, \gamma_k, \lambda_k]$, and (3) implies that for each $k$, $[i_k, \gamma_k, \lambda_k]^* V^* \subseteq \text{nat } U$, so the same is true for $a^*V^*$. The opposite inclusion follows from 1.5. If $a \in \mathfrak{F}_U(V)$ and $b \in \mathfrak{L}(V)$, then $(ab)^*V^* = a^*(b^*V^*) \subseteq \text{nat } U$, and hence $ab \in \mathfrak{F}_U(V)$.

I.2.7 DEFINITION. For $A$ a subsemigroup (subring) of a semigroup (ring) $B$, let $i_B(A)$ be the idealizer of $A$ in $B$, i.e., the largest subsemigroup (subring) of $B$ having $A$ as an ideal. It is easy to see that in both cases

$$i_B(A) = \{b \in B \mid ba, ab \in A \text{ for all } a \in A\}.$$

I.2.8 PROPOSITION.

i) $i_{\mathcal{S}(V)}(\mathfrak{F}_{2,U}(V)) = \mathcal{S}_U(V)$. ii) $i_{\mathcal{L}(V)}(\mathfrak{F}_U(V)) = \mathcal{L}_U(V)$.

PROOF. i) By 1.3, $\mathfrak{F}_{2,U}(V)$ is an ideal of $\mathcal{S}_U(V)$ which implies that $\mathcal{S}_U(V) \subseteq i_{\mathcal{S}(V)}(\mathfrak{F}_{2,U}(V))$. Let $a \in i_{\mathcal{S}(V)}(\mathfrak{F}_{2,U}(V))$, $0 \neq u \in U$, and identify $U$ with its natural image in $V^*$. It suffices to show that $a^*u \in U$. Since $u \neq 0$, there exists $v_\lambda \in V$ such that $v_\lambda u \neq 0$. Then $u = u_i\sigma$ for some $i \in U$, $\sigma \in \Delta^-$; letting $\gamma = \sigma(v_\lambda u)^{-1}$, we obtain $[i,\gamma,\lambda] \in \mathfrak{F}_{2,U}(V)$ and

$$[i,\gamma,\lambda]^*u = u_i\gamma(v_\lambda u) = (u_i\sigma)(v_\lambda u)^{-1}(v_\lambda u) = u_i\sigma = u$$

so that

$$a^*u = a^*([i,\gamma,\lambda]^*u) = (a^*[i,\gamma,\lambda]^*)u = (a[i,\gamma,\lambda])^*u \in U$$

since by hypothesis we have $a[i,\gamma,\lambda] \in \mathfrak{F}_{2,U}(V)$. Thus $a \in \mathcal{S}_U(V)$ and hence $i_{\mathcal{S}(V)}(\mathfrak{F}_{2,U}(V)) \subseteq \mathcal{S}_U(V)$.

ii) The proof is an easy modification of the proof of part i) and is left as an exercise.

I.2.9 DEFINITION. Let $n$ be a positive integer and $S$ a subset of $\mathcal{L}(V)$. Then $S$ is said to be a n-fold transitive if for any linearly independent set $x_1,\ldots,x_n$ and any set $y_1,\ldots,y_n$ in $V$, there exists $a \in S$ such that $x_ia = y_i$ for $1 \leq i \leq n$. Instead of 1-fold transitive, we say transitive, and instead of 2-fold transitive, we say doubly transitive. Further, $S$ is dense if it is n-fold transitive for every positive integer $n$.

The next proposition is due to Gluskin [4].

I.2.10 PROPOSITION. For $n \geq 1$, $\mathfrak{F}_{n+1,U}(V)$ is n-fold transitive.

PROOF. If $\dim V < n$, then $\mathfrak{F}_{n+1,U}(V)$ is n-fold transitive by default. So suppose $\dim V \geq n$ and let $x_1,\ldots,x_n$ be a linearly independent set and $y_1,\ldots,y_n$ be any set of vectors in $V$. Then $y_k = \sigma_k v_{\lambda_k}$ for some $\sigma_k \in \Delta$, $\lambda_k \in I_V$ and $1 \leq k \leq n$. There exist $u_{i_p} \in U$ such that $(x_k, u_{i_p}) = \delta_{kp}\tau_k$ for some $\tau_k \neq 0$ and $1 \leq k,p \leq n$. Let $a = \sum_{k=1}^{n}[i_k, \tau_k^{-1}\sigma_k, \lambda_k]$; then $Va$ is contained in the subspace of $V$ generated by $v_{\lambda_1},\ldots,v_{\lambda_n}$ so that rank $a \leq n$ and hence $a \in \mathfrak{F}_{n+1,U}(V)$. For $1 \leq k \leq n$, we obtain $x_ka = (x_k, u_{i_k})\tau_k^{-1}\sigma_k v_{\lambda_k} = y_k$ and $\mathfrak{F}_{n+1,U}(V)$ is n-fold transitive.

I.2.11 COROLLARY. $\mathfrak{F}_U(V)$ is a dense ring of linear transformations.

PROOF. This follows immediately from 2.6 and 2.10

The next theorem "locates" all transitive subrings of $\mathcal{L}(V)$ containing a linear transformation of rank 1; it appears to be new.

1.2.12 THEOREM. Let $\mathfrak{R}$ be a transitive subring of $\mathfrak{L}(\Delta,V)$ containing a linear transformation of rank 1. Then

$$U = \{u_i\tau \in V^* \mid \tau \in \Delta,\ [i,\gamma,\lambda] \in \mathfrak{R} \text{ for some } \gamma \in \Delta^-,\ \lambda \in I_V\}$$

(where $[i,\gamma,\lambda]$ is written relative to the dual pair $(V^*,\Delta,V)$) is the unique t-subspace of $V^*$ with the property $\mathfrak{F}_U(V) = \mathfrak{R} \cap \mathfrak{F}(V) \subseteq \mathfrak{R} \subseteq \mathfrak{L}_U(V)$.

PROOF. By hypothesis $U \neq 0$. Let $[i,\gamma,\lambda] \in \mathfrak{R}$ with $\gamma \neq 0$. For any $\mu \in I_V$ and $\delta \in \Delta$, by transitivity there exists $a \in \mathfrak{R}$ such that $(\gamma v_\lambda)a = \delta v_\mu$. Hence for any $x \in V$, we have

$$x([i,\gamma,\lambda]a) = (xu_i)\gamma(v_\lambda a) = (xu_i)\delta v_\mu = x[i,\delta,\mu]$$

so that $[i,\delta,\mu] = [i,\gamma,\lambda]a \in \mathfrak{R}$. Hence $[i,\gamma,\lambda] \in \mathfrak{R}$ with $\gamma \neq 0$ implies that $[i,\delta,\mu] \in \mathfrak{R}$ for any $\delta \in \Delta$, $\mu \in I_V$.

Let $u_i\sigma, u_j\tau \in U$ and $t = u_i\sigma + u_j\tau$. If $t = 0$, then $t \in U$ since $U$ contains 0. Hence suppose $t \neq 0$, so $t = u_k\gamma$ for some $k \in I_{V^*}$, $\gamma \in \Delta^-$. We have $[i,\sigma,\lambda][j,\tau,\lambda] \in \mathfrak{R}$ for every $\lambda \in I_V$. For any $x \in V$, we obtain

$$\begin{aligned} x([i,\sigma,\lambda] + [j,\tau,\lambda]) &= (xu_i)\sigma v_\lambda + (xu_j)\tau v_\lambda \\ &= (x(u_i\sigma + u_j\tau))v_\lambda \\ &= (x(u_k\gamma))v_\lambda = x[k,\gamma,\lambda] \end{aligned}$$

and thus $[k,\gamma,\lambda] = [i,\sigma,\lambda] + [j,\tau,\lambda] \in \mathfrak{R}$. But then $t = u_k\gamma \in U$ which proves that $U$ is a subspace of $V^*$ since it is obviously closed under scalar multiplication.

Let $0 \neq x \in V$ and $y \in V$. We know that there exists $k$ such that $[k,\delta,\mu] \in \mathfrak{R}$ for all $\delta \in \Delta$, $\mu \in I_V$. There exist $z \in V$ such that $zu_k \neq 0$ and $a \in \mathfrak{R}$ such that $xa = z$. Further, letting $y = \gamma v_\mu$ and $\mathfrak{R}_2 = \mathfrak{R} \cap \mathfrak{F}_{2,V^*}(V)$, we obtain

$$x(a[k,(zu_k)^{-1}\gamma,\mu]) = (zu_k)(zu_k)^{-1}\gamma v_\mu = y,$$

where $a[k,(zu_k)^{-1}\gamma,\mu] \in \mathfrak{R}_2$ by 1.3. Hence $\mathfrak{R}_2$ is a transitive semigroup. In particular, if $y \neq 0$, there exists $[i,\gamma,\lambda] \in \mathfrak{R}_2$ such that $x[i,\gamma,\lambda] = y$ which implies that $xu_i \neq 0$. Since $u_i \in U$, it follows that $U$ is a t-subspace of $V^*$.

From the definition of $U$ and 2.4 it follows that

$$\mathfrak{R}_2 = \{[i,\gamma,\lambda] \in \mathfrak{F}_{2,V^*}(V) \mid u_i \in U\} = \mathfrak{F}_{2,U}(V). \tag{7}$$

Since $\mathfrak{F}_{2,U}(V)$ is an ideal of $\mathfrak{S}_U(V)$, we have that $\mathfrak{R}_2$ is an ideal of $\mathfrak{M}\mathfrak{R}$. Hence by 2.8, part i), we have $\mathfrak{M}\mathfrak{R} \subseteq i_{\mathfrak{S}(V)}(\mathfrak{F}_{2,U}(V)) = \mathfrak{S}_U(V)$ whence $\mathfrak{R} \subseteq \mathfrak{L}_U(V)$.

By 2.6 and (7), we have $\mathfrak{F}_U(V) = \mathfrak{C}(\mathfrak{F}_{2,U}(V)) = \mathfrak{C}(\mathfrak{R}_2) \subseteq \mathfrak{R}$ since $\mathfrak{R}$ is a ring and hence $\mathfrak{F}_U(V) \subseteq \mathfrak{R} \cap \mathfrak{F}(V)$. Conversely, let $0 \neq a \in \mathfrak{R} \cap \mathfrak{F}(V)$. Then $a = \sum_{k=1}^{n} [i_k,\gamma_k,\lambda_k]$ and we may suppose that $n$ is minimal. By 2.5, the set $v_{\lambda_1},\ldots,v_{\lambda_n}$ is linearly independent and hence there exist $u_{j_\ell} \in U$ with the property $(v_{\lambda_k}, u_{j_\ell}) = \delta_{k\ell}\sigma_k$ for some $\sigma_k \neq 0$ and $1 \leq k,\ell \leq n$. Consequently $[j_k, \sigma_k^{-1}, \lambda_k] \in \mathfrak{R}$ for $1 \leq k \leq n$ and thus the element

$$a[j_k, \sigma_k^{-1}, \lambda_k] = [i_k, \gamma_k(v_{\lambda_k}, u_{j_k})\sigma_k^{-1}, \lambda_k] = [i_k, \gamma_k, \lambda_k]$$

is also in $\mathfrak{R}$, so by (7), we have

$$[i_k, \gamma_k, \lambda_k] \in \mathfrak{R} \cap \mathfrak{F}_2(V) = \mathfrak{R}_2 = \mathfrak{F}_{2,U}(V)$$

for $1 \leq k \leq n$. But then $a \in C(\mathfrak{F}_{2,U}(V)) = \mathfrak{F}_U(V)$ by 2.6. Therefore $\mathfrak{R} \cap \mathfrak{F}(V) \subseteq \mathfrak{F}_U(V)$.

Let $U'$ be a t-subspace of $V^*$ which also satisfies the requirements of the theorem. Then $\mathfrak{F}_U(V) = \mathfrak{F}_{U'}(V)$ and thus also $\mathfrak{F}_{2,U}(V) = \mathfrak{F}_{2,U'}(V)$. Hence if $u_i\sigma \in U$, then $[i,\sigma,\lambda] \in \mathfrak{F}_{2,U}(V)$, so $[i,\sigma,\lambda] \in \mathfrak{F}_{2,U'}(V)$ and thus $u_i\sigma \in U'$. Consequently $U \subseteq U'$ and by symmetry also $U' \subseteq U$, which proves uniqueness of $U$.

We are now able to characterize the rings in the preceding theorem in several ways. The equivalence of parts i), ii) and iii) is due to Jacobson [3], part iv) is new.

I.2.13 THEOREM. <u>The following conditions on a subring $\mathfrak{R}$ of $\mathfrak{L}(\Delta,V)$ are equivalent.</u>

i) $\mathfrak{F}_U(V) \subseteq \mathfrak{R} \subseteq \mathfrak{L}_U(V)$ <u>for some</u> t-<u>subspace</u> $U$ <u>of</u> $V^*$.

ii) $\mathfrak{R}$ <u>is a dense ring containing a nonzero transformation of finite rank.</u>

iii) $\mathfrak{R}$ <u>is a doubly transitive ring containing a transformation of finite rank.</u>

iv) $\mathfrak{R}$ <u>is a transitive ring containing a transformation of rank</u> 1.

<u>PROOF</u>. i) ⇒ ii). This follows from 2.11 since a ring containing a dense ring is itself dense.

ii) ⇒ iii). Trivially.

iii) ⇒ iv). Let $n = \min\{\text{rank } a \mid 0 \neq a \in \mathfrak{R}\}$. It suffices to show that $n = 1$. Suppose $n > 1$ and let $a \in \mathfrak{R}$ be such that rank $a = n$. Let $x_1, \ldots, x_n$ be a basis of $Va$ and let $y_{n-1}$ be any vector in $V$ with the property $y_{n-1}a = x_{n-1}$. By double transitivity there exists $b \in \mathfrak{R}$ such that $x_n b = x_{n-1}b = y_{n-1}$, let $x_i b = y_i$ for $1 \leq i \leq n-2$. Then for any $x \in V$, we obtain

$$xa = \alpha_1 x_1 + \ldots + \alpha_n x_n \quad \text{for some } \alpha_i \in \Delta,$$

$$\begin{aligned} xab &= \alpha_1(x_1 b) + \ldots + \alpha_{n-2}(x_{n-2}b) + \alpha_{n-1}(x_{n-1}b) + \alpha_n(x_n b) \\ &= \alpha_1 y_1 + \ldots + \alpha_{n-2}y_{n-2} + (\alpha_{n-1} + \alpha_n)y_{n-1} \end{aligned}$$

which implies that rank $(ab) \leq n-1$ and $ab \in \mathfrak{R}$. Since

$$y_{n-1}(ab) = x_{n-1}b = y_{n-1} \neq 0$$

we have $ab \neq 0$. But this contradicts minimality of the rank of $a$.

iv) ⇒ i). This follows from 2.12.

I.2.14 Exercises.

i) Let $\Delta$ be a finite field having $m$ elements and $(\Delta, V)$ be a vector space of finite dimension $n$. Show that both $V$ and $V^*$ have $m^n$ elements, and both $I_V$ and $I_{V^*}$ have $\frac{m^n-1}{m-1}$ elements.

ii) Let $a = \sum_{k=1}^{n} [i_k, \gamma_k, \lambda_k]$ and $b = \sum_{t=1}^{m} [j_t, \delta_t, \mu_t]$, with $n$ and $m$ minimal, be nonzero elements of $\mathfrak{F}_U(\Delta, V)$. Show that $a = b$ if and only if $m = n$ and there exist invertible $n \times n$ matrices $A$ and $B$ over $\Delta$ such that

a) $(u_{i_1} \; \ldots \; u_{i_n}) = (u_{j_1} \; \ldots \; u_{j_n})A$,

b) $$\begin{bmatrix} v_{\lambda_1} \\ \vdots \\ v_{\lambda_n} \end{bmatrix} = B \begin{bmatrix} v_{\mu_1} \\ \vdots \\ v_{\mu_n} \end{bmatrix},$$

c) $$A \begin{bmatrix} \gamma_1 & & & 0 \\ & \gamma_2 & & \\ & & \ddots & \\ 0 & & & \gamma_n \end{bmatrix} B = \begin{bmatrix} \delta_1 & & & 0 \\ & \delta_2 & & \\ & & \ddots & \\ 0 & & & \delta_n \end{bmatrix}.$$

The entries of the row and column matrices above are vectors, the multiplication with square matrices over $\Delta$ is performed according to the usual row by column rule and the entries of the resulting matrices are again vectors.

iii) A set $E$ of nonzero idempotents of a ring $\mathfrak{R}$ is said to be orthogonal if $ef = 0$ for any $e, f \in E$, $e \neq f$. Show that the sum of a finite number of orthogonal idempotents is again an idempotent. For $a = \sum_{k=1}^{n} [i_k, \gamma_k, \lambda_k] \in \mathfrak{F}_U(V)$ with $n$ minimal, find necessary and sufficient conditions in order that $a$ be an idempotent. Show that for any $n > 0$, $e \in \mathfrak{F}_U(V)$ is an idempotent of rank $n$ if and only if there exists a set $e_1, \ldots, e_n$ of orthogonal idempotents in $\mathfrak{F}_{2,U}(V)$ such that $e = e_1 + e_2 + \ldots + e_n$.

iv) Prove that every nonzero ideal of a transitive subsemigroup of $\mathfrak{S}(V)$ is itself transitive.

v) Prove that every nonzero ideal of a dense subring of $\mathfrak{L}(V)$ is itself dense.

vi) Show that an element $a$ of $\mathfrak{L}(V)$ is idempotent if and only if $a$ leaves $Va$ elementwise fixed.

vii) For $a, b \in \mathfrak{J}(V)$, prove that

$$|\text{rank } a - \text{rank } b| \leq \text{rank } (a+b).$$

viii) Let $(U,\Delta,V)$ be a dual pair, $I$ be a nonempty subset of $I_U$ and

$$S = \{u_i\tau \mid i \in I,\ \tau \in \Delta\},\quad T = \{[i,\gamma,\lambda] \in \mathfrak{F}_U(V) \mid i \in I\}.$$

Show that $S$ is a subspace of $U$ if and only if $\mathfrak{C}(T) \cap \mathfrak{F}_U(V) = T$.

ix) The least positive integer $n$ for which $nr = 0$ for all elements $r$ of a ring $\mathfrak{R}$, if such exists, is the <u>characteristic</u> of $\mathfrak{R}$; otherwise $\mathfrak{R}$ is said to have characteristic zero. Show that $\Delta$ and $\mathfrak{F}_U(\Delta,V)$ have the same characteristic.

For further results on dense rings of linear transformations see Dieudonné [2], [3], Gluskin [4], Jacobson [3], [4], ([6], Chapter IX, Section 10), ([7], Chapter IV, Section 15), Ribenboim ([1], Chapter III, Section 2). Some related material can be found in Erdos [1], Rosenberg [1], Rosenberg and Zelinsky [1], Zelinsky [3].

## I.3 ONE-SIDED IDEALS OF $\mathfrak{F}_U(V)$

We fix a pair of dual vector spaces $(U,\Delta,V)$, and for $U'$ a subspace of $U$, $V'$ a subspace of $V$, we use the notation

$$\mathfrak{F}_{U'}(V') = \{a \in \mathfrak{F}_U(V) \mid Va \subseteq V',\ a^*U \subseteq U'\}.$$

It is clear that

$$\mathfrak{F}_{U'}(V') = \mathfrak{F}_{U'}(V) \cap \mathfrak{F}_U(V'). \tag{1}$$

We will sometimes write $\mathfrak{F}$ instead of $\mathfrak{F}_U(V)$ and $\mathfrak{F}_2$ instead of $\mathfrak{F}_{2,U}(V)$. Clearly $\mathfrak{F}_{U'}(V)$ is a right ideal and $\mathfrak{F}_U(V')$ is a left ideal of $\mathfrak{L}_U(V)$ and thus also of $\mathfrak{F}$. We will consider only right ideals and subspaces of $U$. The formulation and proof of dual statements for left ideals and subspaces of $V$ will be left as an exercise; for this two different proofs are possible: direct by retaining the notation already used, or using 1.2 one may consider the isomorphic copy of $\mathfrak{F}$ in $\mathfrak{L}(U)$ and then dualize the proofs to be presented here.

I.3.1 <u>LEMMA</u>. <u>We have</u>

$$\mathfrak{F}_{U'}(V') = \{\sum_{k=1}^{n} [i_k,\gamma_k,\lambda_k] \mid u_{i_k} \in U',\ \gamma \in \Delta,\ v_{\lambda_k} \in V'\}.$$

<u>PROOF</u>. By symmetry and using (1), it suffices to show that

$$\mathfrak{F}_{U'}(V) = \{\sum_{k=1}^{n} [i_k,\gamma_k,\lambda_k] \mid u_{i_k} \in U',\ \gamma_k \in \Delta,\ v_{\lambda_k} \in V\}. \tag{2}$$

Let $0 \neq a \in \mathfrak{F}_{U'}(V)$. Then $a^*U \subseteq U'$ and $a = \sum_{k=1}^{n} [i_k,\gamma_k,\lambda_k]$ where $n$ is minimal so that $v_{\lambda_k},\ldots,v_{\lambda_n}$ are linearly independent by 2.5. There exist $u_{j_p} \in U$ such that $(v_{\lambda_k},u_{j_p}) = \delta_{kp}\sigma_k$ for some $\sigma_k \neq 0$ and $1 \leq k,p \leq n$. Hence for $1 \leq k \leq n$, we have

$$[i_k,\gamma_k,\lambda_k] = a[j_k,\sigma_k^{-1},\lambda_k] \in \mathfrak{F}_{U'}(V)$$

since $\mathfrak{F}_{U'}(V)$ is a right ideal of $\mathfrak{F}$. But then $[i_k,\gamma_k,\lambda_k]^*U \subseteq U'$ which implies that $u_{i_k} \in U'$ and thus $a$ is in the right hand side of (2). The opposite inclusion is obvious.

The next result is new.

I.3.2 PROPOSITION. If $(U',\Delta,V')$ is a dual pair with the induced bilinear form, then $\mathfrak{J}_{U'}(V') \cong \mathfrak{F}_{U'}(V')$.

PROOF. Suppose that $(U',\Delta,V')$ is a dual pair, i.e., the restriction of the bilinear form to $V' \times U'$ is nondegenerate. Define the function $\psi$ by

$$\psi : a \to a|_{V'} \qquad (a \in \mathfrak{J}_{U'}(V')).$$

From the definition of $\mathfrak{J}_{U'}(V')$ it follows that $a\psi \in \mathfrak{F}_{U'}(V')$, where the adjoint of $a\psi$ in $U'$ is equal to $a^*|_{U'}$, and it is clear that $\psi$ is a homomorphism.

If $b \in \mathfrak{F}_{U'}(V')$, then by 2.6 we have $b = \sum_{k=1}^{n} [i_k,\gamma_k,\lambda_k]$. Fixing this representation of $b$, we extend $b$ to a function $a$ on all of $V$ by defining $xa = \sum_{k=1}^{n} (x,u_{i_k})\gamma_k v_{\lambda_k}$ $(x \in V)$. Then the function $c$ defined by $cu = \sum_{k=1}^{n} u_{i_k}\gamma_k(v_{\lambda_k},u)$ $(u \in U)$ is evidently the adjoint of $a$ in $U$. Consequently $a\psi = b$ and $\psi$ maps $\mathfrak{J}_{U'}(V')$ onto $\mathfrak{F}_{U'}(V')$.

To show that the kernel of $\psi$ is zero, let $0 \neq a \in \mathfrak{J}_{U'}(V')$. Then by 3.1, we have $a = \sum_{k=1}^{n} [i_k,\gamma_k,\lambda_k]$ with $u_{i_k} \in U'$, $v_{\lambda_k} \in V'$ for $1 \leq k \leq n$, and we may suppose that $n$ is minimal. Since $(U',V')$ is a dual pair, and by 2.5 we have that $u_{i_1},\ldots,u_{i_n}$ are linearly independent, there exists $v \in V'$ such that $(v,u_{i_k}) = \delta_{k1}$ for $1 \leq k \leq n$. Hence

$$va = \sum_{k=1}^{n} (v,u_{i_k})\gamma_k v_{\lambda_k} = (v,u_{i_1})\gamma_1 v_{\lambda_1} \neq 0$$

and $a\psi \neq 0$. Consequently the kernel of $\psi$ is zero and $\psi$ is an isomorphism.

The next lemma will be quite useful.

I.3.3 LEMMA. For any $a \in \mathfrak{L}_U(V)$, if $\mathfrak{F}_2 a = 0$ or $a\mathfrak{F}_2 = 0$, then $a = 0$.

PROOF. If $a \neq 0$, then $xa \neq 0$ for some $0 \neq x \in V$. Hence $x = \gamma v_\lambda$ and thus for any $i \in I_U$ and $y \in V$, we obtain

$$y([i,\gamma,\lambda]a) = (y,u_i)\gamma(v_\lambda a) = (y,u_i)(xa).$$

Since for every $i \in I_U$, there exists $y \in V$ such that $(y,u_i) \neq 0$ and we have $xa \neq 0$, it follows that $[i,\gamma,\lambda]a \neq 0$. Further, there exists $j \in I_U$ such that $(xa,u_j) \neq 0$. Hence for any $\tau \in \Delta^-$, $\lambda \in I_V$, we obtain $x(a[j,\tau,\lambda])$ $= (xa,u_j)\tau v_\lambda \neq 0$ so that $a[j,\tau,\lambda] \neq 0$.

Recall that a lattice isomorphism of a lattice A onto a lattice B is a one-to-one correspondence which preserves both meet and join (it suffices that it preserves either meet or join or the partial order). For matters concerning lattices, consult Szász [1]. The next result is due to Jacobson [5].

I.3.4 THEOREM. *The function*

$$\chi: U' \to \mathfrak{J}_{U'}(V) \qquad (U' \text{ is a subspace of } U)$$

*is a lattice isomorphism of the lattice of all subspaces of* U *onto the lattice of all right ideals of* $\mathfrak{F}_U(V)$.

PROOF. We have already observed that $\mathfrak{J}_{U'}(V)$ is a right ideal of $\mathfrak{F}$. If $U'$ and $U''$ are distinct subspaces of $U$, we may suppose that $u_i \in U' \setminus U''$. Hence for any $v_\lambda \in V$, $\gamma \in \Delta^-$, we have $[i,\gamma,\lambda]^* U \subseteq U'$ and $[i,\gamma,\lambda]^* U \not\subseteq U''$, so that $[i,\gamma,\lambda] \in \mathfrak{J}_{U'}(V) \setminus \mathfrak{J}_{U''}(V)$. Consequently $\chi$ is one-to-one.

For any subspaces $U'$ and $U''$ of $V$, we have

$$\mathfrak{J}_{U'}(V) \cap \mathfrak{J}_{U''}(V) = \mathfrak{J}_{U' \cap U''}(V)$$

which implies that $\chi$ is a meet homomorphism. Hence to complete the proof, it suffices to show that $\chi$ is onto, i.e., that every right ideal of $\mathfrak{F}$ is of the form $\mathfrak{J}_{U'}(V)$ for some $U'$. Note that $0 = \mathfrak{J}_0(V)$.

Let $\mathfrak{R}$ be a nonzero right ideal of $\mathfrak{F}$, and let

$$U' = \{u_i\sigma \in U \mid \sigma \in \Delta,\ [i,\gamma,\lambda] \in \mathfrak{R} \text{ for some } \gamma \in \Delta^-,\ \lambda \in I_V\}.$$

If $0 \neq a \in \mathfrak{R}$, then $a\mathfrak{F}_2 \subseteq \mathfrak{R} \cap \mathfrak{F}_2$ since $\mathfrak{R}$ is a right ideal and $\mathfrak{F}_2$ is an ideal of $\mathfrak{F}$. By 3.3, we also have $a\mathfrak{F}_2 \neq 0$, which implies that $\mathfrak{R}_2 = \mathfrak{R} \cap \mathfrak{F}_2$ is a nonzero right ideal of $\mathfrak{F}_2$. It follows that $U' \neq 0$. If $[i,\gamma,\lambda] \in \mathfrak{R}_2$, $\delta \in \Delta$, $\mu \in I_V$, then there exists $u_j \in U$ such that $(v_\mu, u_j) = \sigma \neq 0$, so that

$$[i,\delta,\mu] = [i,\gamma,\lambda][j,\sigma^{-1}\gamma^{-1}\delta,\mu] \in \mathfrak{R}_2.$$

Hence $[i,\gamma,\lambda] \in \mathfrak{R}_2$ with $\gamma \neq 0$ implies that $[i,\delta,\mu] \in \mathfrak{R}_2$ for any $\delta \in \Delta$, $\mu \in I_V$. As in the proof of 2.12, it follows that $U'$ is a subspace of $U$ and

$$\mathfrak{R}_2 = \mathfrak{R} \cap \mathfrak{F}_2 = \{[i,\gamma,\lambda] \mid u_i \in U'\}$$

so that $\mathfrak{J}_{U'}(V) = C(\mathfrak{R}_2) \subseteq \mathfrak{R}$. Let $0 \neq a = \sum_{k=1}^{n} [i_k,\gamma_k,\lambda_k]$ be in $\mathfrak{R}$ with $n$ minimal. There exist $u_{j_p} \in U$ such that $(v_{\lambda_k}, u_{j_p}) = \delta_{kp}\sigma_k$ for some $\sigma_k \neq 0$ and $1 \leq k,p \leq n$. Hence

$$[i_k,\gamma_k,\lambda_k] = a[j_k,\sigma_k^{-1}\gamma_k,\lambda_k] \in \mathfrak{R}$$

since $\mathfrak{R}$ is a right ideal. Thus $u_{i_k} \in U'$ which by 3.1 implies that $a \in \mathfrak{J}_{U'}(V)$. Therefore $\mathfrak{R} = \mathfrak{J}_{U'}(V)$.

I.3.5 DEFINITION. For $\mathfrak{R}$ a ring, let

$$\mathfrak{R}^2 = \{\sum_{i=1}^{n} a_i b_i \mid a_i, b_i \in \mathfrak{R}\}.$$

It is clear that $\mathfrak{R}^2$ is an ideal, and that $\mathfrak{R}^2 = 0$ if and only if $ab = 0$ for all $a,b \in \mathfrak{R}$. Such a ring is called a zero ring.

A ring $\mathfrak{R}$ is called simple if 0 and $\mathfrak{R}$ are its only ideals and $\mathfrak{R}^2 \neq 0$. An element $a$ of a semigroup (or a ring) $S$ is said to be regular if $a = aba$ for some $b \in S$. A semigroup (or a ring) $S$ is regular if all its elements are regular. This concept was introduced by von Neuman [1]; regular rings are also known as von Neumann (regular) rings.

I.3.6 LEMMA. $\mathfrak{F}_U(V)$ is a regular simple ring.

PROOF. Let $0 \neq a = \sum_{k=1}^{n} [i_k, \gamma_k, \lambda_k] \in \mathfrak{F}_U(V)$ with $n$ minimal. Then both $\{u_{i_1}, \ldots, u_{i_n}\}$ and $\{v_{\lambda_1}, \ldots, v_{\lambda_n}\}$ are linearly independent by 2.5. There exist vectors $v_{\mu_k} \in V$ and $u_{j_p} \in U$ such that $(v_{\mu_k}, u_{i_p}) = \delta_{kp}\sigma_k$, $(v_{\lambda_k}, u_{j_p}) = \delta_{kp}\tau_k$ for some $\sigma_k \neq 0$, $\tau_k \neq 0$, and $1 \leq k,p \leq n$. Hence

$$(\sum_{k=1}^{n} [i_k, \gamma_k, \lambda_k])(\sum_{p=1}^{n} [j_p, \tau_p^{-1}\gamma_p^{-1}\sigma_p^{-1}, \mu_p])(\sum_{k=1}^{n} [i_k, \gamma_k, \lambda_k]) = \sum_{k=1}^{n} [i_k, \gamma_k, \lambda_k]$$

and thus $a$ is regular.

If $I$ is a nonzero ideal of $\mathfrak{F}_U(V)$, then by 3.4 we have $I = \mathfrak{J}_{U'}(V)$ for some nonzero subspace $U'$ of $U$, and by the dual of 3.4, also $I = \mathfrak{J}_U(V')$ for some nonzero subspace $V'$ of $V$. Consequently, $I = \mathfrak{J}_{U'}(V) = \mathfrak{J}_U(V')$ so that $U' = U$, $V' = V$ and thus $I = \mathfrak{F}_U(V)$. Since $\mathfrak{F}_U(V)$ is regular, it is not a zero ring, and hence must be simple.

I.3.7 DEFINITION. For $\mathfrak{R}$ any ring, the set

$$\mathfrak{A}_\ell(\mathfrak{R}) = \{r \in \mathfrak{R} \mid ra = 0 \text{ for all } a \in \mathfrak{R}\}$$

is the left annihilator of $\mathfrak{R}$,

$$\mathfrak{A}_r(\mathfrak{R}) = \{r \in \mathfrak{R} \mid ar = 0 \text{ for all } a \in \mathfrak{R}\}$$

is the right annihilator of $\mathfrak{R}$.

It is clear that both $\mathfrak{A}_\ell(\mathfrak{R})$ and $\mathfrak{A}_r(\mathfrak{R})$ are ideals of $\mathfrak{R}$. The next result appears to be new.

1.3.8 THEOREM. The following statements concerning a subspace $U'$ of $U$ are equivalent.

i) $(U',V)$ is a dual pair.

ii) $\mathfrak{I}_{U'}(V)$ is a simple ring.

iii) $\mathfrak{A}_{\ell}(\mathfrak{I}_{U'}(V)) = 0$.

iv) $\mathfrak{I}_{U'}(V)$ is transitive.

v) $\mathfrak{I}_{U'}(V)$ is a regular ring.

If any of these conditions holds, then $\mathfrak{I}_{U'}(V) \cong \mathfrak{F}_{U'}(V)$.

PROOF. i) $\Rightarrow$ ii). Since $(U',V)$ is a dual pair, 3.2 implies that $\mathfrak{I}_{U'}(V) \cong \mathfrak{F}_{U'}(V)$. This proves the last statement of the theorem and by 3.6 implies that $\mathfrak{I}_{U'}(V)$ is simple.

ii) $\Rightarrow$ iii). By simplicity of $\mathfrak{I}_{U'}(V)$, we must have $\mathfrak{A}_{\ell}(\mathfrak{I}_{U'}(V)) = 0$.

iii) $\Rightarrow$ iv). Let $x,y \in V$ with $x \neq 0$. If $y = 0$, then $x0 = y$ where $0$ is the zero function. So suppose $y \neq 0$. Hence $x = \sigma v_{\lambda}$, $y = \tau v_{\mu}$ where $\sigma \neq 0$, $\tau \neq 0$. For any $i \in I_{U'}$, we have $[i,\sigma,\lambda] \in \mathfrak{I}_{U'}(V)$ and the hypothesis implies that $[i,\sigma,\lambda]a \neq 0$ for some $a \in \mathfrak{I}_{U'}(V)$. But then $[i,\sigma,\lambda][j,\gamma,\nu] \neq 0$ for some $[j,\gamma,\nu] \in \mathfrak{I}_{U'}(V)$. It follows that $j \in I_{U'}$ and $(v_{\lambda},u_j) \neq 0$. Thus

$$x[j,(v_{\lambda},u_j)^{-1}\sigma^{-1}\tau,\mu] = \sigma(v_{\lambda},u_j)(v_{\lambda},u_j)^{-1}\sigma^{-1}\tau v_{\mu} = y$$

where $[j,(v_{\lambda},u_j)\sigma^{-1}\tau,\mu] \in \mathfrak{I}_{U'}(V)$. Hence $\mathfrak{I}_{U'}(V)$ is transitive.

iv) $\Rightarrow$ v). Since $\mathfrak{I}_{U'}(V)$ is a nonzero right ideal, it contains a linear transformation of rank 1. Thus by 2.12, there exists a t-subspace $U''$ of $V^*$ such that $\mathfrak{I}_{U'}(V) = \mathfrak{F}_{U''}(V)$ since all elements of $\mathfrak{I}_{U'}(V)$ are of finite rank. But $\mathfrak{F}_{U''}(V)$ is regular by 3.6, so $\mathfrak{I}_{U'}(V)$ is regular.

v) $\Rightarrow$ i). Let $u_i \in U'$, $\gamma \in \Delta^{-}$, $v_{\lambda} \in V$. Then $[i,\gamma,\lambda] \in \mathfrak{I}_{U'}(V)$, and by regularity, there exists $a \in \mathfrak{I}_{U'}(V)$ such that $[i,\gamma,\lambda] = [i,\gamma,\lambda]a[i,\gamma,\lambda]$. Hence $[i,\gamma,\lambda]a \neq 0$ and thus there exists $u_j \in U'$ such that $(v_{\lambda},u_j) \neq 0$, which implies that $(U',V)$ is a dual pair.

We consider next principal ideals.

1.3.9 THEOREM. For any $a \in \mathfrak{F}$, the principal right ideal generated by $a$ equals $a\mathfrak{F} = \mathfrak{I}_{a^*U}(V)$.

PROOF. If $a = 0$, the assertion is trivial, so let $0 \neq a = \sum_{k=1}^{n} [i_k,\gamma_k,\lambda_k]$ with $n$ minimal. Then a typical element of $a\mathfrak{F}$ is of the form

$$a\sum_{t=1}^{m}[j_t,\delta_t,\mu_t] = (\sum_{k=1}^{n}[i_k,\gamma_k,\lambda_k])(\sum_{t=1}^{m}[j_t,\delta_t,\mu_t])$$
$$= \sum_{k=1}^{n}\sum_{t=1}^{m}[i_k,\gamma_k(v_{\lambda_k},u_{j_t})\delta_t,\mu_t] \tag{3}$$

with $j_t \in I_U$, $\delta_t \in \Delta$, $\mu_t \in I_V$. On the other hand, a typical element of $\mathfrak{I}_{a^*U}(V)$ is of the form $\sum_{q=1}^{p}[\ell_q,\sigma_q,\nu_q]$ with $u_{\ell_q} \in a^*U$. Hence for $1 \leq q \leq p$, we have $u_{\ell_q} = a^*(u_{s_q}\tau_q)$ for some $u_{s_q} \in U$, $\tau_q \in \Delta$ so that

$$u_{\ell_q} = (a^*u_{s_q})\tau_q = \sum_{k=1}^{n}[i_k,\gamma_k,\lambda_k]^*u_{s_q}\tau_q = \sum_{k=1}^{n}u_{i_k}\gamma_k(v_{\lambda_k},u_{s_q})\tau_q.$$

Consequently, for any $x \in V$, we have

$$x(\sum_{q=1}^{p}[\ell_q,\sigma_q,\nu_q]) = \sum_{q=1}^{p}(x,u_{\ell_q})\sigma_q v_{\nu_q} = \sum_{q=1}^{p}\sum_{k=1}^{n}(x,u_{i_k})\gamma_k(v_{\lambda_k},u_{s_q})\tau_q\sigma_q v_{\nu_q}$$
$$= x(\sum_{q=1}^{p}\sum_{k=1}^{n}[i_k,\gamma_k(v_{\lambda_k},u_{s_q})\tau_q\sigma_q,\nu_q])$$

so that

$$\sum_{q=1}^{p}[\ell_q,\sigma_q,\nu_q] = \sum_{k=1}^{n}\sum_{q=1}^{p}[i_k,\gamma_k(v_{\lambda_k},u_{s_q})\tau_q\sigma_q,\nu_q] \tag{4}$$

with $s_q \in I_U$, $\tau_q \in \Delta^-$, $\nu_q \in I_V$. A comparison of (3) and (4) quickly shows that $a\mathfrak{F} = \mathfrak{I}_{a^*U}(V)$.

The set $a\mathfrak{F}$ is indeed the principal right ideal generated by $a$ since $a\mathfrak{F}$ is clearly a right ideal and $a \in a\mathfrak{F}$ since $\mathfrak{F}$ is regular by 3.6.

The next corollary is a supplement to 3.4.

I.3.10 <u>COROLLARY</u>. <u>The isomorphism</u> $\chi: U' \rightarrow \mathfrak{I}_{U'}(V)$ <u>maps the lattice of all finite dimensional subspaces of</u> $U$ <u>onto the lattice of all principal right ideals of</u> $\mathfrak{F}_U(V)$.

<u>PROOF</u>. Let $U'$ be an n-dimensional subspace of $U$ where $n > 0$. Then $U'$ has a basis of the form $u_{i_1},\ldots,u_{i_n}$. Let $v_{\lambda_k} \in V$ be such that $(v_{\lambda_k},u_{i_p}) = \delta_{kp}\sigma_k$ for some $\sigma_k \neq 0$ and $1 \leq k,p \leq n$, and let $a = \sum_{k=1}^{n}[i_k,\sigma_k^{-1},\lambda_k]$. It is clear that $a^*U \subseteq U'$. On the other hand, if $u \in U'$, then $u = \sum_{k=1}^{n}u_{i_k}\gamma_k$ and thus

$$a^*u = \sum_{k=1}^{n}u_{i_k}\sigma_k^{-1}(v_{\lambda_k},u_{i_k})\gamma_k = \sum_{k=1}^{n}u_{i_k}\gamma_k = u.$$

Hence $U' \subseteq a^*U$ and thus $a^*U = U'$. By 3.4 and 3.9, we have

$$U'\chi = \mathfrak{J}_{U'}(V) = \mathfrak{J}_{a^*U}(V) = a\mathfrak{J}.$$

Conversely, if $\mathfrak{R}$ is a principal right ideal of $\mathfrak{J}$, then $\mathfrak{R} = a\mathfrak{J}$ for some $a \in \mathfrak{J}$. Thus by 3.9, we obtain $(a^*U)\chi = \mathfrak{J}_{a^*U}(V) = \mathfrak{R}$.

I.3.11 REMARK. It follows from 3.10 that to 1-dimensional subspaces of U correspond minimal right ideals, and from its proof, that they are all the sets of the form

$$R_i = \{[i,\gamma,\lambda] \mid \gamma \in \Delta,\ \lambda \in I_V\} \qquad (i \in I_U).$$

Similarly all minimal left ideals are given by

$$L_\lambda = \{[i,\gamma,\lambda] \mid i \in I_U,\ \gamma \in \Delta\} \qquad (\lambda \in I_V).$$

Further, $R_i = [i,\sigma^{-1},\lambda]\mathfrak{J}$, $L_\lambda = \mathfrak{J}[i,\sigma^{-1},\lambda]$ where $(v_\lambda,u_i) = \sigma \neq 0$ and $[i,\sigma^{-1},\lambda]$ is an idempotent.

I.3.12 PROPOSITION. If V is a finite dimensional vector space, then $V^*$ has no proper t-subspaces.

PROOF. Let $\dim V = n$, let $x_1,\ldots,x_n$ be a linearly independent set of vectors in V, and let U be a t-subspace of $V^*$. There exist $f_j \in U$ such that $x_i f_j = \delta_{ij}$ for $1 \leq i,j \leq n$ and $f_1,\ldots,f_n$ are linearly independent. Hence $\dim U \geq n$ and since U is a subspace of $V^*$ and $\dim V^* = n$, it follows that $U = V^*$.

I.3.13 COROLLARY. If (U,V) is a dual pair and either U or V is of finite dimension n, then the other is also and each can be identified with the dual of the other.

For $\mathfrak{R}$ any ring and n a positive integer, by $\mathfrak{R}_n$ denote the ring of all $n \times n$ matrices over $\mathfrak{R}$; also write d.c.c. instead of "descending chain condition".

I.3.14 COROLLARY. If $\mathfrak{R}$ is a subring of $\mathfrak{L}_U(\Delta,V)$ containing $\mathfrak{J}_U(\Delta,V)$ and $\mathfrak{R}$ satisfies d.c.c. for either left or right ideals, then both U and V are of the same finite dimension, $\mathfrak{R} = \mathfrak{L}(\Delta,V) \cong \Delta_{\dim V}$, $\mathfrak{R}$ satisfies d.c.c. for both left and right ideals, and all its left and right ideals are principal.

PROOF. Let $U_1 \supseteq U_2 \supseteq \ldots$ be a descending chain of subspaces of U. Then by 3.4 we have that $\mathfrak{J}_{U_1}(V) \supseteq \mathfrak{J}_{U_2}(V) \supseteq \ldots$ is a descending chain of right ideals of $\mathfrak{J}_U(V)$. Since each $\mathfrak{J}_{U_i}(V)$ is also a right ideal of $\mathfrak{R}$ and $\mathfrak{R}$ satisfies d.c.c. for right ideals, there exists n such that $\mathfrak{J}_{U_i}(V) = \mathfrak{J}_{U_n}(V)$ for all $i \geq n$ and thus by 3.4, $U_i = U_n$ for all $i \geq n$. Consequently by 3.4, or its dual, either U or V satisfies d.c.c. for subspaces, so either U or V is finite dimensional, and hence by 3.13 both are of the same finite dimension, and we may write $V^* = U$. Using 1.4 we obtain $\mathfrak{L}_U(V) = \mathfrak{L}(V)$ and hence $\mathfrak{R} = \mathfrak{L}(V) = \mathfrak{F}(V)$. Again by 3.4 and its dual, $\mathfrak{R}$ satisfies d.c.c. for both left and right ideals. The last statement follows from 3.10.

I.3.15 COROLLARY. *$\mathfrak{F}_U(V)$ satisfies d.c.c. for principal left and principal right ideals; it satisfies d.c.c. for right ideals (or left ideals) if and only if $V$ (and thus also $U$) is finite dimensional.*

PROOF. The first statement follows from 3.10 since a vector space evidently satisfies d.c.c. for finite dimensional subspaces. The direct part of the second statement follows from 3.14, and the converse from 3.10 and the first statement. Here, as before, the "left" part of the statements follows from dual statements.

I.3.16 LEMMA. *For $a,b \in \mathfrak{L}(U)$, if $aU \subseteq bU$, then $a = bc$ for some $c \in \mathfrak{L}(U)$.*

PROOF. Let $A$ be a basis of $aU$. For every $u \in A$ choose an element $\bar{u}$ of $U$ such that $b\bar{u} = u$ (possible since $A \subseteq aU \subseteq bU$). Extend $A$ to a basis $B$ of $U$. On $A$ define a function $d$ by letting $du = \bar{u}$, let $d$ map $B \setminus A$ onto zero, and extend $d$ to a linear transformation on $U$. For any $x \in U$, we have $ax = \sum_{k=1}^{n} u_k\sigma_k$ for some $u_k \in A$, $\sigma_k \in \Delta$, and thus

$$bd(ax) = bd(\sum_{k=1}^{n} u_k\sigma_k) = b[\sum_{k=1}^{n} (du_k)\sigma_k] = b(\sum_{k=1}^{n} \bar{u}_k\sigma_k)$$
$$= \sum_{k=1}^{n} (b\bar{u}_k)\sigma_k = \sum_{k=1}^{n} u_k\sigma_k = ax.$$

Letting $c = da$, we get $a = bc$ with $c \in \mathfrak{L}(U)$.

For $S$ a semigroup (or ring) let $E_S$ be the set of all idempotents of $S$ partially ordered by

$$e \leq f \text{ if and only if } e = ef = fe.$$

Let $\mathfrak{P}(U,V)$ be the set of all dual pairs $(U',V')$, where $U'$ is a subspace of $U$, $V'$ is a subspace of $V$, $\dim V'$ is finite, under the induced bilinear form, including the pair $(0,0)$, partially ordered by

$$(U',V') \leq (U'',V'') \text{ if and only if } U' \subseteq U'',\ V' \subseteq V''.$$

If $(A,\leq)$ and $(B,\leq)$ are partially ordered sets, then an *order isomorphism* of $(A,\leq)$ onto $(B,\leq)$ is a one-to-one function $\varphi$ mapping $A$ onto $B$ such that $a \leq a'$ if and only if $a\varphi \leq a'\varphi$.

The next result is new.

I.3.17 THEOREM. *The function $\varphi$ defined by*

$$\varphi : e \to (e^*U, Ve) \qquad (e \in E_{\mathfrak{F}_U(V)})$$

*is an order isomorphism of $E_{\mathfrak{F}_U(V)}$ onto $\mathfrak{P}(U,V)$.*

PROOF. Let $0 \neq e = \sum_{k=1}^{n} [i_k, \gamma_k, \lambda_k] \in E_{\mathfrak{F}}$ with $n$ minimal Then

$$\sum_{k=1}^{n} [i_k, \gamma_k, \lambda_k] = e = e^2 = \sum_{k=1}^{n} \sum_{p=1}^{n} [i_k, \gamma_k (v_{\lambda_k}, u_{i_p}) \gamma_p, \lambda_p].$$

There exist $x_k \in V$ such that $(x_k, u_{i_p}) = \delta_{kp}$ for $1 \leq k,p \leq n$ so that

$$\gamma_k v_{\lambda_k} = x_k e = x_k e^2 = \sum_{p=1}^{n} \gamma_k (v_{\lambda_k}, u_{i_p}) \gamma_p v_{\lambda_p}.$$

By 2.5, the vectors $v_{\lambda_1}, \ldots, v_{\lambda_n}$ are linearly independent so we must have $(v_{\lambda_k}, u_{i_p}) = \delta_{kp} \gamma_k^{-1}$. Since by 2.5, $v_{\lambda_1}, \ldots, v_{\lambda_n}$ form a basis of $Ve$ and $u_{i_1}, \ldots, u_{i_n}$ form a basis of $e^*U$, it follows that $(e^*U, Ve)$ is a dual pair.

Let $(U', V')$ be an element of $\mathfrak{P}(U,V)$ and let $v_{\lambda_1}, \ldots, v_{\lambda_n}$ be a basis of $V'$. There exist $u_{i_p} \in U'$ such that $(v_{\lambda_k}, u_{i_p}) = \delta_{kp} \tau_k$ for some $\tau_k \neq 0$ and $1 \leq k,p \leq n$, and $u_{i_1}, \ldots, u_{i_n}$ are linearly independent. Since $\dim V' = n$, by 3.13 also $\dim U' = n$ and hence the set $u_{i_1}, \ldots, u_{i_n}$ forms a basis of $U'$. Letting $e = \sum_{k=1}^{n} [i_k, \tau_k^{-1}, \lambda_k]$, we see that $e^2 = e$ since $(v_{\lambda_k}, u_{i_p}) = \delta_{kp} \tau_k$. Evidently $Ve \subseteq V'$. If $x \in V'$, then $x = \sum_{k=1}^{n} \sigma_k v_{\lambda_k}$ for some $\sigma_k \in \Delta$ and thus

$$xe = \sum_{k=1}^{n} (x, u_{i_k}) \tau_k^{-1} v_{\lambda_k} = \sum_{k=1}^{n} \left(\sum_{p=1}^{n} \sigma_p v_{\lambda_p}, u_{i_k}\right) \tau_k^{-1} v_{\lambda_k} = \sum_{k=1}^{n} \sigma_k v_{\lambda_k} = x$$

so that $Ve = V'$. Similarly $e^*U = U'$ and therefore $e\varphi = (U', V')$.

Let $e, f \in E_{\mathfrak{F}}$ and assume that $e\varphi = f\varphi$. Then $Ve = Vf$ and $e^*U = f^*U$ which by 3.16 and its dual yields $f = ae$ and $e^* = f^*b$ for some $a \in \mathfrak{L}(V)$ and $b \in \mathfrak{L}(U)$. Hence $f = ae = (ae)e = fe$ and $e^* = f^*b = f^*(f^*b) = f^*e^*$. The last equation implies $e = fe$ which together with the first gives $e = f$.

Let $e, f \in E_{\mathfrak{F}}$. The relation $e \leq f$ means that $e = ef = fe$ which can be written as $e^* = f^*e^*$, $e = ef$ which is in turn equivalent to: $e^* = f^*c$, $e = df$ for some $c \in \mathfrak{L}(U)$, $d \in \mathfrak{L}(V)$. By 3.16 and its dual, the last statement is equivalent to: $e^*U \subseteq f^*U$, $Ve \subseteq Vf$ which means that $e\varphi \leq f\varphi$.

I.3.18 PROPOSITION. *For any* $0 \neq e \in E_{\mathfrak{F}}$, *we have* $e\mathfrak{F}e \cong \Delta_{\text{rank } e}$.

PROOF. Using 3.9 and its dual, (1), 3.17, 3.2, 3.13, and 1.4 we obtain

$$e\mathfrak{F}e = e\mathfrak{F} \cap \mathfrak{F}e = \mathfrak{I}_{e^*U}(V) \cap \mathfrak{I}_U(Ve) = \mathfrak{I}_{e^*U}(Ve)$$

$$\cong \mathfrak{F}_{e^*U}(Ve) = \mathfrak{L}_{(Ve)^*}(Ve) = \mathfrak{L}(Ve) \cong \Delta_{\text{rank } e}.$$

I.3.19 DEFINITION. A ring $\mathfrak{R}$ is a locally matrix ring over a ring $\mathfrak{T}$ if for every finite subset F of $\mathfrak{R}$ there exists a subring $\mathfrak{S}$ of $\mathfrak{R}$ containing F and isomorphic to $\mathfrak{T}_n$ for some positive integer n.

I.3.20 THEOREM. $\mathfrak{F}_U(\Delta,V)$ is a locally matrix ring over $\Delta$.

PROOF. By 3.18, it suffices to show that for every finite set $a_1,a_2,\ldots,a_n$ of elements of $\mathfrak{F}$, there exists an idempotent $e \in \mathfrak{F}$ such that $a_1,\ldots,a_n \in e\mathfrak{F}e$.

Let $U'$ be the subspace of U generated by the set $\bigcup_{i=1}^{n} a_i^*U$. Then $U'$ is finite dimensional and the linear transformation a constructed in the first part of the proof of 3.10 is an idempotent in $\mathfrak{F}$ with the property $a\mathfrak{F} = \mathfrak{I}_{U'}(V)$. It follows from 3.9 that $a_i \in \mathfrak{I}_{U'}(V)$ and hence $a_i = ab_i$ for some $b_i \in \mathfrak{F}$. Consequently $aa_i = a(ab_i) = ab_i = a_i$ for $1 \leq i \leq n$. Next let $V'$ be the subspace of V generated by $(\bigcup_{i=1}^{n} Va_i) \cup Va$. By a dual argument, we conclude that there exists an idempotent $b \in \mathfrak{F}$ such that $\mathfrak{F}b = \mathfrak{I}_U(V')$, $a_i = a_ib$ for $1 \leq i \leq n$ and $a = ab$. Letting $e = a+b-ba$, we obtain $e^2 = e$, $ea = a$, $be = a$, and hence

$$a_i = aa_ib = (ea)a_i(be) \in e\mathfrak{F}e \quad \text{for} \quad 1 \leq i \leq n.$$

The above theorem in conjunction with II.2.8 implies Litoff's theorem, see Jacobson ([7], Chapter IX, Section 15, Theorem 3). Different proofs of Litoff's theorem can be found in Faith and Utumi [1] and Steinfeld [1]. We will prove a generalization in II.2.19 due to Hotzel.

I.3.21 EXERCISES.

i) Find all zero rings without proper ideals.

ii) Show that for any semigroup (or ring) S, the following are equivalent.

a) S is regular,

b) every principal right ideal is generated by idempotent,

c) every principal left ideal is generated by an idempotent.

iii) Let $\mathfrak{R}$ be a ring and $\mathfrak{S}$ be a right ideal of $\mathfrak{R}$ which is a regular ring. Show that every right ideal of $\mathfrak{S}$ is a right ideal of $\mathfrak{R}$.

iv) Prove that $\mathfrak{S}(V)$ is a regular semigroup.

v) Show that dense subrings of $\mathfrak{F}_U(V)$ are precisely the simple right ideals of $\mathfrak{F}_U(V)$.

vi) Let S be a semigroup and e be an idempotent of S. Show that

$$G_e = \{x \in S \mid x = ex = xe,\ e \in xS \cap Sx\}$$

is a subgroup of S containing every subgroup of S having e as its identity; $G_e$ is called a maximal subgroup of S.

vii) Let $e \in E_{\mathfrak{L}(V)}$. Prove that $1 - e$ is an idempotent, $N_e = V(1 - e)$, $V = N_e \oplus Ve$, $G_e = \{a \in \mathfrak{L}(V) \mid N_a = N_e, Va = Ve\}$ is the maximal subgroup of $\mathfrak{S}(V)$ having $e$ as its identity and $G$ is isomorphic to the group of invertible linear transformations on $Ve$.

viii) For $a \in \mathfrak{F}_U(V)$, show that the following are equivalent.

a) $a$ is contained in a subgroup of $\mathfrak{M}\mathfrak{F}_U(V)$,

b) $(a^*U, Va)$ is a dual pair,

c) $V = N_a \oplus Va$.

ix) For any vector space $(\Delta, V)$ and $x \in V$, let $\hat{x}: V^* \to \Delta$ be defined by $\hat{x}f = xf$ $(f \in V^*)$. Prove that the mapping $\pi: x \to \hat{x}$ is an isomorphism of $V$ into $V^{**}$; $\pi$ is called the <u>natural isomorphism</u> of $V$ into $V^{**}$.

x) Prove that the following statements concerning a vector space $V$ are equivalent.

a) $V$ is finite dimensional.

b) $V^*$ is finite dimensional.

c) $V^*$ has no proper t-subspaces.

d) $\pi: x \to \hat{x}$ maps $V$ onto $V^{**}$.

e) $\mathfrak{L}(V)$ is a simple ring.

(Hint: If $V$ is infinite dimensional, show that c), d), e) fail.)

xi) Let $\Delta$ be a division ring and $n$ be a positive integer. For the ring $\Delta_n$

a) find all left (right) ideals,

b) which of them are minimal, maximal, principal, simple?

c) show that $\Delta_n$ is a simple, regular ring satisfying d.c.c. and a.c.c. for both left and right ideals.

Use the results in the text but also give a direct proof whenever you can.

xii) Let $e$ and $f$ be idempotents in a ring $\mathfrak{R}$. Show that $e \leq f$ if and only if $f = e + g$ for some idempotent $g$ in $\mathfrak{R}$ orthogonal to $e$.

xiii) If $\mathfrak{R}$ is a dense subring of $\mathfrak{L}(V)$ containing a nonzero transformation of finite rank, prove that $\mathfrak{R} \cap \mathfrak{F}(V)$ is an ideal of $\mathfrak{R}$ contained in every nonzero ideal of $\mathfrak{R}$.

For more information on this subject see Behrens ([1], Chapter II, Section 3), Dieudonné [2], Gluskin [4], Jacobson [5], ([6], Chapter VIII, Sections 3,4), ([7], Chapter IV, Sections 15,16), Ribenboim ([1], Chapter III, Sections 2,3), Rosenberg [1].

## I.4 IDEALS OF $\mathcal{S}(V)$ AND PRINCIPAL FACTORS OF $\mathfrak{M}\mathfrak{J}_U(V)$

I.4.1 <u>DEFINITION</u>. A nonempty subset $I$ of a semigroup $S$ is a <u>left</u> <u>(right)</u> <u>ideal</u> if $x \in S$, $y \in I$ implies $xy \in I$ $(yx \in I)$. (Hence $I$ is an ideal of $S$ if and only if it is both a left and a right ideal of $S$.)

If $I$ is an ideal of $S$ and $I \neq S$, then the set $Q = (S \setminus I) \cup 0$, where $0$ is not an element of $S \setminus I$, with the multiplication: for $x,y \in S \setminus I$,

$$x * y = \begin{cases} xy & \text{if } xy \notin I \\ 0 & \text{if } xy \in I \end{cases}$$

$$x * 0 = 0 * x = 0 * 0 = 0$$

is a semigroup called the <u>Rees quotient</u> of $S$ relative to the ideal $I$ and is denoted by $S/I$.

Let $J(a)$ denote the principal ideal generated by an element $a$ of a semigroup $S$ (i.e., the smallest ideal of $S$ containing $a$). It is easy to see that

$$J(a) = a \cup aS \cup Sa \cup SaS,$$

and if $S$ has an identity, then $J(a) = SaS$.

Let $I(a) = \{b \in J(a) \mid J(b) \neq J(a)\}$. One verifies easily that $I(a)$ is either empty or is an ideal of $S$. The Rees quotient $J(a)/I(a)$, where $I(a)$ is regarded as an ideal of $J(a)$, is called the <u>principal</u> <u>factor</u> of $S$ containing $a$ (if $I(a) = \emptyset$, let $J(a)/I(a) = J(a)$).

For $\mathfrak{c}$ a cardinal number, let $\mathfrak{c}^+$ denote the least cardinal number greater than $\mathfrak{c}$, and for a $\mathfrak{c}$-dimensional vector space $(\Delta,V)$, let

$$\mathcal{S}_{\mathfrak{a}}(\Delta,V) = \{b \in \mathcal{S}(\Delta,V) \mid \operatorname{rank} b < \mathfrak{a}\}$$

together with the multiplication inherited from $\mathcal{S}(\Delta,V)$, where $\mathfrak{a}$ is a cardinal number, $0 < \mathfrak{a} \leq \mathfrak{c}^+$. As usual, we will omit $\Delta$ if there is no need for stressing the division ring. The next result appears to be new.

I.4.2 <u>THEOREM</u>. <u>The set</u> $\{\mathcal{S}_{\mathfrak{a}}(V) \mid 0 < \mathfrak{a} \leq \mathfrak{c}^+\}$ <u>constitutes the set of all ideals of</u> $\mathcal{S}(V)$ <u>for a</u> $\mathfrak{c}$<u>-dimensional vector space</u> $V$, <u>and</u> $\mathcal{S}_{\mathfrak{a}}(V) \neq \mathcal{S}_{\mathfrak{b}}(V)$ <u>if</u> $\mathfrak{a} \neq \mathfrak{b}$. <u>In addition</u>, $\mathcal{S}_{\mathfrak{a}}(V)$ <u>is a principal ideal if and only if</u> $\mathfrak{a}$ <u>is not a limit cardinal</u>.

<u>PROOF</u>. That $\mathcal{S}_{\mathfrak{a}}(V)$ is an ideal of $\mathcal{S}(V)$ follows from 1.3. We will first find principal ideals. Let $b \in \mathcal{S}(V)$ and let $\operatorname{rank} b = \mathfrak{b}$. The principal ideal $J(b)$ generated by $b$ consists of elements of the form $cbd$ where $c,d \in \mathcal{S}(V)$ since $\mathcal{S}(V)$ has an identity. Again by 1.3, such an element is contained in $\mathcal{S}_{\mathfrak{b}^+}(V)$.

Conversely, let $a \in \mathcal{S}_{\mathfrak{b}^+}(V)$, then $\operatorname{rank} a \leq \mathfrak{b}$. Let $A$ be a basis of $Va$ and extend $A$ to a basis $B$ of $V$. Let $\theta$ be a one-to-one function mapping $A$ into $Vb$ such that the set $A\theta$ is linearly independent (possible since $\operatorname{rank} a \leq \operatorname{rank} b$ by hypothesis). For each $x \in A\theta$, choose an element $\bar{x} \in V$ such that $\bar{x}b = x$ (possible since $A\theta \subseteq Vb$).

Now define a function $d$ by letting $xd = \overline{x\theta}$ for each $x \in A$, define $d$ arbitrarily on $B \setminus A$, and extend it to a linear transformation on $V$. Extend $A\theta$ to a basis $C$ of $V$ Define a function $c$ by letting $xc = x\theta^{-1}$ if $x \in A\theta$, define $c$ arbitrarily on the set $C \setminus A\theta$, and extend it to a linear transformation on $V$. Then for any $x \in V$, we have $xa = \sum_{i=1}^{n} \sigma_i y_i$ for some $\sigma_i \in \Delta$, $y_i \in A$, and thus

$$(xa)dbc = (\sum_{i=1}^{n} \sigma_i (y_i d))bc = (\sum_{i=1}^{n} \sigma_i (\overline{y_i \theta} b))c = \sum_{i=1}^{n} \sigma_i (y_i \theta)c = \sum_{i=1}^{n} \sigma_i y_i = xa.$$

Consequently, $a = adbc \in J(b)$ and thus $J(b) = \mathcal{S}_{\mathfrak{b}^+}(V)$.

Next let $J$ be any ideal of $\mathcal{S}(V)$, and let $\mathfrak{a}$ be the least cardinal number larger than $\operatorname{rank} b$ for all $b \in J$. If $\mathfrak{a}$ is not a limit cardinal, then $\mathfrak{b}^+ = \mathfrak{a}$ for some $\mathfrak{b}$, and hence $J = J(b)$ for any $b$ with $\operatorname{rank} b = \mathfrak{b}$ by the above. If $\mathfrak{a}$ is a limit cardinal and $\operatorname{rank} b < \mathfrak{a}$, then there exists $c \in J$ such that $\operatorname{rank} b \leq \operatorname{rank} c$. But then $b \in J(c) \subseteq J$ which proves that $\mathcal{S}_{\mathfrak{a}}(V) \subseteq J$. Since the opposite inclusion is trivial, we have that $J = \mathcal{S}_{\mathfrak{a}}(V)$. The proof of the remaining assertions of the theorem is left as an exercise.

Let $\mathcal{L}_{\mathfrak{a}}(V)$ denote the semigroup $\mathcal{S}_{\mathfrak{a}}(V)$ provided with the addition inherited from $\mathcal{L}(V)$. The next corollary can be found in Baer ([1], Chapter V, Appendix 1), Jacobson ([6], Chapter IX, Section 9),([7], Chapter IV, Section 17).

I.4.3 COROLLARY. *For a $\mathfrak{c}$-dimensional vector space $V$, the set*

$$\{\mathcal{L}_{\mathfrak{a}}(V) \mid \mathfrak{a} = \mathfrak{c}^+ \text{ or } \mathfrak{a} \leq \mathfrak{c}^+ \text{ and } \mathfrak{a} \text{ is an infinite cardinal}\}$$

*constitutes the set of all ideals of $\mathcal{L}(V)$.*

PROOF. By 4.2 and 1.3, these are ideals of $\mathcal{L}(V)$. Since by 2.6, $C(\mathcal{F}_{2,V^*}(V)) = \mathcal{F}(V)$, no $\mathcal{S}_{\mathfrak{a}}(V)$ for $\mathfrak{a} < \mathfrak{c}^+$ and $\mathfrak{a}$ finite is closed under addition. So by 4.2, $\mathcal{L}(V)$ has no other ideals than those listed in the statement of the corollary.

We have seen in 2.2 that the nonzero elements of $\mathcal{F}_{2,U}(V)$ can be uniquely written in the form $[i,\gamma,\lambda]$ where $i \in I_U$, $\gamma \in \Delta^-$, $\lambda \in I_V$, the zero function is written in the form $[i,0,\lambda]$ for any $i \in I_U$, $\lambda \in I_V$, and the multiplication is given by

$$[i,\gamma,\lambda][j,\delta,\mu] = [i,\gamma(v_\lambda,u_j)\delta,\mu].$$

This is a special case of the following construction.

I.4.4 DEFINITION. Let $I$ and $\Lambda$ be nonempty sets, $G$ be a group, $0$ be an element not in $G$, and let $G^0 = G \cup 0$ with $0$ acting as zero. Let $P: \Lambda \times I \to G^0$, with $p_{\lambda i}$ the value at $(\lambda, i)$, be any function such that for every $i \in I$, there exists $\lambda \in \Lambda$ such that $p_{\lambda i} \neq 0$, and for every $\lambda \in \Lambda$ there exists $i \in I$ such that $p_{\lambda i} \neq 0$. By $\mathfrak{M}^0(I,G,\Lambda;P)$ denote set $(I \times G \times \Lambda) \cup 0$ with multiplication

$$(i,g,\lambda)(j,h,\mu) = (i, gp_{\lambda j}h, \mu)$$

where we write $(i,0,\lambda)$ instead of $0$. Then $S = \mathfrak{M}^0(I,G,\Lambda;P)$ is a semigroup called a Rees matrix semigroup. Here $P = (p_{\lambda i})$ can be considered as a matrix and is called the sandwich matrix of $S$.

We will prove next that $\mathfrak{I}_{n+1,U}(V)/\mathfrak{I}_{n,U}(V)$ is isomorphic to a Rees matrix semigroup for all finite $n \leq \dim V$.

I.4.5 DEFINITION. For $\Delta$ a division ring and $n$ a positive integer, the group $L_n(\Delta)$ of $n \times n$ invertible matrices over $\Delta$ (under multiplication) is called the general (or full) linear group of degree $n$ over $\Delta$.

The following result is new.

I.4.6 THEOREM. *Let $(\Delta, V)$ be a vector space, $U$ a t-subspace of $V^*$, $n$ a positive integer such that $n \leq \dim V$. Let*

$$n_V = \{B \mid B \text{ is an } n\text{-dimensional subspace of } V\},$$

$$n_U = \{A \mid A \text{ is an } n\text{-dimensional subspace of } U\}.$$

*For every $B \in n_V$ fix a basis $v_1^B, \ldots, v_n^B$ of $B$, and for every $A \in n_U$ fix a basis $u_1^A, \ldots, u_n^A$ of $A$. Let*

$$G_{n,U}(\Delta,V) = \mathfrak{M}^0(n_U, L_n(\Delta), n_V; P)$$

*where $P = (p_{BA})$ with*

$$p_{BA} = \begin{bmatrix} v_1^B \\ \vdots \\ v_n^B \end{bmatrix} [u_1^A \ldots u_n^A] = \begin{bmatrix} v_1^B u_1^A & v_1^B u_2^A & \cdots & v_1^B u_n^A \\ \vdots & \vdots & & \\ v_n^B u_1^A & v_n^B u_2^A & \cdots & v_n^B u_n^A \end{bmatrix}$$

*if this matrix is invertible, and $p_{BA} = 0$ otherwise. Finally, let*

$$Q_{n,U}(\Delta,V) = \mathfrak{I}_{n+1,U}(\Delta,V)/\mathfrak{I}_{n,U}(\Delta,V),$$

*let $\theta$ be the function defined by: $\theta$ maps the zero of $Q_{n,U}(\Delta,V)$ onto the zero of $G_{n,U}(\Delta,V)$ and*

$$\theta: b \to (b^*U, [\beta_{ij}], Vb) \qquad (b \in Q_{n,U}(\Delta,V) \setminus 0)$$

<u>where</u> $b^*$ <u>is the conjugate of</u> $b$ <u>and</u>

$$xb = \sum_{j=1}^{n} (xf_j)v_j^{Vb}, \quad f_j = \sum_{i=1}^{n} u_i^{b^*U}\beta_{ij} \quad \text{for } 1 \leq j \leq n \quad (x \in V). \tag{1}$$

<u>Then</u> $\theta$ <u>is an isomorphism of</u> $Q_{n,U}(\Delta,V)$ <u>onto</u> $G_{n,U}(\Delta,V)$.

<u>PROOF</u>. 1. Let $G = G_{n,U}(\Delta,V)$, $Q = Q_{n,U}(\Delta,V)$, and write

$$[v_i^B u_j^A] = \begin{bmatrix} v_1^B \\ \vdots \\ v_n^B \end{bmatrix} [u_1^A \dots u_n^A].$$

In order to prove that $\theta$ maps $Q$ into $G$, we must show that for $b \in Q \setminus 0$, we have that $b^*U \in n_U$ and $[\beta_{ij}]$ is invertible. By 1.5 and 2.5, respectively, we have $b^*U \subseteq U$ and $\dim(b^*U) = \operatorname{rank} b = n$ so that $b^*U \in n_U$.

Define the functions $f_j$ by the first formula in (1). We will show that $f_1,\dots,f_n$ form a basis of $b^*U$. In fact, since $v_1^{Vb},\dots,v_n^{Vb}$ is a basis of $Vb$, for any $x \in V$, $xb$ is a unique linear combination of the vectors $v_1^{Vb},\dots,v_n^{Vb}$. Then $f_j : V \to \Delta$, and from the linearity of $b$, it follows easily that each $f_j$ is linear. Thus $f_j \in V^*$ for $1 \leq j \leq n$. Further for any $x \in V$, $f \in V^*$, using the notation in (1), we obtain

$$x(b^*f) = (xb)f = (\sum_{j=1}^{n} (xf_j)v_j^{Vb})f = \sum_{j=1}^{n} (xf_j)(v_j^{Vb}f) = x(\sum_{j=1}^{n} f_j(v_j^{Vb}f)),$$

so that $b^*f = \sum_{j=1}^{n} f_j(v_j^{Vb}f)$. It follows that $b^*V^* \subseteq T$ where $T$ is the subspace of $V^*$ generated by $f_1,\dots,f_n$. Consequently $b^*U \subseteq b^*V^* \subseteq T$ where $\dim(b^*U) = n$, as we have seen above, and $T$ is generated by the $n$ vectors $f_1,\dots,f_n$. It follows that $b^*U = b^*V = T$ and $f_1,\dots,f_n$ form a basis of $b^*U$. Now the second formula in (1) implies that $[\beta_{ij}] \in L_n(\Delta)$. Therefore $\theta$ maps $Q$ into $G$.

2. Recall that the multiplication in $Q$ is given by: for $b,c \in Q \setminus 0$,

$$bc = \begin{cases} bc & \text{(the product in } \mathcal{S}(\Delta,V)) \text{ if } \operatorname{rank}(bc) = n \\ 0 & \text{otherwise} \end{cases}.$$

Hence let $b,c \in Q \setminus 0$ and $\operatorname{rank}(bc) = n$. Since $Vbc \subseteq Vc$ and $\dim Vbc = \dim Vc = n$, we have $Vbc = Vc$; similarly $b^*c^*U = b^*U$.

For $f_j$ and $[\beta_{ij}]$ as in (1), and

$$xc = \sum_{j=1}^{n} (xg_j)v_j^{Vc}, \quad g_j = \sum_{i=1}^{n} u_i^{c^*U}\gamma_{ij} \quad \text{for } 1 \leq j \leq n \quad (x \in V),$$

we then obtain

$$\begin{aligned}(xb)c &= \sum_{j=1}^{n}[(xb)(\sum_{i=1}^{n} u_i^{c^*U}\gamma_{ij})]v_j^{Vc}\\ &= \sum_{j=1}^{n}\{\sum_{k=1}^{n}[x(\sum_{m=1}^{n} u_m^{b^*U}\beta_{mk})]v_k^{Vb}\}(\sum_{i=1}^{n} u_i^{c^*U}\gamma_{ij})v_j^{Vc}\\ &= \sum_{j=1}^{n}\{x\sum_{m=1}^{n} u_m^{b^*U}[\sum_{k=1}^{n}\beta_{mk}(\sum_{i=1}^{n} v_k^{Vb}u_i^{c^*U})\gamma_{ij}]\}v_j^{Vc}\\ &= \sum_{j=1}^{n} x(\sum_{m=1}^{n} u_m^{b^*c^*U}\sigma_{mj})v_j^{Vbc} = x(bc) \qquad (2)\end{aligned}$$

where we have set

$$[\sigma_{ij}] = [\beta_{ij}][v_i^{Vb}u_j^{c^*U}][\gamma_{ij}].$$

By hypothesis $\operatorname{rank}(bc) = n$, so that (2) together with (1) yields

$$\theta: bc \to (b^*c^*U, [\sigma_{ij}], Vbc)$$

since the linear forms $f_j$ and the matrix $[\beta_{ij}]$ are unique. Consequently $[\sigma_{ij}]$ is invertible and thus also $[v_i^{Vb}u_j^{c^*U}]$ and we obtain

$$\begin{aligned}(b\theta)(c\theta) &= (b^*U,[\beta_{ij}],Vb)(c^*U,[\gamma_{ij}],Vc)\\ &= (b^*U,[\beta_{ij}][v_i^{Vb}u_j^{c^*U}][\gamma_{ij}],Vc)\\ &= (b^*c^*U,[\sigma_{ij}],Vbc) = (bc)\theta.\end{aligned}$$

Suppose next that $\operatorname{rank}(bc) < n$. Then $bc = 0$ in $Q$, so that $(bc)\theta = 0$. For $1 \le i \le n$, let $x_i$ be any vectors in $V$ with the property $x_i b = v_i^{Vb}$. Then the hypothesis $\operatorname{rank}(bc) < n$ implies that the vectors $x_1bc,\dots,x_nbc$ are linearly dependent. Hence, without loss of generality, we may suppose that $x_1bc = \sum_{i=2}^{n}\sigma_i(x_ibc)$ for some $\sigma_i$. For $1 \le j \le n$, let $h_j$ be any vectors in $U$ with the property $c^*h_j = u_j^{c^*U}$. Then for $1 \le i,j \le n$, we obtain

$$v_i^{Vb}u_j^{c^*U} = (x_ib)(c^*h_j) = (x_ibc)h_j$$

and hence

$$v_1^{Vb}u_j^{c^*U} = (x_1bc)h_j = [\sum_{i=2}^{n}\sigma_i(x_ibc)]h_j = \sum_{i=2}^{n}\sigma_i(x_ibc)h_j = \sum_{i=2}^{n}\sigma_i(v_i^{Vb}u_j^{c^*U})$$

which implies that the first row of the matrix $[v_i^{Vb}u_j^{c^*U}]$ is a linear combination of the remaining rows and thus $[v_i^{Vb}u_j^{c^*U}]$ is not invertible. Consequently $(b\theta)(c\theta) = 0$ which completes the proof that $\theta$ is a homomorphism.

3. To show that $\theta$ is one-to-one, let $b,c \in Q\setminus 0$ and suppose that $(b^*U,[\beta_{ij}],Vb) = (c^*U,[\gamma_{ij}],Vc)$ with the notation as above. Then $b^*U = c^*U$, $[\beta_{ij}] = [\gamma_{ij}]$, $Vb = Vc$, which by (1) evidently implies that $b = c$.

4. To show that $\theta$ is onto, let $A \in n_U$, $[\beta_{ij}] \in L_n(\Delta)$, and $B \in n_V$. Define

$$f_j = \sum_{i=1}^{n} u_i^A \beta_{ij} \quad \text{for } 1 \leq j \leq n; \quad xb = \sum_{j=1}^{n} (xf_j)v_j^B \quad (x \in V). \tag{3}$$

Then for $1 \leq j \leq n$, we have that $f_j \in A$, which then implies that $b$ is linear. Since $[\beta_{ij}]$ is invertible, it follows from the first formula in (3) that $f_1,\ldots,f_n$ are linearly independent. Hence there exist $x_j \in V$ such that $x_j f_i = \delta_{ji}$ for $1 \leq j,i \leq n$. But then by (3), $x_j b = v_j^B$ so that $v_j^B \in Vb$. Since clearly $Vb \subseteq B$, we obtain $B = Vb$ and thus $\operatorname{rank} b = n$. It further follows from the first formula in (3) and the calculation in part 1 of the proof that $b^*U = A$. From the statements already proved, the uniqueness of $[\beta_{ij}]$ in (1), and a comparison of (1) and (3), we deduce that $b\theta = (A,[\beta_{ij}],B)$. Therefore $\theta$ maps $Q$ onto $G$.

5. It remains to show that every row and every column of $P$ contains a nonzero entry. For $a \in Q\setminus 0$, we may consider $a$ as an element of $\mathfrak{F}_U(V)$. Hence $a = aba$ for some $b \in \mathfrak{F}_U(V)$ by 3.6. By 1.3, we obtain $\operatorname{rank}(ab) \leq \operatorname{rank} a$ $= \operatorname{rank}(aba) \leq \operatorname{rank}(ab)$ so that $\operatorname{rank}(ab) = n$ and similarly $\operatorname{rank}(ba) = n$. Thus both $ab$ and $ba$ are nonzero elements of $Q$. Since $a$ is arbitrary and $a(ba) = (ab)a = a \in Q\setminus 0$, it follows that the sandwich matrix has the required property.

I.4.7 LEMMA. <u>If $a$ and $b$ are nonzero elements of a Rees matrix semigroup $S$, then there exist $c,d \in S$ such that $a = cbd$.</u>

PROOF. Exercise.

We are now able to find all ideals and all principal factors of $\mathfrak{M}\mathfrak{F}_U(V)$. Recall that an ideal $I$ of a semigroup (or ring) $S$ is *proper* if $I \neq 0$, $I \neq S$.

The next result is new.

I.4.8 THEOREM. *Let $(U,V)$ be a pair of dual vector spaces.*

*i) If $\dim V = m$ is finite, then $\{\mathfrak{F}_{n,V^*}(V) \mid 2 \leq n \leq m\}$ and $\{Q_{n,V^*}(V) \mid 1 \leq n \leq m\}$ are respectively the sets of all (distinct) proper ideals and all principal factors of $\mathfrak{M}\mathfrak{F}_U(V) = \mathfrak{S}_U(V) = \mathfrak{S}(V)$.*

*ii) If $V$ is infinite dimensional, then $\{\mathfrak{F}_{n,U}(V) \mid n = 2,3,\ldots\}$ and $\{Q_{n,U}(V) \mid n = 1,2,\ldots\}$ are respectively the sets of all (distinct) proper ideals and all principal factors of $\mathfrak{M}\mathfrak{F}_U(V)$.*

*Each principal factor different from $Q_{1,U}(V)$ is isomorphic to a Rees matrix semigroup constructed in 4.6.*

PROOF. We will prove part ii); the proof of part i) requires simple modifications of the arguments to be presented and is left as an exercise. That $\mathfrak{F}_{n,U}(V)$ is an ideal of $F = \mathfrak{M}\mathfrak{F}_U(V)$ follows from 1.3, and that $\mathfrak{F}_{n,U}(V) \neq \mathfrak{F}_{m,U}(V)$ if $m \neq n$ is a consequence of 2.4.

Let $I$ be a nonzero ideal of $F$. If $a \in F$, $b \in I$, and $\operatorname{rank} a = \operatorname{rank} b = n > 0$, then both $a$ and $b$ are nonzero elements of $Q_{n,U}(V)$. By 4.6 we know that $Q_{n,U}(V)$ is isomorphic to the Rees matrix semigroup $G_{n,U}(V)$. Hence by 4.7, we have $a = cbd$ for some $c,d \in Q_{n,U}(V)\setminus 0$, and thus $c,d \in F$. Consequently $a = cbd \in I$, which implies that $I$ is the union of some of the sets $\mathfrak{F}_{n+1,U}(V)\setminus\mathfrak{F}_{n,U}(V)$. Next let $b \in I$ and $\operatorname{rank} b = n > m > 0$. Then $b = \sum_{k=1}^{n} [i_k,\gamma_k,\lambda_k]$ with $n$ minimal by 2.4, and $v_{\lambda_1},\ldots,v_{\lambda_n}$ are linearly independent by 2.5. There exist $u_{j_p} \in U$ such that $(v_{\lambda_k},u_{j_p}) = \sigma_k\delta_{kp}$ for some $\sigma_k \neq 0$ and $1 \leq k,p \leq n$. Letting $a = \sum_{p=1}^{m} [j_p,\sigma_p^{-1},\lambda_p]$, we obtain

$$ba = \Big(\sum_{k=1}^{n} [i_k,\gamma_k,\lambda_k]\Big)\Big(\sum_{p=1}^{m} [j_p,\sigma_p^{-1},\lambda_p]\Big)$$

$$= \sum_{k=1}^{n}\sum_{p=1}^{m} [i_k,\gamma_k(v_{\lambda_k},u_{j_p})\sigma_p^{-1},\lambda_p] = \sum_{p=1}^{m} [i_p,\gamma_p,\lambda_p] \in I$$

where both sets $u_{i_1},\ldots,u_{i_m}$ and $v_{\lambda_1},\ldots,v_{\lambda_m}$ are linearly independent. Thus by 2.5, we have that $\operatorname{rank}(ba) = m$ and hence $I$ contains an element of $\operatorname{rank} m$. By the above, it follows that $\mathfrak{F}_{m+1,U}(V)\setminus\mathfrak{F}_{m,U}(V) \subseteq I$. Since $m < n$ is arbitrary, we must have $\mathfrak{F}_{n+1,U}(V) \subseteq I$. Consequently, if the ranks of elements of $I$ are unbounded, then we have $I = F$, otherwise $I = \mathfrak{F}_{n,U}(V)$ for some $n \geq 2$.

Now let $a \in F$, $\operatorname{rank} a = n > 0$. Then $J(a)$ is an ideal of $F$ with the property: $J(a)$ contains an element of $\operatorname{rank} n$ and no element of higher rank (since $J(a) = FaF$). By the part already established, the only ideal of $F$ having this property is $\mathfrak{F}_{n+1,U}(V)$ and hence $J(a) = \mathfrak{F}_{n+1,U}(V)$. It follows that for any $b \in F$, $J(a) = J(b)$ if and only if $\operatorname{rank} b = n$. Consequently $I(a) = \mathfrak{F}_{n,U}(V)$ which in turn implies that the principal factor of $F$ containing $a$ is given by

$$J(a)/I(a) = \mathfrak{F}_{n+1,U}(V)/\mathfrak{F}_{n,U}(V) = Q_{n,U}(V).$$

The last statement about isomorphisms now follows from 4.6.

In light of 2.6 and 4.8, we can view $\mathfrak{M}\mathfrak{F}_U(V)$ as a tower of principal factors $Q_{n,U}(V)$ with zero removed except for $n = 1$. Each of these factors is isomorphic to a Rees matrix semigroup constructed in 4.6, so we can think of each of the "layers" as a modified replica of the preceding one.

I.4.6 EXERCISES.

i) Find all ideals of $\mathcal{S}_{\alpha}(V)$ and of $\mathcal{L}_{\alpha}(V)$.

ii) Characterize all one-sided ideals of a Rees matrix semigroup.

iii) For a dual pair $(U,V)$, show that the sets of all minimal right ideals of the following semigroups and rings coincide: a) $\mathfrak{F}_{n,U}(V)$ for any $n > 1$, b) $\mathfrak{M}\mathfrak{F}_U(V)$, c) $\mathfrak{F}_U(V)$, d) $\mathcal{S}_U(V)$, e) $\mathcal{L}_U(V)$.

iv) For any subspace $V'$ of $V$, let $\mathcal{K}_{V'}(V) = \{b \in \mathfrak{L}(V) \mid Vb \subseteq V'\}$. For $a \in \mathfrak{L}(V)$, show that $\mathfrak{L}(V)a = \mathcal{K}_{Va}(V)$. Prove that the mapping

$$\psi : V' \to \mathcal{K}_{V'}(V) \qquad (V' \text{ is a subspace of } V)$$

is an isomorphism of the lattice of all subspaces of $V$ onto the lattice of all principal left ideals of $\mathfrak{L}(V)$. Deduce that the lattice of all left ideals of $\mathfrak{F}(V)$ is isomorphic to the lattice of all principal left ideals of $\mathfrak{L}(V)$.

## I.5 SEMILINEAR ISOMORPHISMS

Throughout this section, let $(U,\Delta,V)$ and $(U',\Delta',V')$ be pairs of dual vector spaces. Our purpose now is to express every isomorphism of a ring $\mathfrak{R}$ onto a ring $\mathfrak{R}'$, where

$$\mathfrak{F}_U(\Delta,V) \subseteq \mathfrak{R} \subseteq \mathcal{L}_U(\Delta,V), \quad \mathfrak{F}_{U'}(\Delta',V') \subseteq \mathfrak{R}' \subseteq \mathcal{L}_{U'}(\Delta',V'),$$

in terms of mappings of respective pairs of dual vector spaces.

I.5.1 DEFINITION. A semilinear transformation of $(\Delta,V)$ into $(\Delta',V')$ is an ordered pair $(\omega,a)$ where $\omega$ is an isomorphism of $\Delta$ onto $\Delta'$ and $a$ is an additive homomorphism (written on the right) of $V$ into $V'$ such that

$$(\sigma v)a = (\sigma\omega)(va) \qquad (\sigma \in \Delta,\ v \in V). \tag{1}$$

If $a$ is one-to-one and maps $V$ onto $V'$, $(\omega,a)$ is called a semilinear isomorphism (semilinear automorphism if also $\Delta = \Delta'$, $V = V'$), and then $(\Delta,V)$ and $(\Delta',V')$ are said to be isomorphic, to be denoted by $(\Delta,V) \cong (\Delta',V')$.

In the case that $\Delta = \Delta'$ and $\omega$ is the identity automorphism, the semilinear transformation $(\omega,a)$ is a linear transformation. We will usually write $a$ instead of $(\omega,a)$ when there is no danger of confusion. This is also justified by the following statement.

I.5.2 LEMMA. *Let $a$ be a nonzero additive homomorphism of $V$ into $V'$. Then there exists at most one function $\omega:\Delta \to \Delta'$ satisfying (1), and if such an $\omega$ exists, it is a ring homomorphism. If $a$ is one-to-one, then $\omega$ is also.*

PROOF. Let $x \in V$ be such that $xa \neq 0$. If both $\omega$ and $\omega'$ satisfy (1), then for any $\sigma \in \Delta$, we have $(\sigma\omega)(xa) = (\sigma\omega')(xa)$ so that $\sigma\omega = \sigma\omega'$ and thus $\omega = \omega'$. Let $\sigma,\tau \in \Delta$, then

$$[(\sigma\tau)\omega](xa) = [(\sigma\tau)x]a = [\sigma(\tau x)]a = (\sigma\omega)[(\tau x)a] = (\sigma\omega)(\tau\omega)(xa)$$

so that $(\sigma\tau)\omega = (\sigma\omega)(\tau\omega)$, and

$$[(\sigma+\tau)\omega](xa) = [(\sigma+\tau)x]a = (\sigma x+\tau x)a = (\sigma x)a + (\tau x)a$$
$$= (\sigma\omega)(xa) + (\tau\omega)(xa) = (\sigma\omega+\tau\omega)(xa)$$

and hence $(\sigma+\tau)\omega = \sigma\omega + \tau\omega$. Thus $\omega$ is a ring homomorphism. If $a$ is one-to-one and $\sigma\omega = \tau\omega$, then $(\sigma x)a = (\sigma\omega)(xa) = (\tau\omega)(xa) = (\tau x)a$ so $\sigma x = \tau x$ which implies that $\sigma = \tau$; hence $\omega$ is one-to-one.

It follows from 5.2 that it suffices to require that $\omega$ be a one-to-one function of $\Delta$ onto $\Delta'$ (the case where $a$ is the zero transformation is a trivial exception), and in the case that $a$ is one-to-one, the one-to-oneness of $\omega$ can also be omitted. This lemma is useful when we want to show that a given transformation is semilinear. The definition of a semilinear transformation of right vector spaces $(U,\Delta)$ into $(U',\Delta')$ is dual, in particular (1) becomes

$$b(u\tau) = (bu)(\tau\omega) \qquad (u \in U,\ \tau \in \Delta) \tag{2}$$

and we write $(b,\omega)$ or $b$ (written on the left).

Let $(U,\Delta,V;\mathfrak{B})$ and $(U',\Delta',V';\mathfrak{B}')$ be our dual pairs and write $(v,u)$ instead of $\mathfrak{B}(v,u)$, $(v',u')'$ instead of $\mathfrak{B}'(v',u')$.

I.5.3 <u>DEFINITION</u>. Let $(\omega,a)$ be a semilinear transformation of $(\Delta,V)$ into $(\Delta',V')$. Then a function $b:U' \to U$ is an <u>adjoint</u> of $(\omega,a)$ if

$$(v,bu')\omega = (va,u')' \qquad (v \in V,\ u' \in U'). \tag{3}$$

Observe that this is a generalization of the definition of an adjoint given in (2) of Section 1.

I.5.4 <u>LEMMA</u>. <u>If</u> $(\omega,a)$ <u>has an adjoint</u> $b$, <u>it is unique and</u> $(b,\omega^{-1})$ <u>is a semilinear transformation of</u> $(U',\Delta')$ <u>into</u> $(U,\Delta)$. <u>Moreover, if</u> $a$ <u>is onto then</u> $b$ <u>is one-to-one, and if</u> $b$ <u>is onto then</u> $a$ <u>is one-to-one</u>.

PROOF. Let $c$ be another adjoint of $(\omega,a)$. Then for any $u' \in U'$, we have

$$(v,bu')\omega = (va,u')' = (v,cu')\omega \qquad (v \in V)$$

so that $(v,bu') = (v,cu')$ for all $v \in V$, which implies that $bu' = cu'$. Hence $b = c$ and the adjoint is unique. Further,

$$(v,b(u'+s'))\omega = (va,(u'+s'))' = (va,u')' + (va,s')'$$
$$= (v,bu')\omega + (v,bs')\omega = [(v,bu')+(v,bs')]\omega$$
$$= (v,bu'+bs')\omega$$

which implies that $(v,b(u'+s')) = (v,bu'+bs')$ for all $v \in V$ and thus $b(u'+s') = bu'+bs'$. Also

$$(v,b(u'\sigma))\omega = (va,u'\sigma)' = (va,u')'\sigma = [(v,bu')\omega][\sigma\omega^{-1}\omega]$$
$$= [(v,bu')(\sigma\omega^{-1})]\omega = (v,(bu')(\sigma\omega^{-1}))\omega$$

which implies that $(v,b(u'\sigma)) = (v,(bu')(\sigma\omega^{-1}))$ for all $v \in V$ and thus $b(u'\sigma) = (bu')(\sigma\omega^{-1})$. Therefore $(b,\omega^{-1})$ is a semilinear transformation.

If $a$ maps $V$ onto $V'$ and $bu' = bs'$, then for any $v \in V$, we have $(v,bu') = (v,bs')$ so that $(va,u')' = (va,s')'$ which then implies that $u' = s'$. A similar proof shows that if $b$ is onto, then $a$ must be one-to-one.

This lemma shows that by dualizing the concept of adjoint, we may say that $a$ is also an adjoint of $(b,\omega^{-1})$, or simply that $a$ and $b$ are adjoints of each other.

I.5.5 LEMMA. *If $a$ is an additive homomorphism of $V$ onto $V'$, $\omega$ is a one-to-one function of $\Delta$ onto $\Delta'$, $b$ is a function mapping $U'$ onto $U$, such that*

$$(\sigma v)a = (\sigma\omega)(va) \qquad (\sigma \in \Delta,\ v \in V), \tag{4}$$

$$(v,bu')\omega = (va,u')' \qquad (v \in V,\ u' \in U'), \tag{5}$$

*then $(\omega,a)$ is a semilinear isomorphism of $(\Delta,V)$ onto $(\Delta',V')$ with adjoint $b$.*

PROOF. This follows from 5.2 and 5.4.

In this lemma the hypotheses on $a$ and $b$ may be interchanged. The following lemma is a special case of the homomorphism theorem for Rees matrix semigroups, see Clifford and Preston ([1], Chapter 3, Section 4).

I.5.6 LEMMA. *Let $\theta$ be an isomorphism of the Rees matrix semigroup $S = \mathfrak{M}^0(I,G,\Lambda;P)$ onto the Rees matrix semigroup $S' = \mathfrak{M}^0(I',G',\Lambda';P')$. Then there exist functions:*

$\pi: I' \to I$, one-to-one and onto (written on the left),
$c: I' \to G$, with $c: i' \to c_{i'}$,
$\omega: G \to G'$, onto isomorphism and $0\omega = 0'$,
$d: \Lambda \to G'$, with $d: \lambda \to d_\lambda$,
$\eta: \Lambda \to \Lambda'$, one-to-one and onto (written on the right)

*such that*

$$(i,g,\lambda)\theta = (\pi^{-1}i, (c_{\pi^{-1}i}^{-1}g)\omega d_\lambda, \lambda\eta) \qquad ((i,g,\lambda) \in S), \tag{6}$$

$$d_\lambda p'_{\lambda\eta,i'} = (p_{\lambda,\pi i'} c_{i'})\omega \qquad (\lambda \in \Lambda,\ i' \in I'). \tag{7}$$

PROOF. First note that the nonzero idempotents of $S$ are the elements of the form $(i,p_{\lambda i}^{-1},\lambda)$ where $p_{\lambda i} \neq 0$. Fix an element in $\Lambda$ and an element in $I$, both to be denoted $1$, such that $p_{11} \neq 0$. Then $(1,p_{11}^{-1},1)$ is an idempotent and hence $(1,p_{11}^{-1},1)\theta = (k,p'^{-1}_{\tau k},\tau)$ for some $k \in I'$, $\tau \in \Lambda'$. Any element $(i,g,\lambda)$ of $S$ can be written in the form

$$(i,g,\lambda) = (i,e,1)(1,p_{11}^{-1}g,1)(1,p_{11}^{-1},\lambda) \tag{8}$$

where $e$ is the identity of $G$. There exists $\mu \in \Lambda$ such that $p_{\mu i} \neq 0$, so

$$\begin{aligned}(k,p'^{-1}_{\tau k},\tau) &= (1,p_{11}^{-1},1)\theta = [(1,p_{11}^{-1}g^{-1}p_{\mu i}^{-1},\mu)(i,g,1)]\theta \\ &= (1,p_{11}^{-1}g^{-1}p_{\mu i}^{-1},\mu)\theta(i,g,1)\theta\end{aligned}$$

which shows that $(i,g,1)\theta$ is of the form $(\ ,\ ,\tau)$; similarly $(1,g,\lambda)\theta$ is of the form $(k,\ ,\ )$. Using this and motivated by (8), we define the functions $\xi$, $q{:}i \to q_i$, $\omega$, $d{:}\lambda \to d_\lambda$, $\eta$ by

$$\begin{aligned}(i,e,1)\theta &= (\xi i, q_i, \tau) && (i \in I),\\ (1,p_{11}^{-1}g,1)\theta &= (k,p'^{-1}_{\tau k}(g\omega),\tau) && (g \in G^0),\\ (1,p_{11}^{-1},\lambda)\theta &= (k,p'^{-1}_{\tau k}d_\lambda,\lambda\eta) && (\lambda \in \Lambda).\end{aligned}$$

Then by (8), we obtain

$$\begin{aligned}(i,g,\lambda)\theta &= (i,e,1)\theta(1,p_{11}^{-1}g,1)\theta(1,p_{11}^{-1},\lambda)\theta \\ &= (\xi i,q_i,\tau)(k,p'^{-1}_{\tau k}(g\omega),\tau)(k,p'^{-1}_{\tau k}d_\lambda,\lambda\eta) \\ &= (\xi i, q_i(g\omega)d_\lambda, \lambda\eta).\end{aligned} \tag{9}$$

Further, by the definition of $\omega$, we have $\omega{:}G \to G'$ and $0\omega = 0'$; also

$$\begin{aligned}(k,p'^{-1}_{\tau k}(gh)\omega,\tau) &= (1,p_{11}^{-1}gh,1)\theta = [(1,p_{11}^{-1}g,1)(1,p_{11}^{-1}h,1)]\theta \\ &= (1,p_{11}^{-1}g,1)\theta(1,p_{11}^{-1}h,1)\theta \\ &= (k,p'^{-1}_{\tau k}(g\omega),\tau)(k,p'^{-1}_{\tau k}(h\omega),\tau) \\ &= (k,p'^{-1}_{\tau k}(g\omega)(h\omega),\tau)\end{aligned}$$

which implies $(gh)\omega = (g\omega)(h\omega)$. If $g\omega = e'$, the identity of $G'$, then by the definition of $\omega$, we have

$$(1,p_{11}^{-1}g,1)\theta = (k,p'^{-1}_{\tau k}(g\omega),\tau) = (k,p'^{-1}_{\tau k},\tau) = (1,p_{11}^{-1},1)\theta$$

so that $g = e$, and $\omega$ is one-to-one.

Using (9), we see that for any $a \in G'$, there exists $g \in G$ such that $(k,q_1ad_1,\tau) = (1,g,1)\theta$. Hence $(k,q_1ad_1,\tau) = (k,q_1(g\omega)d_1,\tau)$ so $a = g\omega$ and $\omega$ maps $G$ onto $G'$.

For $(j,e,\mu) \in S'$, let $(i,h,\lambda) = (j,e,\mu)\theta^{-1}$. We obtain by (9),

$$(j,e,\mu) = (i,h,\lambda)\theta = (\xi i,q_i(h\omega)d_\lambda,\lambda\eta)$$

which implies that both $\xi$ and $\eta$ are onto.

If $\xi i = \xi j$, then again by (9), we have

$$(i,e,1)\theta = (\xi i,q_i,\tau) = (\xi j,q_j(q_j^{-1}q_i),\tau) = (j,(q_j^{-1}q_id_1^{-1})\omega^{-1},1)\theta$$

which implies that $i = j$. Consequently $\xi$ is one-to-one, similarly $\eta$ is also.

Further, using (9) we obtain

$$(i,e,\lambda)^2\theta = (i,p_{\lambda i},\lambda)\theta = (\xi i,q_i(p_{\lambda i}\omega)d_\lambda,\lambda\eta),$$

$$[(i,e,\lambda)\theta]^2 = (\xi i,q_id_\lambda,\lambda\eta)^2 = (\xi i,q_id_\lambda p'_{\lambda\eta,\xi i}q_id_\lambda,\lambda\eta),$$

so by the homomorphism property, we have

$$p_{\lambda i}\omega = d_\lambda p'_{\lambda\eta,\xi i}q_i. \tag{10}$$

Finally, letting $\pi = \xi^{-1}$ and $c_{i'} = q_{\pi i'}^{-1}\omega^{-1}$ for all $i' \in I'$, we obtain the functions $\pi$ and $c$ as in the statement of the lemma, and (6) and (7) follow from (9) and (10), respectively.

I.5.7 DEFINITION. A semilinear isomorphism $(\omega,a):(\Delta,V) \to (\Delta',V')$ is called a semilinear isomorphism of $(U,\Delta,V)$ onto $(U',\Delta',V')$ if $(\omega,a)$ has an adjoint $b$ which maps $U'$ onto $U$. The dual pairs $(U,\Delta,V)$ and $(U',\Delta',V')$ are then said to be isomorphic, and we write $(U,\Delta,V) \cong (U',\Delta',V')$.

Strictly speaking we should write $(b,\omega,a)$ instead of $(\omega,a)$, but this is actually not necessary since all we need is the existence of $b$ which is then unique by 5.4. We will also write $a$ instead of $(\omega,a)$ if there is no danger of confusion. Again by 5.4, we know that then $(b,\omega^{-1})$ is a semilinear isomorphism of $(U',\Delta')$ onto $(U,\Delta)$, and hence isomorphism of dual pairs is an equivalence relation.

I.5.8 LEMMA. Let $(\omega,a)$ be a semilinear isomorphism of $(U,\Delta,V)$ onto $(U',\Delta',V')$. Then $(\omega^{-1},a^{-1})$ is a semilinear isomorphism of $(U',\Delta',V')$ onto $(U,\Delta,V)$, the function $\zeta_{(\omega,a)}$ defined by

$$c\zeta_{(\omega,a)} = a^{-1}ca \qquad (c \in \mathfrak{L}_U(\Delta,V))$$

is an isomorphism of $\mathfrak{L}_U(\Delta,V)$ onto $\mathfrak{L}_{U'}(\Delta',V')$, and $\operatorname{rank} c = \operatorname{rank}(c\zeta_{(\omega,a)})$.

PROOF. Exercise.

We wish to show that every isomorphism of a semigroup $S$ onto a semigroup $S'$, for which

$$\mathfrak{F}_{2,U}(\Delta,V) \subseteq S \subseteq \mathfrak{S}_U(\Delta,V), \quad \mathfrak{F}_{2,U'}(\Delta',V') \subseteq S' \subseteq \mathfrak{S}_{U'}(\Delta',V')$$

with $\dim V > 1$, is the restriction to $S$ of an isomorphism $\zeta_{(\omega,a)}$ for some semilinear isomorphism $(\omega,a)$ of $(U,\Delta,V)$ onto $(U',\Delta',V')$. We will then establish several important consequences of this result and show that it is in general false for $\dim V = 1$. The next result is crucial for the treatment of the general case; it is due to Gluskin [4].

I.5.9 THEOREM. <u>Let</u> $\theta$ <u>be an isomorphism of</u> $S = \mathfrak{F}_{2,U}(\Delta,V)$ <u>onto</u> $S' = \mathfrak{F}_{2,U'}(\Delta',V')$, <u>fix vectors</u> $u_i \in U$, $v_\lambda \in V$, $u_i' \in U'$, $v_\lambda' \in V'$ <u>as at the beginning of Section</u> 2, <u>consider both</u> $S$ <u>and</u> $S'$ <u>as Rees matrix semigroups</u> $S = \mathfrak{M}^0(I_U,\mathfrak{M}\Delta^-,I_V;P)$ <u>with</u> $p_{\lambda i} = (v_\lambda,u_i)$ <u>and</u> $S' = \mathfrak{M}^0(I_{U'},\mathfrak{M}\Delta'^-,I_{V'};P')$ <u>with</u> $p'_{\lambda' i'} = (v'_\lambda,u'_i)'$, <u>and let</u> $\pi,c,\omega,d,\eta$ <u>be as in</u> 5.6. <u>Suppose that</u> $\dim V > 1$, <u>and define functions</u> $a$ <u>and</u> $b$ <u>by</u>

$$(\gamma v_\lambda)a = (\gamma\omega)d_\lambda v'_{\lambda\eta} \qquad (\gamma v_\lambda \in V), \tag{11}$$

$$b(u_i'\tau) = u_{\pi i}c_i(\tau\omega^{-1}) \qquad (u_i'\tau \in U'). \tag{12}$$

<u>Then</u> $(\omega,a)$ <u>is a semilinear isomorphism of</u> $(U,\Delta,V)$ <u>onto</u> $(U',\Delta',V')$ <u>with adjoint</u> $b$ <u>and</u> $\zeta_{(\omega,a)}|_{\mathfrak{F}_{2,U}(\Delta,V)} = \theta$.

<u>PROOF</u>. The proof of the first statement consists of a verification of the hypotheses in 5.5.

1. Since the representation of nonzero elements of $V$ in the form $\gamma v_\lambda$ is unique, and $0\omega = 0'$ implies $0a = 0'$, $a$ is single valued. If $(\gamma v_\lambda)a = (\delta v_\mu)a \neq 0$, then $(\gamma\omega)d_\lambda v'_{\lambda\eta} = (\delta\omega)d_\mu v'_{\mu\eta} \neq 0'$. It follows that $\lambda\eta = \mu\eta$ and hence $\lambda = \mu$. But then also $\gamma\omega = \delta\omega$ so that $\gamma = \delta$. We thus have $\gamma v_\lambda = \delta v_\mu$ and $a$ is one-to-one. Further for $0' \neq \gamma v_\lambda' \in V'$, we obtain

$$[(\gamma d^{-1}_{\lambda\eta^{-1}})\omega^{-1}v_{\lambda\eta^{-1}}]a = [(\gamma d^{-1}_{\lambda\eta^{-1}})\omega^{-1}\omega]d_{\lambda\eta^{-1}}v'_{\lambda\eta^{-1}\eta} = \gamma v'_\lambda \tag{13}$$

which proves that $a$ maps $V$ onto $V'$.

Now $b$ is defined in a way completely analogous to that of $a$, so by a similar argument, one sees that $b$ is a one-to-one function of $U'$ onto $U$.

2. For any $\sigma \in \Delta$, $\gamma v_\lambda \in V$, we obtain by (11),

$$[\sigma(\gamma v_\lambda)]a = [(\sigma\gamma)v_\lambda]a = (\sigma\gamma)\omega d_\lambda v'_{\lambda\eta} = (\sigma\omega)[(\gamma\omega)d_\lambda v'_{\lambda\eta}] = (\sigma\omega)[(\gamma v_\lambda)a]$$

which verifies (4).

For any $\gamma v_\lambda \in V$ and $u_i'\tau \in U'$, we obtain by (7), (11), and (12),

$$(\gamma v_\lambda, b(u_i'\tau))\omega = (\gamma v_\lambda, u_{\pi i}c_i(\tau\omega^{-1}))\omega = (\gamma\omega)[(v_\lambda, u_{\pi i})c_i]\omega\tau$$
$$= (\gamma\omega)d_\lambda(v_{\lambda\pi}', u_i')'\tau = ((\gamma\omega)d_\lambda v_{\lambda\pi}', u_i'\tau)' = ((\gamma v_\lambda)a, u_i'\tau)'$$

which verifies (5).

We will now show that $a$ is additive and will do this in several steps.

3. Let $x_1, x_2, x_3$ be linearly independent vectors in $V$. There exist $g_1, g_2, g_3 \in U$ such that $(x_i, g_j) = \delta_{ij}$ for $1 \leq i,j \leq 3$. By (5), we have $(x_i a, b^{-1}g_j)' = (x_i, g_j)\omega = \delta_{ij}\omega = \delta_{ij}$ for $1 \leq i,j \leq 3$. If $\sum_{i=1}^{3} \sigma_i(x_i a) = 0$, then for $1 \leq k \leq 3$, we have

$$\sigma_k = \sigma_k(x_k a, b^{-1}g_k) = \sum_{i=1}^{3} \sigma_i(x_i a, b^{-1}g_k) = (\sum_{i=1}^{3} \sigma_i(x_i a), b^{-1}g_k) = 0$$

and hence $x_1 a, x_2 a, x_3 a$ are linearly independent.

4. Now let $x$ and $y$ be linearly independent vectors in $V$. The vectors $xa, ya, xa + ya$ are linearly dependent so by part 3 we have that the vectors $xaa^{-1}, yaa^{-1}, (xa + ya)a^{-1}$ are linearly dependent. Since $x$ and $y$ are linearly independent, we must have $(xa + ya)a^{-1} = \sigma x + \tau y$ for some $\sigma, \tau \in \Delta$, whence

$$xa + ya = (\sigma x + \tau y)a. \tag{14}$$

Using (14) and (5), we obtain for any $u' \in U'$,

$$(x, bu')\omega + (y, bu')\omega = (xa, u')' + (ya, u')' = (xa + ya, u')'$$
$$= ((\sigma x + \tau y)a, u')' = (\sigma x + \tau y, bu')\omega = [\sigma(x, bu') + \tau(y, bu')]\omega. \tag{15}$$

There exists $u \in U$ such that $(x,u) = 1$, $(y,u) = 0$; letting $u' = b^{-1}u$, (15) yields $1\omega = \sigma\omega$. Consequently $\sigma = 1$ and analogously $\tau = 1$. Formula (14) then becomes

$$(x + y)a = xa + ya \qquad (x, y \in V) \tag{16}$$

for the case in xhich $x$ and $y$ are linearly independent.

5. Again let $x, y \in V$. If $x = 0$ or $y = 0$, (16) is trivial. If $x$ and $y$ are linearly dependent but both nonzero, let $z$ be a vector in $V$ such that $y$ and $z$ are linearly independent (here we are using the hypothesis that $\dim V > 1$). It follows easily that if $x + y \neq 0$, then both sets $x, y + z$ and $x + y, z$ are linearly independent. Using (16) several times, we obtain

$$xa + ya + za = xa + (y + z)a = [x + (y + z)]a = [(x + y) + z]a = (x + y)a + za$$

so that $xa + ya = (x + y)a$, which implies that (16) is valid for linearly dependent vectors $x, y$. Therefore $a$ is additive.

By 5.5, $(\omega,a)$ is a semilinear isomorphism of $(U,\Delta,V)$ onto $(U',\Delta',V')$ with adjoint $b$.

6. It remains to verify the last statement of the theorem. For $[i,\gamma,\lambda] \in \mathfrak{F}_{2,U}(\Delta,V)$ and $\sigma v'_{\mu} \in V'$, using (13), (11), (7), (6), we obtain

$$\begin{aligned}
&(\sigma v'_{\mu})(a^{-1}[i,\gamma,\lambda]a) = \{(\sigma v'_{\mu})a^{-1}\}[i,\gamma,\lambda]a \\
&= \{(\sigma d^{-1}_{\mu\eta^{-1}})\omega^{-1}v_{\mu\eta^{-1}}\}[i,\gamma,\lambda]a = \{(\sigma d^{-1}_{\mu\eta^{-1}})\omega^{-1}(v_{\mu\eta^{-1}},u_i)\gamma v_{\lambda}\}a \\
&= \sigma d^{-1}_{\mu\eta^{-1}}\{(v_{\mu\eta^{-1}},u_i)\gamma\}\omega d_{\lambda}v'_{\lambda\eta} = \sigma d^{-1}_{\mu\eta^{-1}}(p_{\mu\eta^{-1},i}\,\omega)(\gamma\omega)d_{\lambda}v'_{\lambda\eta} \\
&= \sigma d^{-1}_{\mu\eta^{-1}}d_{\mu\eta^{-1}}p'_{\mu,\pi^{-1}i}(c^{-1}_{\pi^{-1}i}\gamma)\omega d_{\lambda}v'_{\lambda\eta} = \sigma(v'_{\mu},u'_{\pi^{-1}i})'(c^{-1}_{\pi^{-1}i}\gamma)\omega d_{\lambda}v'_{\lambda\eta} \\
&= (\sigma v'_{\mu})[\pi^{-1}i,(c^{-1}_{\pi^{-1}i}\gamma)\omega d_{\lambda},\lambda\eta] = (\sigma v'_{\mu})\theta
\end{aligned}$$

which shows that $\zeta_{(\omega,a)}|_{\mathfrak{F}_{2,U}(\Delta,V)} = \theta$ and completes the proof.

I.5.10 <u>DEFINITION</u>. Let $S$ be a subsemigroup of $\mathfrak{S}_U(\Delta,V)$ and $S'$ be a subsemigroup of $\mathfrak{S}_{U'}(\Delta',V')$. An isomorphism $\psi$ of $S$ onto $S'$ is said to be <u>induced</u> by a semilinear isomorphism $(\omega,a)$ of $(U,\Delta,V)$ onto $(U',\Delta',V')$ if $\zeta_{(\omega,a)}|_S = \psi$. This same definition is to be used for subrings of $\mathfrak{L}_U(\Delta,V)$ and $\mathfrak{L}_{U'}(\Delta',V')$.

In particular 5.9 says that every isomorphism of $\mathfrak{F}_{2,U}(\Delta,V)$ onto $\mathfrak{F}_{2,U'}(\Delta',V')$ is induced by a semilinear isomorphism. We will show that the conclusion is valid under much less restrictive circumstances.

I.5.11 <u>LEMMA</u>. <u>If</u> $S$ <u>is a subsemigroup of</u> $\mathfrak{S}_U(V)$ <u>containing</u> $\mathfrak{F}_{2,U}(V)$, <u>then</u> $\mathfrak{F}_{2,U}(V)$ <u>is the unique nonzero ideal contained in every nonzero ideal of</u> $S$.

<u>PROOF</u>. Write $\mathfrak{F}_2 = \mathfrak{F}_{2,U}(V)$. By 2.8, part i), the hypothesis implies that $\mathfrak{F}_2 \subseteq S \subseteq i_{\mathfrak{S}(V)}(\mathfrak{F}_2)$ and thus $\mathfrak{F}_2$ is an ideal of $S$. Let $I$ be a nonzero ideal of $S$, and let $a \in I\setminus 0$. By 3.3, there exists $b \in \mathfrak{F}_2$ such $ba \neq 0$; also $ba \in \mathfrak{F}_2 \cap I$. For any $c \in \mathfrak{F}_2$, there exist $g,h \in \mathfrak{F}_2$ such that $c = g(ba)h$ by 4.7. But then $c \in I$ and hence $\mathfrak{F}_2 \subseteq I$.

We are now able to prove the main result of this section, due to Gluskin [4]. For semigroups of matrices, this was first proved by Halezov [2] in a different guise, and then by Gluskin [3].

I.5.12 <u>THEOREM</u>. <u>Let</u> $S$ <u>be a subsemigroup of</u> $\mathfrak{S}_U(\Delta,V)$ <u>containing</u> $\mathfrak{F}_{2,U}(\Delta,V)$, $S'$ <u>be a subsemigroup of</u> $\mathfrak{S}_{U'}(\Delta',V')$ <u>containing</u> $\mathfrak{F}_{2,U'}(\Delta',V')$, <u>suppose that</u> <u>dim</u> $V > 1$, <u>and let</u> $\psi$ <u>be an isomorphism of</u> $S$ <u>onto</u> $S'$. <u>Then</u> $\psi$ <u>is induced by a</u>

semilinear isomorphism $(\omega,a)$ of $(U,\Delta,V)$ onto $(U',\Delta',V')$ and $\zeta_{(\omega,a)}$ is the unique extension of $\psi$ to a homomorphism of $\mathfrak{S}_U(\Delta,V)$ into $\mathfrak{S}_{U'}(\Delta',V')$.

PROOF. First note that by 5.11, we have that $\mathfrak{F}_{2,U}(\Delta,V)$ is the unique nonzero ideal of $S$ contained in every nonzero ideal of $S$. Since the same kind of statement is valid for $\mathfrak{F}_{2,U'}(\Delta',V')$, it is clear that $\psi$ must map $\mathfrak{F}_{2,U}(\Delta,V)$ onto $\mathfrak{F}_{2,U'}(\Delta',V')$. Letting $\theta = \psi|_{\mathfrak{F}_{2,U}(\Delta,V)}$ we see that $\theta$ satisfies the hypotheses of 5.9, so $\theta$ is induced by a semilinear isomorphism $(\omega,a)$.

Let $\varphi$ be a homomorphism of $\mathfrak{S}_U(\Delta,V)$ into $\mathfrak{S}_{U'}(\Delta',V')$ which extends $\theta$. We write an arbitrary element of $V'$ in the form $(\gamma\omega)d_\lambda v'_{\lambda\eta}$ as we may since $a$ maps $V$ onto $V'$, and this representation is unique since $a$ is one-to-one. There exists $i \in I_U$ such that $(v_\lambda,u_i) = \sigma \neq 0$ so that $\gamma v_\lambda = (\gamma v_\lambda)[i,\sigma^{-1},\lambda]$. Hence for any $c \in \mathfrak{S}_U(\Delta,V)$, we obtain

$$\begin{aligned}\{(\gamma\omega)d_\lambda v'_{\lambda\eta}\}a^{-1}ca &= (\gamma v_\lambda)ca = \{(\gamma v_\lambda)[i,\sigma^{-1},\lambda]\}ca \\ &= (\gamma v_\lambda)a\{a^{-1}([i,\sigma^{-1},\lambda]c)a\} = (\gamma v_\lambda)a\{[i,\sigma^{-1},\lambda]c\}\varphi \\ &= (\gamma v_\lambda)a\{([i,\sigma^{-1},\lambda]\varphi)(c\varphi)\} = (\gamma v_\lambda)a\{a^{-1}[i,\sigma^{-1},\lambda]a(c\varphi)\} \\ &= \{(\gamma v_\lambda)[i,\sigma^{-1},\lambda]\}a(c\varphi) = (\gamma v_\lambda)a(c\varphi) = \{(\gamma\omega)d_\lambda v'_{\lambda\eta}\}(c\varphi)\end{aligned}$$

and thus $c\varphi = a^{-1}ca$. Consequently $\varphi = \zeta_{(\omega,a)}$. Since $\psi$ extends $\theta$ to $S$, the same calculation with $\psi$ instead of $\varphi$ and for $c \in S$, shows that $\zeta_{(\omega,a)}|_S = \psi$. Since $\varphi$ is the unique extension of $\theta$ to a homomorphism of $\mathfrak{S}_U(\Delta,V)$ into $\mathfrak{S}_{U'}(\Delta',V')$, it is also the unique extension of $\psi$ to such a homomorphism.

We know by 5.8 that $\zeta_{(\omega,a)}$ is in fact an isomorphism of $\mathfrak{L}_U(\Delta,V)$ onto $\mathfrak{L}_{U'}(\Delta',V')$. Defining $\zeta_{(\omega,a)}$ on all of $\mathfrak{L}(\Delta,V)$ by the same formula as on $\mathfrak{L}_U(\Delta,V)$, we see that $\zeta_{(\omega,a)}$ is an extension of $\psi$ to an isomorphism of $\mathfrak{L}(\Delta,V)$ onto $\mathfrak{L}(\Delta',V')$. For example, if $S$ is contained in $\mathfrak{F}_U(\Delta,V)$, then $S'$ is contained in $\mathfrak{F}_{U'}(\Delta',V')$ since $\zeta_{(\omega,a)}$ preserves rank, and $\zeta_{(\omega,a)}|_{\mathfrak{F}_U(\Delta,V)}$ is the unique extension of $\psi$ to a homomorphism of $\mathfrak{M}\mathfrak{F}_U(\Delta,V)$ into $\mathfrak{M}\mathfrak{F}_{U'}(\Delta',V')$. Furthermore, if $U'$ and $U''$ are t-subspaces of $V^*$, then every isomorphism of a subsemigroup of $\mathfrak{S}_{U'}(\Delta,V)$ containing $\mathfrak{F}_{2,U'}(\Delta,V)$ onto a subsemigroup of $\mathfrak{S}_{U''}(\Delta,V)$ containing $\mathfrak{F}_{2,U''}(\Delta,V)$ can be extended to an (additive and multiplicative) automorphism of $\mathfrak{L}(\Delta,V)$.

I.5.13 COROLLARY. *Let $\mathfrak{R}$ be a subring of $\mathfrak{L}_U(\Delta,V)$ containing $\mathfrak{F}_U(\Delta,V)$, $\mathfrak{R}'$ be a subring of $\mathfrak{L}_{U'}(\Delta',V')$ containing $\mathfrak{F}_{U'}(\Delta',V')$, suppose that $\dim V > 1$, and let $\psi$ be an isomorphism of $\mathfrak{M}\mathfrak{R}$ onto $\mathfrak{M}\mathfrak{R}'$. Then $\psi$ satisfies the conclusions of 5.12 and is thus a ring isomorphism of $\mathfrak{R}$ onto $\mathfrak{R}'$.*

The ring version of 5.12, with the rings $\mathfrak{R}$ and $\mathfrak{R}'$ as in 5.13 and no restriction on the dimension of V, is due to Jacobson [3]; for $\mathfrak{R} = \mathfrak{F}_U(V)$, $\mathfrak{R}' = \mathfrak{F}_{U'}(V)$, the result was established earlier by Dieudonné [2].

In 5.13 we have a class of rings with the remarkable property that every multiplicative isomorphism of a ring in that class is automatically additive, i.e., every multiplicative isomorphism is a ring isomorphism. The hypothesis $\dim V > 1$ cannot be dropped. For if $\dim V = 1$, it is easy to see that $\mathcal{L}(V) \cong \Delta$; similarly $\mathcal{L}(V') \cong \Delta'$. But there exist division rings (even fields) for which there exist multiplicative isomorphisms which are not additive. For example, let $\Delta = \Delta'$ be the field of real numbers and n be any odd integer, $n \neq 1$; then the mapping $x \to x^n$ is a multiplicative automorphism of $\mathfrak{M}\Delta$ but is not additive. It follows that the corresponding automorphism of $\mathcal{L}(V)$ is not induced by a semilinear isomorphism which in turn implies that $\dim V > 1$ cannot be dropped in 5.12 or 5.13.

For $0 < \alpha \leq (\dim V)^+$, let $\mathcal{S}_{\alpha,U}(\Delta,V) = \mathcal{S}_\alpha(\Delta,V) \cap \mathcal{S}_U(\Delta,V)$.

I.5.14 COROLLARY. *Let* $\dim V > 1$ *and* $\alpha$ *and* $\alpha'$ *be cardinal numbers such that* $1 < \alpha,\alpha' < (\dim V)^+$. *Then* $\mathcal{S}_{\alpha,U}(\Delta,V) \cong \mathcal{S}_{\alpha',U'}(\Delta',V')$ *if and only if* $\alpha = \alpha'$ *and* $(U,\Delta,V) \cong (U',\Delta',V')$.

PROOF. If $\psi$ is an isomorphism of $\mathcal{S}_{\alpha,U}(\Delta,V)$ onto $\mathcal{S}_{\alpha',U'}(\Delta',V')$, then by 5.12 we know that $\psi$ is induced by a semilinear isomorphism $(\omega,a)$ of $(U,\Delta,V)$ onto $(U',\Delta',V')$, and since $\zeta_{(\omega,a)}$ preserves rank, we must have $\alpha = \alpha'$.

Conversely if $(\omega,a)$ is a semilinear isomorphism of $(U,\Delta,V)$ onto $(U',\Delta',V')$ and $\alpha = \alpha'$, then $\zeta_{(\omega,a)}|_{\mathcal{S}_{\alpha,U}(\Delta,V)}$ is the required isomorphism.

I.5.15 COROLLARY. *For vector spaces* $(\Delta,V)$ *and* $(\Delta',V')$ *with* $\dim V > 1$, *the following statements are equivalent.*

i) $\Delta \cong \Delta'$, $\dim V = \dim V'$.

ii) $(\Delta,V) \cong (\Delta',V')$.

iii) $\mathcal{S}_\alpha(\Delta,V) \cong \mathcal{S}_{\alpha'}(\Delta',V')$ *for some cardinal numbers* $\alpha,\alpha' > 1$.

iv) $\mathcal{S}(\Delta,V) \cong \mathcal{S}(\Delta',V')$.

v) $\mathcal{L}(\Delta,V) \cong \mathcal{L}(\Delta',V')$.

PROOF. i) $\Rightarrow$ ii). Let $\omega$ be an isomorphism of $\Delta$ onto $\Delta'$ and a be a one-to-one function mapping a basis B of V onto a basis B′ of V′. Extend a to all of V by defining

$$va = \sum_{k=1}^{n} (\sigma_k\omega)(v_k a) \quad \text{if} \quad v = \sum_{k=1}^{n} \sigma_k v_k \qquad (v_k \in B).$$

Verification that $a$ is a semilinear isomorphism of $(\Delta,V)$ onto $(\Delta',V')$ is left as an exercise.

ii) $\Rightarrow$ iii). This follows from 5.14.

iii) $\Rightarrow$ iv). This follows from 5.12 with $U = V^*$, $U' = V'^*$.

iv) $\Rightarrow$ v) and v) $\Rightarrow$ i). This follows from 5.13 with $U = V^*$ and $U' = V'^*$.

This corollary says that under the hypothesis $\dim V > 1$, any of the following determines $(\Delta,V)$ up to a semilinear isomorphism: the division ring $\Delta$ and $\dim V$, the semigroups $\mathfrak{S}_a(\Delta,V)$, $\mathfrak{S}(\Delta,V)$, and the ring $\mathfrak{L}(\Delta,V)$.

We now investigate when two semilinear isomorphisms induce the same semigroup (or ring) isomorphism. If $G$ is a group, we denote by $\varepsilon_g$ the <u>inner automorphism</u> of $G$ induced by $g$, i.e., $x\varepsilon_g = g^{-1}xg$ $(x \in G)$.

I.5.16 <u>DEFINITION</u>. For $\gamma \in \Delta$, the transformation $m_\gamma$ defined by $xm_\gamma = \gamma x$ $(x \in V)$ is called the <u>multiplication induced by</u> $\gamma$, or simply a <u>multiplication</u>.

It is easy to verify that for $\gamma \neq 0$, $m_\gamma = (\varepsilon_{\gamma^{-1}}, m_\gamma)$ is a semilinear automorphism of $(\Delta,V)$. The next result is new, but most of the proof is old.

I.5.17 <u>THEOREM</u>. <u>Let</u> $(\Delta,V)$ <u>be a vector space</u>. <u>If</u> $a$ <u>is any function mapping</u> $V$ <u>into itself which commutes with all elements of</u> $\mathfrak{F}_{2,U}(\Delta,V)$ <u>for some t-subspace</u> $U$ <u>of</u> $V^*$, <u>then</u> $a$ <u>is a multiplication</u>. <u>Conversely, every multiplication commutes with all elements of</u> $\mathfrak{L}(V)$.

<u>PROOF</u>. Let $a$ be as in the statement of the theorem. Let $0 \neq x \in V$. Then $x = \gamma v_\lambda$ and there exists $i \in I_U$ such that $(v_\lambda, u_i) = \sigma \neq 0$. Since $[i,\sigma^{-1},\lambda]$ is in $\mathfrak{F}_{2,U}(V)$, the hypothesis implies

$$[(xa,u_i)\sigma^{-1}\gamma^{-1}]x = (xa,u_i)\sigma^{-1}v_\lambda = x(a[i,\sigma^{-1},\lambda])$$
$$= x([i,\sigma^{-1},\lambda]a) = (\gamma(v_\lambda,u_i)\sigma^{-1}v_\lambda)a = (\gamma v_\lambda)a = xa.$$

It follows that $a$ maps $x$ into the subspace of $V$ generated by $x$, and thus for some scalar $\gamma_x$ we have $xa = \gamma_x x$. If $x$ and $y$ are nonzero vectors, by 2.10 there exists $b \in \mathfrak{F}_{2,U}(V)$ such that $xb = y$. Using the hypothesis, we obtain

$$\gamma_y y = ya = (xb)a = x(ba) = x(ab) = (xa)b = (\gamma_x x)b = \gamma_x(xb) = \gamma_x y$$

and hence $\gamma_x = \gamma_y$. But then $\gamma = \gamma_x$ is a constant and we have $xa = \gamma x$ for all $x \neq 0$. Let $o$ be the zero function on $V$. Then $0 = (0a)o = (0o)a = 0a$. Consequently $a = m_\gamma$.

Conversely, if $b \in \mathfrak{L}(V)$, then for every $x \in V$, we have

$$x(m_\gamma b) = (xm_\gamma)b = (\gamma x)b = \gamma(xb) = (xb)m_\gamma = x(bm_\gamma)$$

and thus $m_\gamma b = bm_\gamma$.

If $S$ is a semigroup (or ring), by $C(S)$ we denote the <u>center</u> of $S$, i.e., $C(S) = \{a \in S \mid xa = ax \text{ for all } x \in S\}$.

I.5.18 <u>LEMMA</u>. <u>For</u> $\gamma \in \Delta$, $m_\gamma$ <u>is linear if and only if</u> $\gamma \in C(\Delta)$.

<u>PROOF</u>. Exercise.

I.5.19 <u>COROLLARY</u>. $C(\mathcal{L}_U(\Delta,V)) = \{m_\gamma \mid \gamma \in C(\Delta)\}$.

<u>PROOF</u>. If $a \in C(\mathcal{L}_U(\Delta,V))$, then $a = m_\gamma$ for some $\gamma \in \Delta$ by 5.17, and since $a \in \mathcal{L}_U(\Delta,V)$, by 5.18 we also have that $\gamma \in C(\Delta)$. Conversely, let $\gamma \in C(\Delta)$. Then by 5.18, we have $m_\gamma \in \mathcal{L}(\Delta,V)$, and by 5.17, $m_\gamma$ commutes with all elements of $\mathcal{L}(\Delta,V)$. Further, for any $x \in V$, $f \in V^*$, we obtain

$$x(m_\gamma^* f) = (xm_\gamma)f = (\gamma x)f = \gamma(xf) = (xf)\gamma = x(f\gamma)$$

which proves that $m_\gamma^* f = f\gamma$ (in particular, the conjugate of a multiplication is again a multiplication). Consequently $m_\gamma^* U \subseteq U$ which implies that $m_\gamma \in \mathcal{L}_U(\Delta,V)$ and therefore $m_\gamma \in C(\mathcal{L}_U(\Delta,V))$.

I.5.20 <u>LEMMA</u>. <u>If</u> $(\omega,a)$ <u>and</u> $(\omega',a')$ <u>are semilinear transformations of</u> $(\Delta,V)$ <u>into</u> $(\Delta',V')$ <u>and of</u> $(\Delta',V')$ <u>into</u> $(\Delta'',V'')$, <u>respectively, then</u> $(\omega\omega',aa')$ <u>is a semilinear transformation of</u> $(\Delta,V)$ <u>into</u> $(\Delta'',V'')$.

<u>PROOF</u>. Exercise.

I.5.21 <u>COROLLARY</u>. <u>Let</u> $(\omega,a)$ <u>and</u> $(\omega',a')$ <u>be semilinear isomorphisms of</u> $(U,\Delta,V)$ <u>onto</u> $(U',\Delta',V')$. <u>Then</u> $\zeta_{(\omega,a)} = \zeta_{(\omega',a')}$ <u>if and only if</u> $a' = m_\gamma a$ <u>for some</u> $\gamma \in \Delta^-$. <u>In such a case</u>, $\omega' = \varepsilon_{\gamma^{-1}}\omega$.

<u>PROOF</u>. The equation $\zeta_{(\omega,a)} = \zeta_{(\omega',a')}$ is equivalent to

$$a^{-1}ca = a'^{-1}ca' \qquad (c \in \mathcal{L}_U(\Delta,V))$$

which in turn can be written as

$$(a'a^{-1})c = c(a'a^{-1}) \qquad (c \in \mathcal{L}_U(\Delta,V)), \tag{17}$$

i.e., $a'a^{-1}$ commutes with all elements of $\mathcal{L}_U(\Delta,V)$. Hence by 5.17, formula (17) is equivalent to $a'a^{-1} = m_\gamma$ for some $\gamma \in \Delta^-$; the last expression is evidently equivalent to $a' = m_\gamma a$. The last statement of the corollary follows from 5.20.

I.5.22 <u>EXERCISES</u>.

A ring $(\mathfrak{R},+,.)$ is said to have <u>unique addition</u> if for any ring $(\mathfrak{R},\oplus,\cdot)$ defined on the same set and under the same multiplication, we have that $+ = \oplus$. This concept was introduced by R.E. Johnson [1].

i) Show that a ring $\mathfrak{R}$ has unique addition if and only if for any ring $\mathfrak{R}'$, every isomorphism of $\mathfrak{M}\mathfrak{R}$ onto $\mathfrak{M}\mathfrak{R}'$ is additive.

ii) A ring $\mathfrak{B}$ all of whose elements are idempotent is called a Boolean ring. Show that a Boolean ring has characteristic 2, is commutative, and has unique addition.

iii) Show that the ring Z of integers has only one automorphism, wheras $\mathfrak{M}Z$ has an infinite number of automorphisms. Hence Z is not a ring with unique addition.

For any ring $\mathfrak{R}$, by $\mathfrak{R}^+$ denote the additive group of $\mathfrak{R}$.

iv) Let $\mathfrak{R}$ be a ring and F be a prime field (i.e., F is either the field of rational numbers or the ring of residue classes of integers mod a prime). Prove that if $\mathfrak{M}\mathfrak{R} \cong \mathfrak{M}F$, then $\mathfrak{R} \cong F$, and if $\mathfrak{R}^+ \cong F^+$, then either $\mathfrak{R} \cong F$ or $\mathfrak{R}^2 = 0$. Give an example of a prime field F with an automorphism of $\mathfrak{M}F$ which is not additive (and hence F does not have unique addition) and an automorphism of $F^+$ which is not multiplicative.

v) Prove that if $\mathfrak{R}$ is a ring for which $\mathfrak{R}^+ \cong Q^+/Z^+$ (the additive group of rational numbers over the additive group of integers), then $\mathfrak{R}^2 = 0$. Give an example of a semigroup S with zero and with the property $S \cong \mathfrak{M}\mathfrak{R}$ for no ring $\mathfrak{R}$.

vi) Show that for a vector space V with $\dim V > 1$, any nonzero ideal of $\mathcal{S}(V)$ and any nonzero ideal of $\mathcal{L}(V)$ have isomorphic automorphism groups.

An algebra $(\Delta,*,A,+,\cdot)$ is a system consisting of a vector space $(\Delta,*,A,+)$ over a field $\Delta$ with scalar multiplication $*$ and addition $+$, and a ring $(A,+,\cdot)$ with the same addition $+$ and multiplication $\cdot$, and satisfying

$$\alpha(ab) = (\alpha a)b = a(\alpha b) \quad (\alpha \in \Delta,\ a,b \in A)$$

where we have denoted both "multiplications" by juxtaposition. The concepts of subalgebra, homomorphism, etc. for an algebra refer to the same notion for both the vector space and the ring structure of A.

vii) Let $(U,\Delta,V)$ be a pair of dual vector spaces where $\Delta$ is a field and $\dim V > 1$. Show that $\mathcal{L}_U(\Delta,V)$ can be made into an algebra over $\Delta$ by defining

$$x(\alpha a) = \alpha(xa) \quad (x \in V,\ \alpha \in \Delta,\ a \in \mathcal{L}_U(\Delta)).$$

Prove that every algebra isomorphism of $\mathcal{L}_U(\Delta,V)$ onto $\mathcal{L}_{U'}(\Delta,V')$ is induced by a linear isomorphism of $(U,\Delta,V)$ onto $(U',\Delta,V')$. Deduce that for $V = V'$, every algebra isomorphism of $\mathcal{L}_U(\Delta,V)$ onto $\mathcal{L}_{U'}(\Delta,V)$ can be extended to an algebra inner automorphism of $\mathcal{L}(\Delta,V)$, and that every algebra automorphism of $\mathcal{L}(\Delta,V)$ is inner.

viii) Prove that if $\Delta$ is a field, then every automorphism of $\mathcal{L}(\Delta,V)$ which leaves the elements of the center of $\mathcal{L}(\Delta,V)$ fixed is an inner automorphism.

ix) Show by an example that a subring of a ring with unique addition need not have a unique addition. Such an example is given by the ring $\mathfrak{R}$ of all matrices of the form $\begin{bmatrix} a_{11} & 0 & 0 \\ a_{21} & 0 & 0 \\ a_{31} & a_{32} & 0 \end{bmatrix}$ over a 2-element field. In fact, show that the mapping

$$\theta: \begin{bmatrix} a_{11} & 0 & 0 \\ a_{11} & 0 & 0 \\ a_{32} & a_{32} & 0 \end{bmatrix} \to \begin{bmatrix} a_{11} & 0 & 0 \\ a_{21} & 0 & 0 \\ a_{31} + a_{11}a_{21} & a_{32} & 0 \end{bmatrix}$$

is an automorphism of $\mathfrak{M}\mathfrak{R}$ but is not additive. Find another addition for $\mathfrak{M}\mathfrak{R}$ which makes it a ring.

x) State and prove the ring version of 5.12, 5.14 and 5.15 without the restriction on the dimension of V.

xi) Show that if $a$ in 5.17 is additive and commutes with all idempotents of $\mathfrak{F}_{2,U}(\Delta,V)$, and $\dim V > 1$, then $a$ must be a multiplication.

For further information on isomorphisms and semilinear transformations, see Baer ([1], Chapter V, Section 4), Behrens ([1], Chapter II, Section 6), Dieudonné [2], Gluskin [4], Jacobson [3], ([6], Chapter IX, Section 11), ([7], Chapter IV, Section 11), and for related material Baer ([1], Chapters III,IV), Eidelheit [1], Fajans [1], [2], Gewirtzman [1],[2], Gluskin [2],[6],[7], Halezov [1],[2], Jacobson [1], ([2], Chapter 2, Sections 7,8 and Chapter 3, Section 12),([6], Chapter IX, Section 12), ([7], Chapter IV, Section 12), Jodeit and Lam [1], R.E. Johnson [1], Johnson and Kiokemeister [1], Kaplansky [3], Mackey [1],[2], Martindale [1], Mihalev [1], Mihalev and Šatalova [1], Morita [1], Rédei and Steinfeld [1], Rickart [1],[2], [3], Rjabuhin [1], Stephenson [1], Wolfson [5],[6].

## I.6 GROUPS OF SEMILINEAR AUTOMORPHISMS

We will discuss only a few properties of certain groups of semilinear automorphisms we have essentially encountered in the preceding section. Throughout this section, we fix the following notation. Multiplication of functions on a set is taken to be their composition; we write all functions on the right.

$(\Delta,V)$ is a fixed vector space with $\dim V > 1$,

$\Sigma$ is the group of all semilinear automorphisms of $V$,

$\Lambda$ is the group of all linear automorphisms of $V$ (invertible linear transformations),

$M$ is the group of all multiplications $(m_\gamma = (\varepsilon_{\gamma^{-1}}, m_\gamma)$ with $\gamma \in \Delta^{\cdot})$,

$I$ is the group of all semilinear inner automorphisms (i.e., $I = \{(\varepsilon_\gamma, a) \in \Sigma \mid \gamma \in \Delta^{\cdot}\}$),

$\mathfrak{L} = \mathfrak{L}(\Delta, V)$.

Furthermore, for $S$ a semigroup (or ring), let

$\mathfrak{G}(S)$ be the group of all automorphisms of $S$,

$\mathfrak{J}(S)$ be the group of all inner automorphisms of $S$,

where an <u>inner automorphism</u> $\varepsilon_g$ of a semigroup or a ring is defined by $x\varepsilon_g = g^{-1}xg$ $(x \in S)$, as in the case of groups, with the provision that $g^{-1}$ exists, i.e., $S$ must have an identity element and $g$ must have an inverse relative to it.

If $H$ is a subset of a group $G$, then the <u>centralizer</u> of $H$ in $G$ is the set $\{g \in G \mid gh = hg$ for all $h \in H\}$. For subsets $H$ and $K$ of $G$, let $HK = \{hk \mid h \in H, k \in K\}$.

I.6.1 <u>PROPOSITION</u>. <u>The mapping</u> $\gamma \to m_{\gamma^{-1}}$ <u>is an isomorphism of</u> $\mathfrak{m}\Delta^-$ <u>onto</u> $M$. <u>Further,</u> $\Lambda$ <u>is the centralizer of</u> $M$ <u>in</u> $\Sigma$ <u>and is a normal subgroup of</u> $\Sigma$, $I$ <u>is a normal subgroup of</u> $\Sigma$, $I = M\Lambda$, <u>and</u> $M \cap \Lambda = \{m_\gamma \mid \gamma \in C(\Delta^-)\} = C(M)$.

<u>PROOF</u>. Exercise.

I.6.2 <u>THEOREM</u>. <u>The mapping</u> $\zeta: (\omega, a) \to \zeta_{(\omega,a)}$ $((\omega,a) \in \Sigma)$, <u>is a homomorphism of</u> $\Sigma$ <u>onto</u> $\mathfrak{G}(\mathfrak{L})$ <u>with kernel</u> $M$, <u>and maps</u> $I$ <u>onto</u> $\mathfrak{J}(\mathfrak{L})$. <u>Hence</u> $M$ <u>is a normal subgroup of</u> $\Sigma$ <u>and of</u> $I$ <u>and</u> $\Sigma/M \cong \mathfrak{G}(\mathfrak{L})$, $I/M \cong \mathfrak{J}(\mathfrak{L})$.

<u>PROOF</u>. By 5.8, $\zeta$ maps $\Sigma$ into $\mathfrak{G}(\mathfrak{L})$, and by 5.13, $\zeta$ maps $\Sigma$ onto $\mathfrak{G}(\mathfrak{L})$. For any $c \in \mathfrak{L}$ and $(\omega,a),(\pi,b) \in \Sigma$, we have

$$c\zeta_{(\omega,a)}\zeta_{(\pi,b)} = (a^{-1}ca)\zeta_{(\pi,b)} = b^{-1}(a^{-1}ca)b = (ab)^{-1}c(ab) = c\zeta_{(\omega\pi,ab)}$$

which implies that $\zeta_{(\omega,a)}\zeta_{(\pi,b)} = \zeta_{(\omega,a)(\pi,b)}$ and hence $\zeta$ is a homomorphism. By 5.21, $M$ is the kernel of $\zeta$. Hence $M$ is a normal subgroup of $\Sigma$ and $\Sigma/M \cong \mathfrak{G}(\mathfrak{L})$.

Since for any $\gamma \in \Delta^-$, we have $m_\gamma = (\varepsilon_{\gamma^{-1}}, m_\gamma)$, it follows that $M \subseteq I$ and thus $M$ is also a normal subgroup of $I$ and the kernel of $\zeta$ restricted to $I$. It remains to show that $\zeta$ maps $I$ onto $\mathfrak{J}(\mathfrak{L})$. Hence let $(\varepsilon_\gamma, a) \in I$. For any $\sigma \in \Delta$, $x \in V$, we obtain

$$(\sigma x)(m_\gamma a) = (\gamma\sigma x)a = \gamma^{-1}(\gamma\sigma)\gamma(xa) = \sigma\gamma(xa) = \sigma\{[(\gamma\varepsilon_{\gamma^{-1}})x]a\}$$
$$= \sigma[(\gamma x)a] = \sigma[(xm_\gamma)a] = \sigma[x(m_\gamma a)]$$

and hence $m_\gamma a$ is linear. This together with 5.20 yields

$$\zeta_{(\varepsilon_\gamma, a)} = \zeta_{m_\gamma a} = \varepsilon_{m_\gamma a} \in \mathfrak{J}(\mathfrak{L}).$$

Finally, let $\zeta_{(\omega,a)} \in \mathfrak{J}(\mathfrak{L})$. Then there exists a linear automorphism $b$ such that $\zeta_{(\omega,a)} = \zeta_{(\iota_\Delta, b)}$. But then 5.21 implies that $a = m_\gamma b$ for some $\gamma \in \Delta^-$ and $\omega = \varepsilon_{\gamma^{-1}}\iota_\Delta = \varepsilon_{\gamma^{-1}} \in \mathfrak{J}(\Delta)$ so that $(\omega,a) \in I$. It follows that $I$ is the complete inverse image of $\mathfrak{J}(\mathfrak{L})$ under $\zeta$. In particular, $I/M \cong \mathfrak{J}(\mathfrak{L})$.

I.6.3 COROLLARY. <u>The following conditions on a semilinear isomorphism</u> $(\omega,a)$ <u>are equivalent</u>: i) $\omega \in \mathcal{J}(\Delta)$, ii) $\zeta_{(\omega,a)} \in \mathcal{J}(\mathfrak{L})$, iii) $m_\gamma a$ <u>is linear for some</u> $\gamma \in \Delta^-$.

PROOF. Exercise.

I.6.4 COROLLARY. <u>The identity mapping is the only automorphism of</u> $\mathfrak{L}$ <u>which leaves fixed every element of</u> $\mathfrak{F}_{2,U}(V)$ <u>for some</u> t-<u>subspace</u> $U$ <u>of</u> $V^*$.

PROOF. Let $\theta$ be such an automorphism. Then $\theta = \zeta_{(\omega,a)}$ for some semilinear automorphism $(\omega,a)$ by 5.13. For every element $b$ of $\mathfrak{F}_{2,U}(V)$, we have $b = b\zeta_{(\omega,a)} = a^{-1}ba$ so that $ba = ab$. By 5.17, we must have $a = m_\gamma$ for some $\gamma \in \Delta^-$. Hence $(\omega,a) \in M$ and by 6.2, we obtain $\theta = \zeta_{(\omega,a)} = \iota_{\mathfrak{L}}$.

I.6.5 COROLLARY. <u>Let</u> $S$ <u>and</u> $S'$ <u>be as in</u> 5.12. <u>If</u> $\psi$ <u>and</u> $\psi'$ <u>are isomorphisms of</u> $S$ <u>onto</u> $S'$ <u>which agree on</u> $\mathfrak{F}_{2,U}(\Delta,V)$, <u>then</u> $\psi = \psi'$.

PROOF. By 5.12, both $\psi$ and $\psi'$ are induced by semilinear isomorphisms, say $(\omega,a)$ and $(\omega',a')$, respectively. Now $\zeta_{(\omega,a)}$ and $\zeta_{(\omega',a')}$ are isomorphisms of $\mathfrak{L}(\Delta,V)$ onto $\mathfrak{L}(\Delta',V')$, so that $\zeta_{(\omega,a)}\zeta^{-1}_{(\omega',a')}$ satisfies the conditions of 6.4, and thus $\zeta_{(\omega,a)}\zeta^{-1}_{(\omega',a')} = \iota_{\mathfrak{L}(\Delta,V)}$. Consequently $\zeta_{(\omega,a)} = \zeta_{(\omega',a')}$, and since these are extensions of $\psi$ and $\psi'$, respectively, we must have $\psi = \psi'$.

For the remainder of this section, we fix a basis $B$ of $V$ and let

$$\Theta = \{(\omega,a) \in \Sigma \mid va = v \quad \text{for all} \quad v \in B\}.$$

I.6.6 PROPOSITION. $\Theta$ <u>is a subgroup of</u> $\Sigma$, <u>the mapping</u>

$$\varphi: (\omega,a) \to \omega \qquad ((\omega,a) \in \Theta)$$

<u>is on isomorphism of</u> $\Theta$ <u>onto</u> $\mathcal{G}(\Delta)$, $\Sigma = \Theta\Lambda$, <u>and</u> $\Theta \cap \Lambda = \iota_V$.

PROOF. Verification that $\Theta$ is a subgroup of $\Sigma$ and that $\varphi$ is a homomorphism of $\Theta$ onto $\mathcal{G}(\Delta)$ with trivial kernel is left as an exercise (for "onto" part use the function a constructed in the proof of 5.15).

For $(\omega,a) \in \Sigma$ define $\bar{a}$ on $V$ as follows. Let $v\bar{a} = va$ for every $v \in B$, and extend $\bar{a}$ to a linear transformation on $V$. Since $a$ maps $B$ onto a basis of $V$, we have that $\bar{a}$ maps $V$ onto itself. Suppose that $v\bar{a} = 0$, where $v = \sum_{i=1}^{n} \sigma_i v_i$ $(v_i \in B)$. Then $0 = v\bar{a} = \sum_{i=1}^{n} \sigma_i(v_i a)$ which implies that $\sigma_1 = \sigma_2 = \dots = \sigma_n = 0$ since the vectors $v_1 a, v_2 a, \dots, v_n a$ are linearly independent. Hence $v = 0$ and $\bar{a}$ is one-to-one. Therefore $\bar{a} \in \Lambda$ and $(\omega,a) = (\omega,a\bar{a}^{-1})(\iota_\Delta,\bar{a})$. Hence $a\bar{a}^{-1}$ maps $V$ onto $V$ and is one-to-one; for $\sigma \in \Delta$, $x \in V$, we have

$$(\sigma x)a\bar{a}^{-1} = [(\sigma x)a]\bar{a}^{-1} = [(\sigma\omega)(xa)]\bar{a}^{-1} = (\sigma\omega)(xa\bar{a}^{-1})$$

and for $v \in B$, $v\bar{a} = va$ implies that $va\bar{a}^{-1} = v$. Thus $(\omega,a\bar{a}^{-1}) \in \Theta$ and $(\iota_\Delta,\bar{a}) \in \Lambda$ which proves that $\Sigma = \Theta\Lambda$.

If $(\omega,a) \in \Theta \cap \Lambda$, then for every $v \in B$, $va = v$ since $(\omega,a) \in \Theta$, which implies that $a = \iota_V$ since $(\omega,a) \in \Lambda$ and $B$ is a basis of $V$.

We now discuss a group theoretic construction which will clarify some of the concepts we have studied, see Hall ([1], Chapter 6, Section 5).

I.6.7 DEFINITION. Let $H$ and $A$ be groups and $\eta: H \to \mathcal{G}(A)$ be a homomorphism, we write $\eta: h \to \eta_h$. The set $H \times A$ together with the multiplication

$$(h,a)(h',a') = (hh',(a\eta_{h'})a')$$

is a semidirect product of $H$ with $A$ (determined by $\eta$).

We say that a group $G$ is a semidirect product of $H$ with $A$ (more precisely "determined by a homomorphism $\eta$") if $G$ is isomorphic to a group constructed as above. The mappings $h \to (h,1)$ and $a \to (1,a)$ are easily seen to be isomorphisms of $H$ and $A$, respectively, into the semidirect product; it is convenient to identify $H$ and $A$ with their respective images, and we do so henceforth.

I.6.8 THEOREM. *If $G$ is a semidirect product of a group $H$ with a group $A$, then $G$ is a group, $A$ is a normal subgroup of $G$, $G = HA$, $H \cap A = 1$. Conversely, if $G$ is a group having a subgroup $H$ and a normal subgroup $A$ such that $G = HA$ and $H \cap A = 1$, then $G$ is a semidirect product of $H$ with $A$.*

PROOF. Let $G$ be a semidirect product of $H$ with $A$. Then

$$[(h,a)(h',a')](h'',a'') = (hh',(a\eta_{h'})a')(h'',a'')$$
$$= (hh'h'',[(a\eta_{h'})a']\eta_{h''}a'') = (hh'h'',(a\eta_{h'}\eta_{h''})(a'\eta_{h''})a'')$$
$$= (hh'h'',(a\eta_{h'h''})(a'\eta_{h''})a'') = (h,a)(h'h'',(a'\eta_{h''})a'')$$
$$= (h,a)[(h',a')(h'',a'')]$$

and the multiplication is associative. Further

$$(h,a)(1,1) = (h1,(a\eta_1)1) = (h,a),$$
$$(1,1)(h,a) = (1h,(1\eta_h)a) = (h,a),$$

so that $(1,1)$ is the identity of $G$. Finally,

$$(h^{-1},a^{-1}\eta_{h^{-1}})(h,a) = (hh^{-1},(a^{-1}\eta_{h^{-1}})\eta_h a) = (1,a^{-1}a) = (1,1)$$

which shows that $(h^{-1},a^{-1}\eta_{h^{-1}})$ is a left inverse of $(h,a)$. It follows that $G$ is a group.

Now we note that

$$(h,a)^{-1}(1,b)(h,a) = (h^{-1},a^{-1}\eta_{h^{-1}})(h,(b\eta_h)a)$$
$$= (1,(a^{-1}\eta_{h^{-1}}\eta_h)(b\eta_h)a) = (1,a^{-1}(b\eta_h)a),$$

so that $A$ is a normal subgroup of $G$. It is clear that $G = HA$ and $H \cap A = 1$.

Conversely, let $G$ be a group with a subgroup $H$ and a normal subgroup $A$ such that $G = HA$ and $H \cap A = 1$. Since $G = HA$, any $g \in G$ can be written in the form $g = ha$ with $h \in H$, $a \in A$. This representation is unique since $ha = h'a'$ implies that $h'^{-1}h = a'a^{-1} \in H \cap A = 1$, so $h = h'$, $a = a'$. Next let $g = ha$, $g' = h'a'$; then

$$gg' = (ha)(h'a') = hh'(h'^{-1}ah')a', \tag{1}$$

where $h'^{-1}ah' \in A$ since $A$ is a normal subgroup of $G$. For $h \in H$, writing $a\eta_h = h^{-1}ah$ $(a \in A)$, we see that $\eta_h \in \mathfrak{G}(A)$. For $h,h' \in H$, we have

$$(a\eta_h)\eta_{h'} = h'^{-1}(h^{-1}ah)h' = (hh')^{-1}a(hh') = a\eta_{hh'}$$

so the mapping $\eta : H \to \mathfrak{G}(A)$ defined by $\eta : h \to \eta_h$ $(h \in H)$ is a homomorphism. We then can write (1) as

$$gg' = (hh')[(a\eta_{h'})a'] \tag{2}$$

or, writing the elements of $G$ as ordered pairs in $H \times A$, (1) becomes by virtue of (2), $(h,a)(h',a') = (hh',(a\eta_{h'})a')$.

Therefore $G$ is a semidirect product of $H$ with $A$.

With the notation of 6.8, let $G/A = B$. The conditions $G = HA$, $H \cap A = 1$ are equivalent to saying that $H$ intersects each coset of $A$ exactly once. It follows easily that $B \cong H$. In terms of (group) extensions, $G$ is an extension of $A$ by $H$; in fact, $G$ is a very special extension called a <u>split extension</u> of $A$ by $H$ (one also says that the extension $G$ <u>splits</u> over $A$). For a discussion of group extensions see Rédei ([1], § 50).

I.6.9 <u>PROPOSITION</u>. <u>Let $G$ be a semidirect product of $H$ with $A$ determined by the homomorphism $\eta : H \to \mathfrak{G}(A)$. Then the following statements are equivalent.</u>

i) <u>$G$ is the direct product of $H$ with $A$.</u>

ii) <u>$\eta$ maps $H$ onto the identity of $\mathfrak{G}(A)$.</u>

iii) <u>Every element of $H$ commutes with every element of $A$.</u>

<u>PROOF</u>. Exercise.

I.6.10 <u>PROPOSITION</u>. <u>$\Sigma$ is a semidirect product of $\mathfrak{G}(\Delta)$ with $\Lambda$ and $I$ is a semidirect product of $\mathfrak{J}(\Delta)$ with $\Lambda$.</u>

<u>PROOF</u>. We have seen in 6.1 that $\Lambda$ is a normal subgroup of $\Sigma$, so that by 6.6 and 6.8, $\Sigma$ is a semidirect product of $\Theta$ with $\Lambda$. Again by 6.6 we have $\Theta \cong \mathfrak{G}(\Delta)$ and hence $\Sigma$ is also a semidirect product of $\mathfrak{G}(\Delta)$ with $\Lambda$.

Now let $\Theta' = \Theta \cap I$. It is left as an exercise to show that $\Theta' \cong \mathfrak{J}(\Delta)$, $I = \Theta'\Lambda$, and $\Theta' \cap I = \iota_V$, $\Lambda$ is a normal subgroup of $I$, so that the proof of the second statement of the proposition follows along the same lines as the proof of the first.

I.6.11 **REMARK**. We can make 6.10 more precise by writing the semidirect product of $\mathfrak{G}(\Delta)$ with $\Lambda$ (and thus also of $\mathfrak{I}(\Delta)$ with $\Lambda$) in terms of ordered pairs and exhibiting the determining homomorphism $\eta$. To this end, we take into account the isomorphism in 6.6 between $\Theta$ and $\mathfrak{G}(\Delta)$, the representation of $(\omega,a) \in \Sigma$ as the product $(\omega, a\bar{a}^{-1})(\iota_\Delta, \bar{a})$ in the proof of 6.6, and the construction of the homomorphism $\eta$ in the proof of 6.8.

Let $\omega \in \mathfrak{G}(\Delta)$ and define $b$ on $V$ by

$$vb = \sum_{i=1}^{n} (\sigma_i\omega)v_i \quad \text{if} \quad v = \sum_{i=1}^{n} \sigma_i v_i \qquad (v \in B).$$

Then $(\omega,b) \in \Sigma$ and we obtain the function $\eta_\omega : a \to b^{-1}ab$ $(a \in \Lambda)$. Let $a \in \Lambda$. For any $v \in B$, we have $va = \sum_{i=1}^{n} \sigma_i v_i$ for some $\sigma_i \in \Delta$ and $v_i \in B$ whence

$$v(b^{-1}ab) = (va)b = (\sum_{i=1}^{n} \sigma_i v_i)b = \sum_{i=1}^{n} (\sigma_i\omega)v_i$$

since $b$ leaves the elements of $B$ fixed. Hence $\eta_\omega : a \to a^\omega$ $(a \in \Lambda)$ where

$$va^\omega = \sum_{i=1}^{n} (\sigma_i\omega)v_i \quad \text{if} \quad va = \sum_{i=1}^{n} \sigma_i v_i \qquad (v \in V,\ v_i \in B). \tag{3}$$

Thus the semidirect product multiplication is given by

$$(\omega,a)(\omega',a') = (\omega\omega', a^{\omega'} \cdot a') \qquad (\omega,\omega' \in \mathfrak{G}(\Delta),\ a,a' \in \Lambda),$$

where $a^\omega$ is given by (3).

We are now able to express automorphisms of $\mathfrak{L}$ in matrix form. If $c \in \mathfrak{L}$, then for any $v \in B$ we have $vc = \sum_{i=1}^{n} \sigma_i v_i$ for some $\sigma_i \in \Delta$ and $v_i \in B$. Hence $c$ can be represented as a $B \times B$ matrix over $\Delta$ having $\sigma_i$ in its $(v,v_i)$-entry for $1 \le i \le n$ and every $v \in B$, and $0$ otherwise. This matrix $C = [\gamma_{ij}]$ is row finite (i.e., every row has at most a finite number of nonzero entries). Conversely, it is clear that to every row finite $B \times B$ matrix $C$ one can associate uniquely an endomorphism of $(\Delta,V)$. It is easy to show that the correspondence $c \to C = [\gamma_{ij}]$ is an isomorphism of $\mathfrak{L}$ onto the set of all row finite $B \times B$ matrices over $\Delta$, where the matrices are added and multiplied as usual ("row by column" multiplication is possible because of row finiteness).

Now let $c \in \mathfrak{L}$ and $(\omega,a) \in \Sigma$. Then, as we have seen above,

$$c\zeta_{(\omega,a)} = a^{-1}ca = \bar{a}^{-1}[(a\bar{a}^{-1})^{-1}c(a\bar{a}^{-1})]\bar{a} \tag{4}$$

where $\bar{a} = (\iota_\Delta, \bar{a}) \in \Lambda$, $a\bar{a}^{-1} = (\omega, a\bar{a}^{-1}) \in \Theta$. For $v \in B$, we have $vc = \sum_{i=1}^{n} \sigma_i v_i$ for some $\sigma_i \in \Delta$, $v_i \in B$, and thus

$$v(a\bar{a}^{-1})^{-1}c(a\bar{a}^{-1}) = (vc)a\bar{a}^{-1} = (\sum_{i=1}^{n}\sigma_i v_i)a\bar{a}^{-1} = \sum_{i=1}^{n}(\sigma_i\omega)v_i$$

using the fact that $a\bar{a}^{-1} \in \mathfrak{G}$ twice. Then if $c$ corresponds to $C = [\gamma_{ij}]$ and $\bar{a}$ to $A = [\alpha_{ij}]$ in matrix form, we obtain by (4),

$$[\gamma_{ij}]\zeta_{(\omega,a)} = A^{-1}[\gamma_{ij}\omega]A.$$

We have thus proved, using 5.13, the following result. Call a matrix $C$ invertible if the corresponding endomorphism in $\mathfrak{L}$ is invertible.

For the finite dimensional vector spaces the next result has been around for some time, see Jacobson ([2], Chapter 2, Section 5); the semigroup version is due to Halezov [1] and Gluskin [3].

I.6.12 THEOREM. Let $(\Delta,V)$ be a vector space with $\dim V > 1$, and write the elements of $\mathfrak{L} = \mathfrak{L}(\Delta,V)$ in matrix form. Let $\omega \in \mathfrak{G}(\Delta)$, $A$ be an invertible matrix in $\mathfrak{L}$, and for $[\gamma_{ij}] \in \mathfrak{L}$, let

$$[\gamma_{ij}]\hat{\omega} = [\gamma_{ij}\omega], \quad [\gamma_{ij}]\varepsilon_A = A^{-1}[\gamma_{ij}]A.$$

Then $\zeta_{(\omega,A)} = \hat{\omega}\varepsilon_A$ is an automorphism of $\mathfrak{L}$, and conversely, every automorphism of $\mathfrak{L}$ is of this form. Moreover, $\zeta_{(\omega,a)} \in \mathfrak{I}(\mathfrak{L}(\Delta,V))$ if and only if $\omega \in \mathfrak{I}(\Delta)$.

We see that every automorphism of $\mathfrak{L}$ is the product of an automorphism of the form $\hat{\omega}$ and an inner automorphism. From the above discussion and 6.10, we see that $\Sigma$ is a semidirect product of $\mathfrak{G}(\Delta)$ with the group of invertible $B \times B$ matrices over $\Delta$ with multiplication

$$(\omega,A)(\omega',A') = (\omega\omega', A^{\omega'} \cdot A')$$

where for $A = [\alpha_{ij}]$, $A^{\omega} = [\alpha_{ij}\omega]$.

I.6.13 EXERCISES.

i) Define a semidirect product of a semigroup $H$ with a semigroup $A$ by using a homomorphism of $H$ into the endomorphism semigroup of $A$, otherwise the definition is to be the same as for groups. Let $(\Delta,V)$ be a vector space, $\mathfrak{P}(V)$ be the semigroup of semilinear transformations on $V$. For $(\omega,a) \in \mathfrak{P}(V)$, show that $Va$ is a subspace of $V$; let $\operatorname{rank}(\omega,a) = \dim Va$. Define

$$\mathfrak{P}_{\mathfrak{a}}(V) = \{(\omega,a) \in \mathfrak{P}(V) \mid \operatorname{rank} a < \mathfrak{a}\}.$$

Find all ideals of $\mathfrak{P}(V)$, show that $\mathfrak{P}_{\mathfrak{a}}(V)$ is a semidirect product of $\mathfrak{G}(\Delta)$ with $\mathfrak{S}_{\mathfrak{a}}(V)$ and that each maximal subgroup of $\mathfrak{P}(V)$ is a semidirect product of $\mathfrak{G}(\Delta)$ with a maximal subgroup of $\mathfrak{S}(V)$ (as groups).

ii) Show that $\mathcal{P}_2(V) \cong \mathfrak{M}^o(I_{V^*}, G, I_V; P)$, where $G$ is the semidirect product of $\mathcal{G}(\Delta)$ with $\mathfrak{M}\Delta^-$ (as groups) determined by $\eta: \mathcal{G}(\Delta) \to \mathcal{G}(\mathfrak{M}\Delta^-)$ with $\eta_\omega = \omega|_{\mathfrak{M}\Delta^-}$ and $p_{\mu i} = (\iota_\Delta, (v_\mu, u_i))$. (Hint: Consider $x\langle i,(\omega,\gamma),\lambda\rangle = (x,u_i)\omega\gamma v_\lambda\cdot$.)

iii) Let $(\Delta,V)$ be a vector space. For $a \in \Lambda$ and $z \in V$, the function $\alpha_{(a,z)}$ defined by $x\alpha_{(a,z)} = xa + z$ $(x \in V)$ is called an affine transformation; for $\gamma \in \Delta^-$, $\alpha_{(m_\gamma, z)}$ is called a homothetic transformation; $\alpha_{(m_1,z)}$ is called a translation. Let $A$, $H$, and $T$ be the groups of all affine transformations, homothetic transformations, and translations, respectively. Show that

a) $T \cong V^+$ (the additive group of $V$), $T$ is a normal subgroup of $A$, $A = \Lambda T$, $\Lambda \cap T = \iota_V$, so that $A$ is a semidirect product of $\Lambda$ with $V^+$; express this explicitly.

b) $M \cong \mathfrak{M}\Delta^-$, $T$ is a normal subgroup of $H$, $H = MT$, $M \cap T = \iota_V$, so that $H$ is a semidirect product of $\mathfrak{M}\Delta^-$ with $V^+$; express this explicitly.

iv) Find the center of the following groups: $\Sigma, \Lambda, I, \Theta, A, H$.

v) Show that "element" in the statement of 6.4 can be substituted by "idempotent". Use this to strengthen the assertion of 6.5.

For information on various (automorphism) groups related to vector spaces, see Baer ([1], Chapter VI), Dieudonné [4], Gluskin [2],[4],[6], Plotkin ([1], Chapter IV).

## I.7 EXTENSIONS OF SEMIGROUPS AND RINGS

I.7.1 DEFINITION. i) A semigroup $K$ is an extension of a semigroup $S$ if $S$ is an ideal of $K$.

ii) A congruence $\sigma$ on a semigroup $S$ is an equivalence relation on $S$ with the property that $a \,\sigma\, b$ implies $ac \,\sigma\, bc$, $ca \,\sigma\, cb$ for all $a,b,c \in S$. If $\varphi: S \to S'$ is a homomorphism, then the relation $\sigma$ defined on $S$ by: $x \,\sigma\, y$ if and only if $x\varphi = y\varphi$ is a congruence, called the congruence induced by $\varphi$. Conversely, if $\sigma$ is a congruence on $S$, then the set $S/\sigma$ of all classes of $\sigma$ can be given a multiplication in a natural way.

iii) An extension $K$ of $S$ is dense if the equality relation on $K$ is the only congruence on $K$ whose restriction to $S$ is the equality relation on $S$. A dense extension $K$ of $S$ is maximal if there exists no dense extension $K'$ of $S$ containing $K$ as a proper subsemigroup.

iv) A semigroup $S$ is weakly reductive if for any $a,b \in S$, $ax = bx$, $xa = xb$ for all $x \in S$, implies that $a = b$.

The main result of this section is the statement that $\mathcal{S}_U(V)$ is a maximal dense extension of each of its nonzero ideals (and the corresponding statement for $\mathcal{L}_U(V)$). For this we need some preparation.

I.7.2 DEFINITION. Let $S$ be a semigroup. A function $\lambda$ on $S$, written on the left, is a left translation if $\lambda(xy) = (\lambda x)y$ $(x,y \in S)$; a function $\rho$ on $S$, written on the right, is a right translation if $(xy)\rho = x(y\rho)$ $(x,y \in S)$. The pair $(\lambda,\rho)$ is a bitranslation if $\lambda$ is a left translation, $\rho$ is a right translation, and $x(\lambda y) = (x\rho)y$ $(x,y \in S)$. The set $\Omega(S)$ of bitranslations of $S$ under the multiplication $(\lambda,\rho)(\lambda',\rho') = (\lambda\lambda',\rho\rho')$ where $(\lambda\lambda')x = \lambda(\lambda'x)$, $x(\rho\rho') = (x\rho)\rho'$ $(x \in S)$, is a semigroup, the translational hull $\Omega(S)$ of $S$. For $a \in S$, the functions $\lambda_a$ and $\rho_a$ defined by: $\lambda_a x = ax$ and $x\rho_a = xa$ $(x \in S)$ are called inner left and inner right translations induced by $a$, respectively; $\pi_a = (\lambda_a,\rho_a)$ is the inner bitranslation induced by $a$. Let $\Pi(S) = \{\pi_a \mid a \in S\}$. We will sometimes write $\omega = (\lambda,\rho) \in \Omega(S)$ with $x\omega = x\rho$, $\omega x = \lambda x$ $(x \in S)$.

I.7.3 LEMMA. *With the notation as in 7.2, we have*

$$\omega\pi_a = \pi_{\omega a}, \quad \pi_a\omega = \pi_{a\omega} \qquad (a \in S,\ \omega \in \Omega(S))$$

*so that $\Pi(S)$ is an ideal of $\Omega(S)$.*

PROOF. Exercise.

I.7.4 LEMMA. *Let $K$ be an extension of a semigroup $S$. Let*

$$\tau = \tau(K:S) : k \to (\lambda^k,\rho^k) \qquad (k \in K)$$

*where $\lambda^k$ and $\rho^k$ are functions on $S$ defined by*

$$\lambda^k x = kx, \quad x\rho^k = xk \qquad (x \in S).$$

*Then $\tau$ is a homomorphism of $K$ into $\Omega(S)$ and maps $S$ onto $\Pi(S)$. Letting $\pi = \tau|_S$, we have $\pi : a \to \pi_a$ and $\pi$ is one-to-one if and only if $S$ is weakly reductive.*

PROOF. That $(\lambda^k,\rho^k) \in \Omega(S)$ and that $\tau$ is a homomorphism follows immediately from the associativity in $K$. For $a,b \in S$, $\pi_a = \pi_b$ if and only if $ax = bx$, $xa = xb$ $(x \in S)$ which implies that $\pi$ is one-to-one if and only if $S$ is weakly reductive.

Hence for a weakly reductive $S$, we may identify $S$ with $\Pi(S)$. The next result is due to Grillet and Petrich [1].

I.7.5 THEOREM. *Let $K$ be an extension of a weakly reductive semigroup $S$ and let $\tau = \tau(K:S)$. Then $K$ is a dense extension of $S$ if and only if $\tau$ is one-to-one, and is a maximal dense extension if and only if $\tau$ also maps $K$ onto $\Omega(S)$. If we identify $S$ with $\Pi(S)$, then dense extensions of $S$ may be identified with all subsemigroups of $\Omega(S)$ containing $\Pi(S)$, and maximal dense extensions with $\Omega(S)$.*

<u>PROOF</u>. Let $S$ and $K$ be as in the statement of the theorem and let $\sigma$ be a congruence on $K$ whose restriction to $S$ is the equality relation. If $a \sigma b$, then for any $x \in S$, $ax \sigma bx$, $xa \sigma xb$, so by hypothesis $ax = bx$, $xa = xb$. Consequently $\lambda^a = \lambda^b$, $\rho^a = \rho^b$, which proves that $\sigma$ is contained in the congruence induced by $\tau$. Recalling that $\tau|_S$ is one-to-one, we conclude that $K$ is a dense extension if and only if $\tau$ is one-to-one.

Let $\sigma$ be a congruence on a subsemigroup $T$ of $\Omega(S)$ containing $\Pi(S)$ whose restriction to $\Pi(S)$ is the equality relation. If $\omega \sigma \omega'$, then for any $\pi_a \in \Pi(S)$, we have $\omega\pi_a \sigma \omega'\pi_a$, $\pi_a\omega \sigma \pi_a\omega'$, so by 7.3 and the hypothesis, we obtain $\pi_{\omega a} = \pi_{\omega' a}$, $\pi_{a\omega} = \pi_{a\omega'}$. Hence $\omega a = \omega' a$ and $a\omega = a\omega'$ for all $a \in S$ since $S$ is weakly reductive. Thus $\omega = \omega'$ and $\sigma$ is the equality relation, so $T$ is a dense extension of $\Pi(S)$.

Let $K$ be a dense extension of $S$ for which $K\tau$ is a proper subsemigroup of $\Omega(S)$. Identifying $K$ with $K\tau$, we see that $\Omega(S)$ is a dense extension of $S$ properly containing $K$ so that $K$ is not a maximal dense extension of $S$. Hence if $K$ is a maximal dense extension of $S$, then $\tau$ must be onto. To prove the converse, it suffices to show that $\Omega(S)$ is a maximal dense extension of $\Pi(S)$. We have already proved that $\Omega(S)$ is a dense extension of $\Pi(S)$.

Let $K$ be an extension of $\Pi(S)$ containing $\Omega(S)$ as a proper subsemigroup, and let $a \in K\backslash\Omega(S)$. Then

$$\lambda^a\pi_x = a\pi_x, \quad \pi_x\rho^a = \pi_x a \qquad (x \in S).$$

Define $\lambda$ and $\rho$ on $S$ by $\pi_{\lambda x} = a\pi_x$, $\pi_{x\rho} = \pi_x a$ $(x \in S)$. It is easy to check that $(\lambda,\rho) \in \Omega(S)$, and writing $\omega = (\lambda,\rho)$ that $\lambda^a = \lambda^\omega$, $\rho^a = \rho^\omega$. Consequently $\tau = \tau(K:\Pi(S))$ has the property $a\tau = \omega\tau$ with $a \neq \omega$ so that $K$ is not a dense extension of $\Pi(S)$. Therefore $\Omega(S)$ is a maximal dense extension of $\Pi(S)$.

For $S$ a semigroup, let $S^2 = \{ab \mid a,b \in S\}$. The next theorem is mentioned in Petrich [3], its proof appears here for the first time.

I.7.6 <u>THEOREM</u>. <u>Let $S$ be a semigroup such that $S^2 = S$ and let $K$ be a maximal dense extension of $S$. Then $K$ is a maximal dense extension of each of its ideals containing $S$.</u>

<u>PROOF</u>. Let $T$ be an ideal of $K$ containing $S$. Since $K$ is a dense extension of $S$, it follows that $K$ is a dense extension of $T$. Let $D$ be a dense extension of $T$ containing $K$ as a subsemigroup. If $s \in S$, then $s = ab$ for some $a,b \in S$ since $S^2 = S$. Hence for any $d \in D$, we obtain $sd = a(bd) \in S$ since $T$ is an ideal of $D$ so $bd \in T$ and $S$ is an ideal of $T$. Consequently $S$ is an ideal of $D$. If $\sigma$ is a congruence on $D$ whose restriction to $S$ is the equality relation, then $\sigma$ restricted to $K$ is the equality relation on $K$ since $K$ is a dense extension of $S$. But then $\sigma$ restricted to $T$ is the

equality relation on T, and since D is a dense extension of T, it follows that $\sigma$ is the equality relation on all of D. Thus D is a dense extension of S which by maximality of K implies that D = K. Therefore K is a maximal dense extension of T.

I.7.7 COROLLARY. For a weakly reductive semigroup S for which $S^2 = S$, $\Omega(S)$ is a maximal dense extension of each of its ideals containing $\Pi(S)$.

PROOF. Exercise.

The next lemma can be found in Petrich [1].

I.7.8 LEMMA. Let $S = \mathfrak{M}^{\circ}(I,G,\Lambda;P)$ be a Rees matrix semigroup and let $0 \neq (\lambda,\rho) \in \Omega(S)$. Then there exist a subset A of I and functions $\alpha: A \to I$, $\varphi: A \to G$ (written on the left) and a subset $\Gamma$ of $\Lambda$ and functions $\beta: \Gamma \to \Lambda$, $\psi: \Gamma \to G$ (written on the right) satisfying

i) $$\lambda(i,a,\mu) = \begin{cases} (\alpha i, (\varphi i)a, \mu) & \text{if } i \in A \\ 0 & \text{otherwise} \end{cases},$$

ii) $$(i,a,\mu)\rho = \begin{cases} (i, a(\mu\psi), \mu\beta) & \text{if } \mu \in \Gamma \\ 0 & \text{otherwise} \end{cases},$$

iii) $i \in A$, $p_{\mu(\alpha i)} \neq 0$ if and only if $\mu \in \Gamma$, $p_{(\mu\beta)i} \neq 0$ and if so, then

$$p_{\mu(\alpha i)}(\varphi i) = (\mu\psi)p_{(\mu\beta)i},$$

for all $(i,a,\mu) \in S \setminus 0$.

PROOF. The condition $(\lambda,\rho) \neq 0$ means that at least one of the functions $\lambda$, $\rho$ is nonzero. But if, e.g., $\lambda x \neq 0$ then since S is a Rees matrix semigroup, there exists $y \in S$ such that $y(\lambda x) \neq 0$ so that $(y\rho)x \neq 0$ and $y\rho \neq 0$. Thus $\lambda \neq 0$ implies $\rho \neq 0$, and conversely, and hence both functions are nonzero.

If $\lambda(i,a,\mu) \neq 0$, then there exists $j \in I$ such that $p_{\mu j} \neq 0$ and

$$\lambda(i,a,\mu) = \lambda[(i,a,\mu)(j,p_{\mu j}^{-1},\mu)] = [\lambda(i,a,\mu)](j,p_{\mu j}^{-1},\mu) \neq 0$$

which implies that $\lambda(i,a,\mu)$ is of the form $( \ , \ ,\mu)$.

Suppose next that $\lambda(i,1,\mu) = (j,b,\mu) \neq 0$ and $\lambda(i,1,\nu) = (k,c,\nu) \neq 0$, where 1 is the identity of G. Then for some $m \in I$, we have $p_{\mu m} \neq 0$ and thus

$$\begin{aligned}(j,b,\nu) &= (j,b,\mu)(m,p_{\mu m}^{-1},\nu) = [\lambda(i,1,\mu)](m,p_{\mu m}^{-1},\nu) \\ &= \lambda[(i,1,\mu)(m,p_{\mu m}^{-1},\nu)] = \lambda(i,1,\nu) = (k,c,\nu) \neq 0.\end{aligned}$$

Consequently $j = k$ and $b = c$ which shows that the value of the first two indices of $\lambda(i,1,\mu)$ depends only upon i.

This makes it possible to define $A = \{i \in I \mid \lambda(i,1,\mu) \neq 0\}$ and the functions $\alpha : A \to I$, $\varphi : A \to G$ by the formula $\lambda(i,1,\mu) = (\alpha i, \varphi i, \mu)$ $(i \in A)$.

If now $\lambda(i,a,\mu) \neq 0$, then for some $j \in I$, $p_{\mu j} \neq 0$ so that

$$\lambda(i,a,\mu) = \lambda[(i,1,\mu)(j,p_{\mu j}^{-1}a,\mu)] = [\lambda(i,1,\mu)](j,p_{\mu j}^{-1}a,\mu)$$
$$= (\alpha i,\varphi i,\mu)(j,p_{\mu j}^{-1}a,\mu) = (\alpha i,(\varphi i)a,\mu)$$

with $i \in A$. The same calculation shows that if $\lambda(i,a,\mu) = 0$, then we must have $\lambda(i,1,\mu) = 0$ so that $i \notin A$. This proves part i) of the lemma; a symmetric argument establishes part ii).

For any $i,j \in I$ and $\mu,\nu \in \Lambda$, we obtain

$$(j,1,\mu)[\lambda(i,1,\nu)] = \begin{cases} (j,1,\mu)(\alpha i,\varphi i,\nu) & \text{if } i \in A \\ (j,1,\mu)0 & \text{otherwise} \end{cases}$$

$$= \begin{cases} (j,p_{\mu(\alpha i)}(\varphi i),\nu) & \text{if } i \in A,\ p_{\mu(\alpha i)} \neq 0 \\ 0 & \text{otherwise} \end{cases} \tag{1}$$

and analogously

$$[(j,1,\mu)\rho](i,1,\nu) = \begin{cases} (j,(\mu\psi)p_{(\mu\beta)i},\nu) & \beta \in \Gamma,\ p_{(\mu\beta)i} \neq 0 \\ 0 & \text{otherwise} \end{cases}. \tag{2}$$

Since by hypothesis, the left hand sides of (1) and (2) are equal, the assertions in part iii) of the lemma follow.

For the remainder of this section, let $(U,\Delta,V)$ be a pair of dual vector spaces. The next result is essentially taken from Petrich [2].

I.7.9 THEOREM. *For* $S = \mathfrak{F}_{2,U}(V)$ *and* $0 \neq (\lambda,\rho) \in \Omega(S)$, *let* $A,\alpha,\varphi,\Gamma,\beta,\psi$ *be as in* 7.8 *and define functions* $a$ *and* $b$ *by*

$$(\gamma v_\mu)a = \begin{cases} \gamma(\mu\psi)v_{\mu\beta} & \text{if } \mu \in \Gamma \\ 0 & \text{otherwise} \end{cases} \qquad (\gamma v_\mu \in V),$$

$$b(u_i\tau) = \begin{cases} u_{\alpha i}(\varphi i)\tau & \text{if } i \in A \\ 0 & \text{otherwise} \end{cases} \qquad (u_i\tau \in U).$$

*Then* $a \in \mathfrak{L}_U(V)$, $b$ *is the adjoint of* $a$ *in* $U$, *and*

$$\lambda r = ar, \quad r\rho = ra \qquad (r \in \mathfrak{F}_{2,U}(V)). \tag{3}$$

PROOF. For any $\gamma v_\mu \in V$ and $u_i\tau \in U$, using 7.8 we obtain

$$((\gamma v_\mu)a, u_i\tau) = \begin{cases} (\gamma(\mu\psi)v_{\mu\beta}, u_i\tau) & \text{if } \mu \in \Gamma \\ (0, u_i, \tau) & \text{otherwise} \end{cases}$$

$$= \begin{cases} \gamma(\mu\psi)(v_{\mu\beta}, u_i)\tau & \text{if } \mu \in \Gamma \\ 0 & \text{otherwise} \end{cases}$$

$$= \begin{cases} \gamma(v_\mu, u_{\alpha i})(\varphi i)\tau & \text{if } i \in A \\ 0 & \text{otherwise} \end{cases}$$

$$= \begin{cases} (\gamma v_\mu, u_{\alpha i}(\varphi i)\tau) & \text{if } i \in A \\ (\gamma v_\mu, 0) & \text{otherwise} \end{cases}$$

$$= (\gamma v_\mu, b(u_i\tau))$$

hence $b$ is an adjoint of $a$ so that $a \in \mathfrak{L}_U(V)$ and 1.2 implies that unique adjoint of $a$ in $U$.

For any $\gamma v_\mu \in V$ and $[j,\sigma,\nu] \in \mathfrak{F}_{2,U}(V)$, using 7.8 we obtain

$$\gamma v_\mu(\lambda[j,\sigma,\nu]) = \begin{cases} \gamma v_\mu[\alpha j, (\varphi j)\sigma, \nu] & \text{if } j \in A \\ 0 & \text{otherwise} \end{cases}$$

$$= \begin{cases} \gamma(v_\mu, u_{\alpha j})(\varphi j)\sigma v_\nu & \text{if } j \in A \\ 0 & \text{otherwise} \end{cases}$$

$$= \begin{cases} \gamma(\mu\psi)(v_{\mu\beta}, u_j)\sigma v_\nu & \text{if } \mu \in \Gamma \\ 0 & \text{otherwise} \end{cases}$$

$$= \begin{cases} \gamma(\mu\psi)v_{\mu\beta}[j,\sigma,\nu] & \text{if } \mu \in \Gamma \\ 0 & \text{otherwise} \end{cases}$$

$$= \gamma v_\mu(a[j,\sigma,\nu])$$

which proves the first part of (3). Finally

$$\gamma v_\mu([j,\sigma,\nu]\rho) = \begin{cases} \gamma v_\mu[j, \sigma(\nu\psi), \nu\beta] & \text{if } \nu \in \Gamma \\ 0 & \text{otherwise} \end{cases}$$

$$= \begin{cases} \gamma(v_\mu, u_j)\sigma(\nu\psi)v_{\nu\beta} & \text{if } \nu \in \Gamma \\ 0 & \text{otherwise} \end{cases}$$

$$= \gamma(v_\mu, u_j)\sigma v_\nu a = \gamma v_\mu([j,\sigma,\nu]a)$$

which proves the second part of (3).

The following result is new; for $\mathcal{S}(V)$ it is due to Gluskin [5].

I.7.10 THEOREM. *$\mathcal{S}_U(V)$ is a maximal dense extension of each of its nonzero ideals.*

PROOF. By 5.11 we know that $S = \mathcal{J}_{2,U}(V)$ is contained in every nonzero ideal of $K = \mathcal{S}_U(V)$. Since $S$ is a Rees matrix semigroup, it follows easily that $S$ is weakly reductive and $S^2 = S$. Thus to prove the theorem, according to 7.5 and 7.6, it suffices to show that $\tau = \tau(K:S)$ is one-to-one and maps $K$ onto $\Omega(S)$. The onto part follows from 7.9 since for a given $(\lambda,\rho) \in \Omega(S)$ we obtain $(\lambda^a,\rho^a) = (\lambda,\rho)$ with the notation of 7.9. If $a,b \in K$ and $a\tau = b\tau$, then $ar = br$ and thus $(a-b)r = 0$ for all $r \in S$ which by 3.3 implies that $a = b$ and thus $\tau$ is one-to-one.

I.7.11 COROLLARY. *For every nonzero ideal $I$ of $\mathcal{S}_U(V)$, we have $\Omega(I) \cong \mathcal{S}_U(V)$.*

PROOF. This follows directly from 7.10 and 7.5.

We will now obtain the corresponding results for nonzero (ring) ideals of $\mathcal{L}_U(V)$. For this we need the concepts and statements for rings which are analogues of those we have seen for semigroups. To this end we substitute semigroups with rings with the following modifications.

In 7.1: i) the definition remains the same; ii) instead of congruences, we speak of ideals; iii) an extension $K$ of a ring $S$ is *essential* if $0$ is the only ideal of $K$ whose intersection with $S$ is $0$ (such an ideal is called *large*); iv) is expressed by saying that the *annihilator*

$$\mathfrak{A}(S) = \{a \in S \mid as = sa = 0 \quad \text{for all} \quad s \in S\}$$

is $0$.

In 7.2 we only add that every left and every right translation also be an additive homomorphism.

In 7.6 the condition $S^2 = S$ can be weakened to $\{\sum_{i=1}^{n} s_i t_i \mid s_i, t_i \in S\} = S$ (i.e., the same kind of condition with $S^2$ defined as in ring theory).

With these modifications, all statements in 7.3-7.7 remain valid; the proof of this assertion is left as an exercise (work with ideals instead of congruences as is usual in ring theory).

I.7.12 LEMMA. *$\mathcal{J}_U(V)$ is contained in every nonzero ideal of $\mathcal{L}_U(V)$.*

PROOF. If $I$ is a nonzero ideal of $\mathcal{L}_U(V)$, then $\mathfrak{M}I$ is a nonzero ideal of $\mathcal{S}_U(V)$, so that by 5.11 we must have $\mathcal{J}_{2,U}(V) \subseteq I$. But then by 2.6 we obtain $\mathcal{J}_U(V) = C(\mathcal{J}_{2,U}(V)) \subseteq I$.

The following result appears to be new.

I.7.13 <u>THEOREM</u>. $\mathfrak{L}_U(V)$ <u>is a maximal essential extension of each of its nonzero ideals</u>.

<u>PROOF</u>. We know by 3.6 that $\mathfrak{F} = \mathfrak{F}_U(V)$ is regular and hence satisfies the (ring and semigroup) condition $\mathfrak{F}^2 = \mathfrak{F}$; by 7.12, $\mathfrak{F}$ is contained in every nonzero ideal of $\mathfrak{L} = \mathfrak{L}_U(V)$. Hence by the ring analogue of 7.6, in order to prove the theorem it suffices to show that $\mathfrak{L}$ is a maximal essential extension of $\mathfrak{F}$. Since $\mathfrak{F}$ is regular, we have $\mathfrak{A}(\mathfrak{F}) = 0$, so by the ring analogue of 7.5, in order to prove that $\mathfrak{L}$ is a maximal essential extension of $\mathfrak{F}$, it suffices to show that $\tau:\mathfrak{L} \to \Omega(\mathfrak{F})$ as in 7.4 is a (ring) isomorphism of $\mathfrak{L}$ onto $\Omega(\mathfrak{F})$. By the ring analogue of 7.4, we know that $\tau$ is a homomorphism. That the kernel of $\tau$ is trivial follows from 3.3.

To prove that $\tau$ maps $\mathfrak{L}$ onto $\Omega(\mathfrak{F})$, let $(\lambda,\rho) \in \Omega(\mathfrak{F})$. Next let $\lambda' = \lambda_{|\mathfrak{F}_2}$, $\rho' = \rho_{|\mathfrak{F}_2}$. Then for any $r \in \mathfrak{F}_2$, there exists $e \in \mathfrak{F}_2$ such that $r = re$ (recall that $\mathfrak{F}_2$ is a Rees matrix semigroup). Thus $\lambda'r = \lambda(re) = (\lambda r)e \in \mathfrak{F}_2$ so that $\lambda'$ maps $\mathfrak{F}_2$ into itself; similarly $\rho'$ maps $\mathfrak{F}_2$ into itself. It follows that $(\lambda',\rho') \in \Omega(\mathfrak{F}_2)$. By 7.9, there exists $a \in \mathfrak{L}$ such that

$$\lambda'r = ar, \quad r\rho' = ra \quad (r \in \mathfrak{F}_2).$$

If $c \in \mathfrak{F}$, then $c = \sum_{k=1}^{n} r_k$ for some $r_k \in \mathfrak{F}_2$, which then implies

$$\lambda c = \lambda(\sum_{k=1}^{n} r_k) = \sum_{k=1}^{n} \lambda r_k = \sum_{k=1}^{n} \lambda' r_k = \sum_{k=1}^{n} ar_k = a\sum_{k=1}^{n} r_k = ac$$

and similarly $c\rho = ca$. Since $a \in \mathfrak{L}$, it follows that $\tau$ maps $a$ onto $(\lambda,\rho)$, which proves that $\tau$ maps $\mathfrak{L}$ onto $\Omega(\mathfrak{F})$.

I.7.14 <u>COROLLARY</u>. <u>For every nonzero ideal</u> I <u>of</u> $\mathfrak{L}_U(V)$, <u>we have</u> $\Omega(I) \cong \mathfrak{L}_U(V)$.

<u>PROOF</u>. This follows directly from 7.13 and the ring analogue of 7.5.

I.7.15 <u>EXERCISES</u>.

i) Show that every regular semigroup is weakly reductive. Deduce that a regular ring has trivial annihilator.

ii) Let $\theta$ be an isomorphism of a semigroup $S$ onto a semigroup $S'$. Prove that $\theta$ induces an isomorphism $\bar{\theta}$ of $\Omega(S)$ onto $\Omega(S')$. If $S$ is weakly reductive and we identify $S$ with $\Pi(S)$, show that $\bar{\theta}$ is the unique extension of $\theta$ to a homomorphism of $\Omega(S)$ into $\Omega(S')$. Do the same for rings. (Hint: Let $(\lambda,\rho)\bar{\theta} = (\bar{\lambda},\bar{\rho})$, where $\bar{\lambda}x = [\lambda(x\theta^{-1})]\theta$, $x\bar{\rho} = [(x\theta^{-1})\rho]\theta$.)

iii) Let $S$ be a semigroup and $\lambda$, $\rho$ be a left and right translation of $S$, respectively. We say that $\lambda$ and $\rho$ are <u>permutable</u> if $(\lambda x)\rho = \lambda(x\rho)$ $(x \in S)$. Now let $(\lambda,\rho) \in \Omega(S)$ be invertible and suppose that $\lambda$ and $\rho$ are permutable. Show that the function

$$\delta_{(\lambda,\rho)}: x \to (\lambda^{-1}x)\rho \qquad (x \in S)$$

is an automorphism of $S$. If $S$ is weakly reductive, show that for any $(\lambda,\rho),(\lambda',\rho') \in \Omega(S)$, $\lambda$ and $\rho'$ are permutable. In such a case, if we identify $S$ with $\Pi(S)$, prove that $\delta_{(\lambda,\rho)} = \varepsilon_{(\lambda,\rho)}|_S$ where $\varepsilon_{(\lambda,\rho)}$ is the inner automorphism of $\Omega(S)$ induced by $(\lambda,\rho)$. Do the same for rings. We call $\delta_{(\lambda,\rho)}$ a <u>generalized inner automorphism</u>.

iv) Let $\mathfrak{R}$ be any ring and $(\lambda,\rho),(\lambda',\rho') \in \Omega(\mathfrak{R})$. Show that for every $r \in R$, $(\lambda r)\rho' - \lambda(r\rho') \in \mathfrak{A}(\mathfrak{R})$. Deduce that if $\mathfrak{A}(\mathfrak{R}) = 0$ and $(\lambda,\rho),(\lambda',\rho') \in \Omega(\mathfrak{R})$, then $\lambda$ and $\rho'$ are permutable.

v) Let $\mathfrak{R}$ be any ring, $\lambda$ and $\rho$ be functions on $\mathfrak{R}$ such that $x(\lambda y) = (x\rho)y$ $(x,y \in \mathfrak{R})$. Prove that for any $r,s \in \mathfrak{R}$,

a) $\lambda(rs) - (\lambda r)s \in \mathfrak{A}_r(\mathfrak{R})$, $\lambda(r+s) - (\lambda r + \lambda s) \in \mathfrak{A}_r(\mathfrak{R})$,

b) $(rs)\rho - r(s\rho) \in \mathfrak{A}_\ell(\mathfrak{R})$, $(r+s)\rho - (r\rho + s\rho) \in \mathfrak{A}_\ell(\mathfrak{R})$.

Deduce that if $\mathfrak{A}_r(\mathfrak{R}) = \mathfrak{A}_\ell(\mathfrak{R}) = 0$, for a pair $(\lambda,\rho)$ of functions on $\mathfrak{R}$ to be a bitranslation it suffices that for all $x,y \in \mathfrak{R}$, $x(\lambda y) = (x\rho)y$. Show that this is the case, e.g., when $\mathfrak{R}$ is a subring of $\mathfrak{L}_U(V)$ containing $\mathfrak{F}_U(V)$ for some dual pair $(U,V)$.

vi) In any ring $\mathfrak{R}$ define the <u>circle composition</u> $\circ$ by

$$x \circ y = x + y + xy \qquad (x,y \in \mathfrak{R}).$$

Show that under the circle composition $\mathfrak{R}$ is a semigroup with identity.

vii) Let $a$ and $a'$ be elements of a ring $\mathfrak{R}$ such that $a \circ a' = a' \circ a = 0$. Define a mapping $\varphi_{(a,a')}$ by

$$\varphi_{(a,a')}: x \to a' \circ x \circ a \qquad (x \in \mathfrak{R}).$$

Find an invertible bitranslation $(\lambda,\rho) \in \Omega(\mathfrak{R})$ for which $\lambda$ and $\rho$ are permutable and $\varphi_{(a,a')} = \delta_{(\lambda,\rho)}$; the automorphism $\varphi_{(a,a')}$ is called <u>quasi-inner</u>. Denoting by $\mathcal{Q}(\mathfrak{R})$ and $\mathcal{G}(\mathfrak{R})$ the sets of all quasi-inner and all generalized inner automorphisms, respectively, prove $\mathfrak{I}(\mathfrak{R}) \subseteq \mathcal{Q}(\mathfrak{R}) \subseteq \mathcal{G}(\mathfrak{R}) \subseteq \mathfrak{G}(\mathfrak{R})$.

viii) For $\mathfrak{F} = \mathfrak{F}_U(\Delta,V)$ with $\dim V > 1$, prove the following statements.

a) $\mathcal{Q}(\mathfrak{F}) = \{\varepsilon_b|_{\mathfrak{F}} \mid b = \iota_V + a,\ a \in \mathfrak{F},\ b$ invertible$\}$.

b) $\mathcal{G}(\mathfrak{F}) = \{\varepsilon_b|_{\mathfrak{F}} \mid b \in \mathfrak{L}_U(\Delta,V)$ and $b$ is invertible$\}$.

c) If $V$ is finite dimensional, then $\emptyset \neq \mathfrak{I}(\mathfrak{F}) = \mathcal{Q}(\mathfrak{F}) = \mathcal{G}(\mathfrak{F})$.

d) If $V$ is infinite dimensional, then $\emptyset = \mathfrak{I}(\mathfrak{F}) \subset \mathcal{Q}(\mathfrak{F}) \subset \mathcal{G}(\mathfrak{F})$.

e) If $\Delta$ is a field with more than two elements, then $\mathcal{Q}(\mathfrak{F})$ has more than one element.

f) $\mathfrak{G}(\mathfrak{F}) = \mathfrak{G}(\mathfrak{F})$ if and only if $\mathfrak{J}(\Delta) = \mathfrak{G}(\Delta)$.

ix) Let B and B′ be Boolean rings and $\varphi$ be a one-to-one function of B onto B′. Prove that $\varphi$ is a ring isomorphism if and only if $\varphi$ is an isomorphism of the corresponding semigroups of B and B′ under the circle composition. Give an example of an isomorphism of the additive groups of two Boolean rings which is not multiplicative.

x) Prove that if $\mathfrak{R}$ is a Boolean ring, then $\Omega(\mathfrak{R})$ is also. Show that a Boolean ring has trivial annihilator, so $\mathfrak{R} \to \Pi(\mathfrak{R}) \subseteq \Omega(\mathfrak{R})$ provides an embedding of a Boolean ring into a Boolean ring with identity. Do the same for a ring without proper zero divisors.

xi) Prove that every extension of a Boolean ring by a Boolean ring is again Boolean.

xii) Let $\mathfrak{R}$ be a commutative ring without proper zero divisors. Let $\mathfrak{Q}$ be the field of quotients of $\mathfrak{R}$, and $\mathfrak{R}'$ be the image of $\mathfrak{R}$ in $\mathfrak{Q}$ in the usual embedding of $\mathfrak{R}$ into $\mathfrak{Q}$. Show that $\Omega(\mathfrak{R}) \cong i_{\mathfrak{Q}}(\mathfrak{R}')$.

xiii) Let a ring $\mathfrak{E}$ be an extension of a ring $\mathfrak{R}$. Show that $\mathfrak{E}$ has no proper zero divisors if and only if $\mathfrak{R}$ has no proper zero divisors and the extension is essential.

xiv) Let $\mathfrak{R}$ be a ring with $\mathfrak{A}(\mathfrak{R}) = 0$ and $\mathfrak{E}$ be an essential extension of $\mathfrak{R}$. Prove that $\mathfrak{R}$ has unique addition if and only if $\mathfrak{E}$ has unique addition, and that the identity automorphism is the only endomorphism of $\mathfrak{E}$ leaving every element of $\mathfrak{R}$ fixed.

xv) A semigroup (or ring) S is said to be <u>uniform</u> if S is a maximal dense (essential) extension of each of its nonzero ideals (cf. 7.10 and 7.13). Show that the following semigroups and rings are uniform:

a) every simple semigroup or ring with identity,

b) $\{x \mid 0 < x \leq 1\}$ under multiplication,

c) $\{x \mid 0 \leq x \leq 1\}$ under multiplication,

d) nonnegative integers under addition,

e) the semigroup of all transformations on a nonempty set,

f) the ring of integers,

g) the ring of polynomials in one indeterminate over a field.

xvi) Let R be the field and topological space of real numbers, R[x] be the ring of polynomials over R in the indeterminate x, C be the ring of continuous functions mapping R into itself. For every nonnegative integer n, let $A_n$ be the set of all fractions $p(x)/q(x)$ such that $p(x), q(x) \in R[x]$, $q(x)$ has no real zeros, and $\deg q(x) - \deg p(x) \geq n$. Show that $i_C(A_n) = A_o$ and that $A_o$ is a maximal essential extension of $A_n$.

For the material concerning ideal extensions and translations in semigroups, see Grillet and Petrich [1], Petrich [1],[4], a systematic discussion can be found in Petrich ([5], Chapter III); in rings, see Everett [1], MacLane [1], Müller and Petrich [1],[2], Rédei ([1],§§52,53,54,85); (c) in both semigroups and rings, see B.E. Johnson [1] and the survey article Petrich [3]. For semigroups and rings of linear transformations related to the present subject, consult Gluskin [4],[5], Lambek ([1], Chapter IV), Petrich [2].

# PART II

## SEMIPRIME RINGS WITH MINIMAL ONE-SIDED IDEALS

This is a study of the class of rings in the title with various additional restrictions such as simplicity, atomicity and maximality relative to certain properties. Prime rings with a nonzero socle are characterized in several ways, one of these being that they are isomorphic to dense rings of linear transformations containing a nonzero transformation of finite rank. Simple rings with minimal one-sided ideals are characterized in several ways and a matrix representation is provided for them. These results are used to give several characterizations of the ring of all linear transformations as well as the ring of all linear transformations of finite rank on an arbitrary or a finite dimensional vector space. The nonzero socle of a semiprime ring is decomposed into homogeneous components and the semigroup socle is decomposed concordantly. Semiprime atomic rings as well as semiprime rings which are essential extensions of their nonzero socles are also characterized in several ways. For a subclass of the latter class of rings, isomorphisms of semigroup socles are extended uniquely to isomorphisms of the entire rings. Isomorphisms of Rees matrix rings are found.

Each of the classes of rings studied here is also characterized by the properties of the multiplicative semigroups of the rings belonging to the given class. This is possible because all these rings have unique addition.

### II.1 PRIME RINGS

The purpose of this section is to give several characterizations of prime rings with a nonzero socle. One of these characterizations says that they are precisely the rings isomorphic to dense rings of linear transformations containing a nonzero linear transformation of finite rank. This latter class was itself characterized in several ways in I.2. The first mentioned characterization can thus be interpreted as an abstract characterization of the class of concrete rings appearing in the second mentioned characterization.

After an introduction of needed concepts and preliminary results, we will collect all the characterizations in a single theorem. We denote by $\mathfrak{R}$ an arbitrary ring unless specified otherwise.

II.1.1 NOTATION. If $A$ and $B$ are nonempty subsets of $\mathfrak{R}$, let $AB$ denote the set of all elements of $\mathfrak{R}$ which are sums of elements of the form $ab$ with $a \in A$, $b \in B$. If $A = \{a\}$, we write $aB$ instead of $\{a\}B$; $Ab$ has a similar meaning. By induction, we define $A^n = (A^{n-1})A$ for $n > 1$, where $A^1 = A$, and see that $A^n$ is the set of all sums of elements of the form $a_1 a_2 \ldots a_n$ where $a_i \in A$.

II.1.2 DEFINITION. An (left, right, two-sided) ideal $I$ of $\mathfrak{R}$ is nilpotent if $I^n = 0$ for some $n$. A ring $\mathfrak{R}$ is semiprime if it has no nonzero nilpotent ideals. For a nonempty subset $B$ of $\mathfrak{R}$, the sets

$$\mathfrak{A}_{\ell,\mathfrak{R}}(B) = \{r \in \mathfrak{R} \mid rb = 0 \text{ for all } b \in B\}$$

$$\mathfrak{A}_{r,\mathfrak{R}}(B) = \{r \in \mathfrak{R} \mid br = 0 \text{ for all } b \in B\}$$

are the left and the right annihilators of $B$ in $\mathfrak{R}$, respectively.

It is clear that both $\mathfrak{A}_\ell(\mathfrak{R})$ and $\mathfrak{A}_r(\mathfrak{R})$ are nilpotent ideals of $\mathfrak{R}$.

II.1.3 LEMMA. A semiprime ring $\mathfrak{R}$ has no one-sided nilpotent ideals.

PROOF. Let $L$ be a left ideal of $\mathfrak{R}$ such that $L^n = 0$ and let $A = L\mathfrak{R}$. Then $A$ is a two-sided ideal of $\mathfrak{R}$ and a typical element of $A^n$ is a sum of elements of the form

$$\left(\sum_{k=1}^{p_1} \ell_{k,1} r_{k,1}\right)\left(\sum_{k=1}^{p_2} \ell_{k,2} r_{k,2}\right) \cdots \left(\sum_{k=1}^{p_n} \ell_{k,n} r_{k,n}\right).$$

But

$$(\ell_1 r_1)(\ell_2 r_2) \cdots (\ell_n r_n) = \ell_1 (r_1 \ell_2) \cdots (r_{n-1}\ell_n) r_n = 0$$

where $\ell_i \in L$, $r_i \in \mathfrak{R}$, since $r_i \ell_{i+1} \in L$ and $L^n = 0$. Hence $A^n = 0$ so $A = 0$ by hypothesis. Thus $L \subseteq \mathfrak{A}_\ell(\mathfrak{R})$ where $\mathfrak{A}_\ell(\mathfrak{R}) = 0$ by hypothesis, so that $L = 0$. The case of right ideals is analogous.

It follows from the lemma that the absence of nilpotent (1) left ideals, (2) right ideals, (3) ideals are equivalent statements. We will be dealing mainly with such rings.

II.1.4 DEFINITION. A (left, right, two-sided) ideal $A$ of $\mathfrak{R}$ is idempotent if $A^2 = A$.

Note that if $A$ is a minimal one-sided ideal of $\mathfrak{R}$, then $A^2 \subseteq A$ and $A^2$ is an ideal on the same side as $A$; so by minimality of $A$ either $A^2 = 0$ or $A^2 = A$. Hence a minimal one-sided ideal is either nilpotent or idempotent.

II.1.5 LEMMA. Every idempotent minimal one-sided ideal of $\mathfrak{R}$ is generated by an idempotent.

PROOF. Let $L$ be an idempotent minimal left ideal of $\mathfrak{R}$. If $La = 0$ for all $a \in L$, then $L^2 = 0$, a contradiction, hence $La \neq 0$ for some $a \in L$ and thus $La = L$ since $L$ is a minimal left ideal. Consequently $ea = a$ for some $e$ in $L$. Define $\varphi : x \to xa$ $(x \in L)$. Then $0\varphi^{-1}$ is a left ideal of $\mathfrak{R}$ contained in $L$, and is different from $L$ since $e \notin \mathfrak{O}\varphi^{-1}$. By minimality of $L$, we have $\mathfrak{O}\varphi^{-1} = 0$. Hence $e^2 - e \in \mathfrak{O}\varphi^{-1}$ implies $e^2 = e$. Thus $Le$ is a nonzero left ideal contained in $L$, so $Le = L$. Hence $\mathfrak{R}e = L$. Again the case of right ideals is analogous.

The following statement will prove quite useful.

II.1.6 COROLLARY. *In a semiprime ring every minimal one-sided ideal is generated by an idempotent.*

II.1.7 NOTATION. If $A$ and $B$ are nonempty subsets of a semigroup $S$, $AB$ denotes the set of all elements of the form $ab$ with $a \in A$, $b \in B$. Defining $A^n$ by induction, we see that $A^n$ is the set of all elements of the form $a_1 a_2 \dots a_n$ with $a_i \in A$. Again the same convention of writing $aB$ if $A = \{a\}$ will be used.

II.1.8 DEFINITION. For a semigroup $S$ with zero, the definition of a nilpotent left, right or two-sided ideal remains the same as in the ring case with the semigroup definition of $A^n$. Further, $S$ is a *semiprime semigroup* if it has no nilpotent ideals. The annihilators $\mathfrak{A}_\ell(S)$ and $\mathfrak{A}_r(S)$ are defined as in the case of a ring.

Note that in spite of the different definitions of $A^n$ in a ring or a semigroup, for a subset $A$ of $\mathfrak{R}$, $A^n = 0$ in $\mathfrak{R}$ if and only if $A^n = 0$ in $\mathfrak{M}\mathfrak{R}$.

II.1.9 LEMMA. *A semiprime semigroup has no one-sided nilpotent ideals.*

PROOF. Exercise.

II.1.10 DEFINITION. In a semigroup $S$ with zero, a nonzero left (right, two-sided) ideal $L$ is *minimal* if it contains no nonzero left (right, two-sided) ideal of $S$ different from $L$ (in semigroup theory such ideals are called 0-minimal).

II.1.11 LEMMA. *Let $S$ be a semigroup or a ring. Then a nonempty subset $L$ of $S$ satisfying $Sa = L$ for every $0 \neq a \in L$ is a minimal left ideal; the converse holds if $S$ is semiprime. The corresponding statement is valid for right ideals.*

PROOF. The following arguments are valid for both a semigroup and a ring $S$. Let $L$ have the property in the statement of the lemma. It is easy to see that $L$ is a left ideal. Let $L'$ be a left ideal of $S$ such that $0 \neq L' \subseteq L$; then for any $0 \neq a \in L'$, we have $L = Sa \subseteq L'$ so that $L' = L$ and $L$ is minimal.

Conversely, suppose that $S$ is semiprime, let $L$ be a minimal left ideal of $S$, and let $0 \neq a \in L$. Then $Sa$ is a left ideal of $S$ contained in $L$ and thus $Sa = L$ or $Sa = 0$. The latter alternative cannot occur since then $a \in \mathfrak{A}_\ell(S)$, and $\mathfrak{A}_\ell(S)$ would be a nonzero nilpotent ideal.

II.1.12 LEMMA. A ring $\mathfrak{R}$ is semiprime if and only if $\mathfrak{M}\mathfrak{R}$ is a semiprime semigroup.

PROOF. Necessity. Let $I$ be an ideal of $\mathfrak{M}\mathfrak{R}$ with $I^n = 0$. Then $\mathfrak{C}(I)$ is an ideal of $\mathfrak{R}$ and $[\mathfrak{C}(I)]^n = 0$, which by hypothesis implies $\mathfrak{C}(I) = 0$. But then $I = 0$ and hence $\mathfrak{M}\mathfrak{R}$ is semiprime.

Sufficiency. Obvious.

II.1.13 COROLLARY. In a semiprime ring $\mathfrak{R}$, minimal one-sided ideals of $\mathfrak{R}$ and $\mathfrak{M}\mathfrak{R}$ coincide.

II.1.14 LEMMA. The following statements concerning an idempotent $e$ of a semiprime ring $\mathfrak{R}$ are equivalent.

i) $\mathfrak{R}e$ is a minimal left ideal.

ii) $e\mathfrak{R}e$ is a division ring.

iii) $e\mathfrak{R}$ is a minimal right ideal.

PROOF. i) $\Rightarrow$ ii). Since $e\mathfrak{R}e = \mathfrak{R}e \cap e\mathfrak{R}$, we have that $e\mathfrak{R}e$ is an additive group. Let $0 \neq a \in e\mathfrak{R}e$, then $0 \neq a \in \mathfrak{R}e$ and 1.11 implies that $\mathfrak{R}a = \mathfrak{R}e$. Hence $e = ra$ for some $r \in \mathfrak{R}$. Consequently

$$e = ra = (er)(ea) = (ere)a$$

where $ere \in e\mathfrak{R}e$, and since $e$ is the identity of $e\mathfrak{R}e$, the nonzero elements of $e\mathfrak{R}e$ form a multiplicative group, so $e\mathfrak{R}e$ is a division ring.

ii) $\Rightarrow$ i). Let $0 \neq a \in \mathfrak{R}e$. Then $a = ae \in a\mathfrak{R}$ so $a\mathfrak{R} \neq 0$ and thus also $(\mathfrak{R}a)^2 \neq 0$ since $\mathfrak{R}$ is semiprime. Hence $aba \neq 0$ for some $b \in \mathfrak{R}$ so that $(ae)ba \neq 0$ and thus $0 \neq eba \in e\mathfrak{R}e$. Since $e\mathfrak{R}e$ is a division ring, there exists $y \in e\mathfrak{R}e$ such that $y(eba) = e$. For any $c \in \mathfrak{R}e$, we obtain $c = ce = cyeba \in \mathfrak{R}a$ which proves that $\mathfrak{R}e \subseteq \mathfrak{R}a$. Conversely, $a \in \mathfrak{R}$ which implies $\mathfrak{R}a \subseteq \mathfrak{R}e$ so that $\mathfrak{R}a = \mathfrak{R}e$ for every $0 \neq a \in \mathfrak{R}e$ and 1.11 implies that $\mathfrak{R}e$ is a minimal left ideal.

The equivalence of ii) and iii) now follows by symmetry.

The proof of 1.14 is almost entirely multiplicative, hence the same statements remain valid in a semigroup by replacing "division ring" by "group with zero". The notion of socle introduced below is due to Dieudonné [2].

II.1.15 DEFINITION. Let $\{A_i\}_{i\in I}$ be a nonempty family of subrings of $\mathfrak{R}$. Then the subring of $\mathfrak{R}$ generated by $\bigcup_{i\in I} A_i$ is called a sum of the subrings $\{A_i\}_{i\in I}$, and is denoted by $\sum_{i\in I} A_i$. A sum of an empty set of subrings is defined to be equal to zero. A sum of all minimal right ideals of $\mathfrak{R}$ is the socle $\mathfrak{S}$ of $\mathfrak{R}$.

Note that all the elements of a nonzero socle $\mathfrak{S}$ are of the form $r_1+r_2+\ldots+r_n$ where $r_i \in R_i$ for some minimal right ideals $R_i$, and that $\mathfrak{S}$ is a right ideal of $\mathfrak{R}$.

II.1.16 LEMMA. In a semiprime ring $\mathfrak{R}$, the socle $\mathfrak{S}$ of $\mathfrak{R}$ coincides with the sum of all minimal left ideals.

PROOF. It follows from 1.14 that $\mathfrak{R}$ has a minimal left ideal if and only if it has minimal right ideal. By symmetry, it suffices to prove that $\mathfrak{S}$ contains every minimal left ideal of $\mathfrak{R}$. Hence let $L$ be a minimal left ideal and $0 \neq a \in L$. By 1.5, there exists an idempotent $e$ such that $\mathfrak{R}e = L$. Further $\mathfrak{R}a = L$ by 1.11, and thus $\mathfrak{R}(a\mathfrak{R})a = (\mathfrak{R}a)^2 = \mathfrak{R}a$ since $\mathfrak{R}$ is semiprime, proving that $a\mathfrak{R} \neq 0$. Let $0 \neq b \in a\mathfrak{R}$, then $b\mathfrak{R} \subseteq a\mathfrak{R}$. We prove next that the opposite inclusion also holds. It follows that for some $x \in \mathfrak{R}$, we have $b = ax = (ae)x = a(ex)$ and hence $ex \neq 0$. Since $\mathfrak{U}_\ell(\mathfrak{R}) = 0$, it follows that $ex\mathfrak{R} \neq 0$ and thus also $(ex\mathfrak{R})^2 \neq 0$. Hence $exye \neq 0$ for some $y \in \mathfrak{R}$, so that $0 \neq exye \in e\mathfrak{R}e$, where $e\mathfrak{R}e$ is a division ring by 1.14. Thus there exists $c \in e\mathfrak{R}e$ such that $(exye)c = e$. Consequently

$$b(yec) = a(exyec) = ae = a$$

which proves that $a \in b\mathfrak{R}$ and therefore $b\mathfrak{R} = a\mathfrak{R}$. By 1.11, $a\mathfrak{R}$ is a minimal right ideal and hence $a \in a\mathfrak{R} \subseteq \mathfrak{S}$.

It follows from the proof that the set theoretic union of all minimal left and of all minimal right ideals coincide.

II.1.17 COROLLARY. The socle of a semiprime ring $\mathfrak{R}$ is an ideal of $\mathfrak{R}$.

The sum of all minimal left ideals is called the left socle, whereas the socle above is called the right socle. Because of 1.16, we need not make the left-right distinction when we are interested in semiprime rings. The following concept is due to Jacobson [4].

II.1.18 DEFINITION. A ring $\mathfrak{R} \neq 0$ is atomic if it coincides with its socle (i.e., $\mathfrak{R}$ is a sum of its minimal right ideals).

II.1.19 LEMMA. If $\mathfrak{R}$ is a simple atomic ring, then $\mathfrak{R} \cong \mathfrak{F}_U(V)$ for some pair $(U,V)$ of dual vector spaces over a division ring $\Delta$.

PROOF. First note that $\mathfrak{R}^2 \neq 0$ implies that $\mathfrak{R}$ is semiprime. Hence 1.6 implies that every minimal right ideal is of the form $V = e\mathfrak{R}$ for some idempotent $e$. By 1.14, $\Delta = e\mathfrak{R}e$ is a division ring and $U = \mathfrak{R}e$ is a minimal left ideal of $\mathfrak{R}$. Letting $\Delta$ act on $V$ on the left and on $U$ on the right by multiplication in $\mathfrak{R}$, we easily see that $(\Delta,V)$ is a left and $(U,\Delta)$ is a right vector space over $\Delta$. Define a form $\mathfrak{B}: V \times U \to \Delta$ by $(er,se) = e(rs)e$ $(r,s \in \mathfrak{R})$. Then $\mathfrak{B}$ is evidently bilinear. If $er \neq 0$, then $er\mathfrak{R} = e\mathfrak{R} = V$ by 1.11, so $ers = e$ for some $s \in \mathfrak{R}$ and hence $erse = e \neq 0$ which shows that $(er,se) \neq 0$. Dually $se \neq 0$ implies $(er,se) \neq 0$ for some $r \in \mathfrak{R}$ and hence $\mathfrak{B}$ is nondegenerate. Consequently $(U,\Delta,V;\mathfrak{B}) = (U,V)$ is a pair of dual vector spaces.

For every $r \in \mathfrak{R}$, let $\overline{r}$ be defined by

$$(es)\overline{r} = e(sr) \qquad (s \in \mathfrak{R});$$

then it follows immediately that $\overline{r} \in \mathfrak{L}(\Delta,V)$. Define $\hat{r}$ by

$$\hat{r}(se) = (rs)e \qquad (s \in \mathfrak{R});$$

then similarly $\hat{r} \in \mathfrak{L}(U,\Delta)$. Since

$$((es)\overline{r},te) = (e(sr),te) = esrte = (es,r(te)) = (es,\hat{r}(te)),$$

it follows that $\hat{r}$ is the adjoint of $\overline{r}$ in U, and hence $\overline{r} \in \mathfrak{L}_U(V)$.

The mapping $\chi: r \to \overline{r}$ $(r \in \mathfrak{R})$ is easily seen to be a homomorphism. Since $\overline{e} \neq 0$, we have $\ker \chi \neq \mathfrak{R}$ and since $\mathfrak{R}$ is simple and $\ker \chi$ is an ideal of $\mathfrak{R}$, it follows that $\ker \chi = 0$. Thus $\chi$ is an isomorphism of $\mathfrak{R}$ into $\mathfrak{L}_U(V)$.

For any $r \in \mathfrak{R}$, we have $(er)\overline{e} = ere = (ere)e$ where $ere \in \Delta$, $e \in V$, and hence rank $\overline{e} = 1$. Since $\mathfrak{F}_U(V)$ is an ideal of $\mathfrak{L}_U(V)$, the set

$$T = \{r \in \mathfrak{R} \mid \overline{r} \in \mathfrak{F}_U(V)\}$$

must be a nonzero ideal of $\mathfrak{R}$. Hence $T = \mathfrak{R}$ and thus $\chi$ maps $\mathfrak{R}$ into $\mathfrak{F}_U(V)$.

For $[i,\gamma,\lambda] \in \mathfrak{F}_{2,U}(V)$, we have $u_i = te$, $\gamma = epe$, $v_\lambda = er$ for some $t,p,r \in \mathfrak{R}$, and thus

$$(es)[i,\gamma,\lambda] = (es)(te)(epe)(er) = e(steper) = (es)\overline{teper}$$

which shows that $(teper)\chi = [i,\gamma,\lambda]$. Since $\mathfrak{F}_{2,U}(V)$ generates $\mathfrak{F}_U(V)$, it follows that $\chi$ maps $\mathfrak{R}$ onto $\mathfrak{F}_U(V)$. Therefore $\mathfrak{R} \cong \mathfrak{F}_U(V)$.

The following concepts will play an important role.

II.1.20 DEFINITION. Let M be an additively written abelian group and $\mathfrak{R}$ be a ring. If $\mathfrak{R}$ acts on M on the right satisfying: for any $x,y \in M$, $a,b \in \mathfrak{R}$,

$$(x+y)a = xa + ya,$$
$$x(a+b) = xa + xb,$$
$$(xa)b = x(ab),$$

then M is a right $\mathfrak{R}$-module or simply an $\mathfrak{R}$-module. An additive subgroup N of M for which $xa \in N$ for all $x \in N$, $a \in \mathfrak{R}$ is an $\mathfrak{R}$-submodule of M. If $\emptyset \neq A \subseteq M$, $\emptyset \neq B \subseteq \mathfrak{R}$, let AB be the set of all elements of M which are sums of elements of the form ab with $a \in A$, $b \in B$.

An $\mathfrak{R}$-module $M$ is <u>irreducible</u> if (1) $mr \neq 0$ for some $m \in M$, $r \in \mathfrak{R}$ and (2) $M$ has no $\mathfrak{R}$-submodule different from $M$ and $0$ (caution: the zero of both $M$ and $\mathfrak{R}$ are denoted by $0$). Further, $M$ is <u>faithful</u> if for any nonzero element $r \in \mathfrak{R}$, there exists an element $m \in M$ such that $mr \neq 0$.

II.1.21 <u>DEFINITION</u>. A ring $\mathfrak{R}$ is <u>primitive</u> if it has a faithful irreducible module. An ideal $I$ of $\mathfrak{R}$ is a <u>prime ideal</u> if for any ideals $A$ and $B$ of $\mathfrak{R}$, $AB \subseteq I$ implies that either $A \subseteq I$ or $B \subseteq I$. Further, $\mathfrak{R}$ is a <u>prime ring</u> if the ideal $0$ is prime. An ideal $I$ of $\mathfrak{R}$ is <u>semiprime</u> if for any ideal $A$ of $\mathfrak{R}$, $A^n \subseteq I$ implies that $A \subseteq I$.

Hence a ring $\mathfrak{R}$ is semiprime if and only if $0$ is a semiprime ideal of $\mathfrak{R}$. This is in conformity with the definition of a semiprime ring and can be taken as a definition. Also note that $\mathfrak{R}$ is a prime ring if and only if for any nonzero ideals $A$ and $B$ of $\mathfrak{R}$, we have $AB \neq 0$.

II.1.22 <u>LEMMA</u>. <u>Every primitive ring is prime</u>.

<u>PROOF</u>. Let $\mathfrak{R}$ be a primitive ring, $A$ and $B$ be nonzero ideals of $\mathfrak{R}$, and $M$ be a faithful irreducible $\mathfrak{R}$-module. Since $M$ is faithful and $A \neq 0$, there exists $m \in M$ such that $mA \neq 0$, so that $mA = M$ since $mA$ is an $\mathfrak{R}$-submodule of $M$ and $M$ is irreducible. If $AB = 0$, then $0 = mAB = MB$ which is impossible since $B \neq 0$ and $M$ is faithful. Consequently $AB \neq 0$ and $\mathfrak{R}$ is a prime ring.

II.1.23 <u>LEMMA</u>. <u>If $S$ is a semigroup with zero and a dense extension of a semigroup $K$ without proper nonzero ideals, then $K$ is contained in every nonzero ideal of $S$</u>.

<u>PROOF</u>. Let $I$ be an ideal of $S$ such that $K \not\subseteq I$. Thus $K \cap I = 0$ since $K$ has no proper nonzero ideals. On $S$ define the following relation: $x \tau y$ if and only if either $x, y \in I$ or $x = y$. It is easy to see that $\tau$ is a congruence on $S$ whose restriction to $K$ is the equality relation. Thus $\tau$ must be the equality relation on $S$ since $S$ is a dense extension of $K$, but then $I = 0$.

II.1.24 <u>DEFINITION</u>. A semigroup is <u>completely 0-simple</u> if it is isomorphic to a Rees matrix semigroup.

This is not a customary definition of a completely 0-simple semigroup in the theory of semigroups, but it turns out to be equivalent to the usual one, see Clifford and Preston ([1], Chapter 3, Section 2). We are now ready to prove the principal result of this section. The implication ii) $\Rightarrow$ v) is a special case of the Jacobson Density Theorem, see Jacobson [3]. The equivalences ii) $\Leftrightarrow$ v) $\Leftrightarrow$ I.2.13 i) are due to Jacobson [4]. The implications i) $\Rightarrow$ I.2.13 i) and vii) $\Rightarrow$ I.2.13 i) are proved in Gluskin [4]. The remaining implications appeared in Petrich [4].

II.1.25 THEOREM. *The following conditions on a ring* $\mathfrak{R}$ *are equivalent.*

i) $\mathfrak{R}$ *is a semiprime ring with a minimal right ideal* $A$ *such that* $\mathfrak{A}_{r,\mathfrak{R}}(A) = 0$.

ii) $\mathfrak{R}$ *is a primitive ring with a nonzero socle.*

iii) $\mathfrak{R}$ *is a prime ring with a nonzero socle.*

iv) $\mathfrak{R}$ *is an essential extension of its simple socle.*

v) $\mathfrak{R}$ *is isomorphic to a dense ring of linear transformations on a vector space containing a nonzero transformation of finite rank.*

vi) $\mathfrak{M}\mathfrak{R}$ *is a dense extension of a completely 0-simple semigroup.*

vii) $\mathfrak{M}\mathfrak{R}$ *has a completely 0-simple ideal contained in every nonzero ideal of* $\mathfrak{M}\mathfrak{R}$.

PROOF. i) ⇒ ii). We consider $A$ as an $\mathfrak{R}$-module. Then its minimality as a right ideal implies that it is irreducible and the condition $\mathfrak{A}_{r,\mathfrak{R}}(A) = 0$ that it is faithful.

ii) ⇒ iii). This is a special case of 1.22.

iii) ⇒ iv). Let $\mathfrak{S}$ be the socle of $\mathfrak{R}$, $I$ be a nonzero ideal of $\mathfrak{S}$, and $A$ be a minimal right ideal of $\mathfrak{R}$. If $\mathfrak{R} I \mathfrak{R} = 0$, then $I\mathfrak{R} \subseteq \mathfrak{A}_{\ell}(\mathfrak{R}) = 0$ and thus $I \subseteq \mathfrak{A}_r(\mathfrak{R}) = 0$, a contradiction. Similarly $\mathfrak{R}A \neq 0$, and thus $(\mathfrak{R} I \mathfrak{R})(\mathfrak{R}A) \neq 0$ since $\mathfrak{R}$ is a prime ring. Hence $bra \neq 0$ for some $b \in I$, $r \in \mathfrak{R}$, $a \in A$. By 1.6, there exists an idempotent $e$ such that $A = e\mathfrak{R}$. Hence $a = ea$ and thus $b(re)a \neq 0$ where $re \in \mathfrak{S}$ by 1.17, which implies that $b(re) \in I$. Letting $c = bre$, we see that $0 \neq c \in I$ and $ce \neq 0$. The hypothesis implies that $(\mathfrak{R}ce\mathfrak{R})^2 \neq 0$ so that $esc \neq 0$ for some $s \in \mathfrak{R}$. Letting $d = esc$, we deduce that $0 \neq d \in A \cap I$ since $es \in \mathfrak{S}$ again by 1.17. Further, 1.11 together with the hypothesis yields $0 \neq A^2 = (d\mathfrak{R})(d\mathfrak{R})$, and hence $d(ud) \neq 0$ for some $u \in \mathfrak{R}$. It follows that $d\mathfrak{S}$ is a nonzero right ideal of $\mathfrak{R}$ contained in $A$ and thus $d\mathfrak{S} = A$. On the other hand, $d \in I$ implies that $d\mathfrak{S} \subseteq I$ so that $A \subseteq I$. This shows that $I$ contains every minimal right ideal of $\mathfrak{R}$, which implies that $I = \mathfrak{S}$. Since $\mathfrak{S}$ contains a nonzero idempotent, we have $\mathfrak{S}^2 \neq 0$ and thus $\mathfrak{S}$ is simple.

If $J$ is an ideal of $\mathfrak{R}$ such that $J \cap \mathfrak{S} = 0$, then also $J\mathfrak{S} = 0$, and since $\mathfrak{R}$ is prime we must have $J = 0$. Thus $\mathfrak{R}$ is an essential extension of $\mathfrak{S}$.

iv) ⇒ v). Letting $\mathfrak{S}$ be the socle of $\mathfrak{R}$, by 1.19 there exists a pair $(U,V)$ of dual vector spaces such that $\mathfrak{S} \cong \mathfrak{F}_U(V)$. By I.7.13, $\mathfrak{L}_U(V)$ is a maximal essential extension of $\mathfrak{F}_U(V)$. By the ring version of I.7.5, $\mathfrak{R}$ is isomorphic to a subring $\mathfrak{R}'$ of $\mathfrak{L}_U(V)$ containing $\mathfrak{F}_U(V)$. Finally, I.2.13 asserts that $\mathfrak{R}'$ is a dense ring of linear transformations.

v) $\Rightarrow$ vi). By I.2.13, we have $\mathfrak{R} \cong \mathfrak{R}'$, where $\mathfrak{F}_U(V) \subseteq \mathfrak{R}' \subseteq \mathfrak{L}_U(V)$ for some pair (U,V) of dual vector spaces. By I.7.10, we have that $\mathfrak{S}_U(V) = \mathfrak{M}\mathfrak{L}_U(V)$ is a maximal dense extension of $\mathfrak{F}_{2,U}(V)$. Hence $\mathfrak{M}\mathfrak{R}'$ is a dense extension of the completely 0-simple semigroup $\mathfrak{F}_{2,U}(V)$.

vi) $\Rightarrow$ vii). Since a Rees matrix semigroup has no nonzero proper ideals (simple calculation), the same holds for a completely 0-simple semigroup. The desired conclusion now follows from 1.23.

vii) $\Rightarrow$ i). The hypothesis implies that the completely 0-simple ideal $\mathfrak{K}$ in question is unique. Letting $I = C(\mathfrak{K})$, we see that I is an ideal of $\mathfrak{R}$ contained in every nonzero ideal of $\mathfrak{R}$. Since $\mathfrak{K}$ contains nonzero idempotents, the hypothesis implies that $\mathfrak{R}$ cannot have nilpotent ideals and is thus semiprime. Let A be a minimal right ideal of $\mathfrak{K}$. By a simple calculation in a Rees matrix semigroup, we see that for every $a \in A$ there exists $e \in \mathfrak{K}$ such that $a = ae$. Hence for any $r \in \mathfrak{R}$, we have $ar = (ae)r = a(er) \in A$ since $er \in \mathfrak{K}$. Consequently A is also a right ideal of $\mathfrak{M}\mathfrak{R}$. If B is a right ideal of $\mathfrak{M}\mathfrak{R}$ contained in A, then B is also a right ideal of $\mathfrak{K}$ which by minimality implies that $B = A$ or $B = 0$. Consequently A is a minimal right ideal of $\mathfrak{M}\mathfrak{R}$, and thus also of $\mathfrak{R}$ by 1.13 since $\mathfrak{R}$ is semiprime. Further, since A is a right ideal of $\mathfrak{R}$, $J = \mathfrak{A}_{r,\mathfrak{R}}(A)$ is an ideal of $\mathfrak{R}$. If $J \neq 0$, then $I \subseteq J$ and thus $0 \neq A^2 \subseteq AI \subseteq AJ = 0$, a contradiction. Therefore $\mathfrak{A}_{r,\mathfrak{R}}(A) = 0$.

A <u>left $\mathfrak{R}$-module</u> M is defined in a dual manner to a (right) $\mathfrak{R}$-module by making $\mathfrak{R}$ act on M on the left. The definition of a "left $\mathfrak{R}$-submodule", "irreducible", "faithful", "left primitive ring" should be clear. Now note that conditions vi) and vii) contain no left or right modifier, which implies that in the remaining conditions left and right can be interchanged. Indeed 1.16 says that left and right socles coincide in this case. In condition i) we may take a minimal left ideal A with $\mathfrak{A}_{\ell,\mathfrak{R}}(A) = 0$; in ii) a left primitive ring; in v) the linear transformations can be taken on a left or right vector space.

II.1.26 <u>DEFINITION</u>. The set theoretic union of minimal right ideals of $\mathfrak{R}$ is the <u>semigroup socle</u> $\mathfrak{K}$ of $\mathfrak{R}$.

Exercise 1.28,v) below implies that $\mathfrak{K}$ is an ideal of $\mathfrak{M}\mathfrak{R}$. We have remarked after 1.16 that $\mathfrak{K}$ agrees with the set theoretic union of all minimal left ideals in a semiprime ring.

II.1.27 <u>COROLLARY</u>. <u>A ring $\mathfrak{R}$ is isomorphic to $\mathfrak{L}_U(V)$ for some pair (U,V) of dual vector spaces if and only if $\mathfrak{R}$ satisfies the conditions of 1.25 and is maximal relative to ii) or iii) or iv) for a fixed socle or vi) or vii) for a fixed semigroup socle</u>.

PROOF. This follows from 1.25, I.2.13, I.7.10 and I.7.13.

II.1.28 EXERCISES.

i) Let $\mathfrak{R}$ be a ring and $I$ be a nonzero ideal of $\mathfrak{R}$. Prove the following statements.

(a) $\mathfrak{A}(I) = 0$ and $\mathfrak{R}$ is an essential extension of $I$ if and only if $\mathfrak{A}_{\mathfrak{R}}(I) = 0$ where $\mathfrak{A}_{\mathfrak{R}}(I) = \{r \in \mathfrak{R} \mid rI = Ir = 0\}$.

(b) $\mathfrak{A}_r(I) = 0$ and $\mathfrak{R}$ is an essential extension of $I$ if and only if $\mathfrak{A}_{r,\mathfrak{R}}(I) = 0$.

ii) Show that a ring $\mathfrak{R}$ for which $\mathfrak{M}\mathfrak{R}$ is completely 0-simple must be a division ring.

iii) Show that a prime ring with a nonzero socle and without nonzero nilpotent elements must be a division ring (an element $r$ is nilpotent if $r^n = 0$ for some positive integer $n$).

iv) Let $\mathfrak{K}$ and $\mathfrak{S}$ denote the semigroup socle and the socle, respectively, of the ring $\mathfrak{R}$ in 1.25. Prove the following statements.

(a) $\mathfrak{S} = \mathfrak{C}(\mathfrak{K})$ and is contained in every nonzero ideal of $\mathfrak{R}$.

(b) The sets of minimal (left) right ideals of $\mathfrak{K}$, $\mathfrak{S}$, and $\mathfrak{R}$ coincide.

(c) Every (left) right ideal of $\mathfrak{S}$ is also a (left) right ideal of $\mathfrak{R}$.

v) Let $A$ be a minimal right ideal and $a$ be an element of a ring $\mathfrak{R}$. Show that $aA$ is either $0$ or a minimal right ideal of $\mathfrak{R}$.

vi) Let $A$ be an idempotent minimal right ideal of a ring $\mathfrak{R}$ and let $0 \neq a \in A$. Show that $aA = a\mathfrak{R} = A$.

The main original references for this section are Dieudonné [2],[3], Gluskin [4] and Jacobson [3],[4],[5]. Further material concerning primitive rings and rings of linear transformations can be found in Behrens ([1], Chapter II, Section 1), Jacobson ([7], Chapter II, Section 2 and Chapter IV, Sections 9, 15), Kaplansky ([4], Part II, Section 6), McCoy ([1], Sections 20,21,23-27), Ribenboim ([1], Chapter III, Section 2), Wolfson [1],[3],[4].

## II.2 SIMPLE RINGS

After some preliminaries, we will characterize simple rings with minimal one-sided ideals in several ways. We will supplement these results by establishing further properties of dual pairs of vector spaces in connection with these rings.

II.2.1 <u>DEFINITION</u>. If $\lambda$ and $\rho$ are binary relations on a set X, their <u>product</u> $\lambda \circ \rho$ is the binary relation on X given by: $a \lambda \circ \rho b$ if and only if $a \lambda x$ and $x \rho b$ for some $x \in X$.

II.2.2 <u>DEFINITION</u>. On a semigroup S, define the following relations.

i) $a \mathcal{L} b \Leftrightarrow a \cup Sa = b \cup Sb$.

ii) $a \mathcal{R} b \Leftrightarrow a \cup aS = b \cup bS$.

iii) $\mathcal{D} = \mathcal{L} \circ \mathcal{R}$.

Then $\mathcal{L}$, $\mathcal{R}$, $\mathcal{D}$ are called <u>Green's relations</u>.

Clearly $\mathcal{L}$ and $\mathcal{R}$ are equivalence relations. The set $L(a) = a \cup Sa$ is the principal left ideal of S generated by a, so that $a \mathcal{L} b$ if and only if a and b generate the same principal left ideal. If a has a left identity, say $a = ea$, then $L(a) = Sa$. The corresponding statements hold for right ideals.

II.2.3 <u>LEMMA</u>. <u>In a Rees matrix semigroup, any two nonzero elements are $\mathcal{D}$-related</u>.

<u>PROOF</u>. For $S = \mathfrak{M}^{o}(I,G,M;P)$ and $(i,a,\mu),(j,b,\nu) \in S$ where $a \neq 0$, $b \neq 0$, the verification that $(i,a,\mu) \mathcal{L} (j,a,\mu)$ and $(j,a,\mu) \mathcal{R} (j,b,\nu)$ is left as an exercise.

II.2.4 <u>DEFINITION</u>. If A and B are $\mathfrak{R}$-modules (respectively left $\mathfrak{R}$-modules), a function $\varphi$ mapping A into B is an $\mathfrak{R}$-<u>homomorphism</u> if $\varphi$ is a homomorphism of abelian groups and $(\varphi a)r = \varphi(ar)$ (respectively $r(a\varphi) = (ra)\varphi$) for all $a \in A$, $r \in \mathfrak{R}$. If $\varphi$ is one-to-one and onto it is called an $\mathfrak{R}$-<u>isomorphism</u> and A and B are said to be $\mathfrak{R}$-<u>isomorphic</u>, to be denoted by "$A \cong B$ as $\mathfrak{R}$-modules" (respectively as left $\mathfrak{R}$-modules).

II.2.5 <u>LEMMA</u>. <u>The following conditions on idempotents</u> e <u>and</u> f <u>of a ring</u> $\mathfrak{R}$ <u>are equivalent</u>.

i) $e\mathfrak{R} \cong f\mathfrak{R}$ <u>as $\mathfrak{R}$-modules</u>.

ii) $\mathfrak{R}e \cong \mathfrak{R}f$ <u>as left $\mathfrak{R}$-modules</u>.

iii) <u>There exist</u> $a,b \in \mathfrak{R}$ <u>such that</u> $ab = e$, $ba = f$.

iv) $e \mathcal{D} f$.

<u>PROOF</u>. ii) $\Rightarrow$ iii). Let $\varphi: \mathfrak{R}e \to \mathfrak{R}f$ be an $\mathfrak{R}$-isomorphism, and $a = e\varphi$, $b = f\varphi^{-1}$. Then $a \in \mathfrak{R}f$, $b \in \mathfrak{R}e$ and

$$e = e\varphi\varphi^{-1} = a\varphi^{-1} = (af)\varphi^{-1} = a(f\varphi^{-1}) = ab,$$

$$f = f\varphi^{-1}\varphi = b\varphi = (be)\varphi = b(e\varphi) = ba,$$

since $\varphi^{-1}$ is also an $\mathfrak{R}$-isomorphism.

iii) $\Rightarrow$ iv). Letting $c = fbe$, we obtain $c \in \mathfrak{R}e$ and

$$e = ab = a(ba)b = a(fbe) = ac \in \mathfrak{R}c$$

which implies that $e \mathcal{L} c$. One shows similarly that $c \mathcal{R} f$ so that $e \mathcal{D} f$.

iv) $\Rightarrow$ ii). Let $e \mathcal{L} b$, $b \mathcal{R} f$. Then $e = xb$ and $f = by$ for some $x,y \in \mathfrak{R}$ and also $b = fb = be$. Letting $a = xf$, we obtain

$$ab = (xf)b = x(fb) = xb = e,$$
$$ba = b(xf) = bx(by) = b(xb)y = bey = by = f,$$
$$a = af.$$

Then the mappings $\varphi: r \to ra$ $(r \in \mathfrak{R}e)$, $\psi: r \to rb$ $(r \in \mathfrak{R}f)$, are $\mathfrak{R}$-homomorphisms of $\mathfrak{R}e$ into $\mathfrak{R}f$ and $\mathfrak{R}f$ into $\mathfrak{R}e$, respectively. For any $r \in \mathfrak{R}e$ we have $r\varphi\psi = rab = re = r$ and similarly $r\psi\varphi = r$ for any $r \in \mathfrak{R}f$, which proves that both $\varphi$ and $\psi$ are $\mathfrak{R}$-isomorphisms.

The equivalence of i) and iii) now follows by symmetry.

II.2.6 <u>DEFINITION</u>. For a division ring $\Delta$ and nonempty sets $I$ and $\Lambda$, any function $P = (p_{\lambda i})$ mapping $\Lambda \times I$ into $\Delta$ is a $\Lambda \times I$-<u>matrix over</u> $\Delta$; $P$ is <u>left row-independent</u> if for any finite set $\lambda_1, \lambda_2, \ldots, \lambda_n \in \Lambda$ and scalars $\sigma_1, \sigma_2, \ldots, \sigma_n \in \Delta$, the relation $\sum_{j=1}^{n} \sigma_j p_{\lambda_j i} = 0$ for every $i \in I$ implies $\sigma_1 = \sigma_2 = \ldots = \sigma_n = 0$; <u>right column-independence</u> is defined analogously.

II.2.7 <u>DEFINITION</u> (cf. I.4.4). Let $I$ and $\Lambda$ be nonempty sets, $\Delta$ be a division ring, $P = (p_{\lambda i})$ be a left row-independent and right column-independent $\Lambda \times I$-matrix over $\Delta$; by $\mathfrak{M}(I,\Delta,\Lambda;P)$ denote the set of all $I \times \Lambda$-matrices over $\Delta$ having only a finite number of nonzero entries, together with addition by entries $(a_{i\lambda}) + (b_{i\lambda}) = (a_{i\lambda} + b_{i\lambda})$ and multiplication $*$ given by $A * B = APB$, where the multiplication on the right is the usual row-by-column multiplication of matrices (which is possible because of the hypothesis on $A$ and $B$). It is left as an exercise to show that we obtain a ring in this way: we call it the <u>Rees matrix ring over</u> $\Delta$ <u>with sandwich matrix</u> $P$, or simply a <u>Rees matrix ring</u>.

<u>Caution</u>. An unrelated class of rings is called "Rees rings".

We are finally ready for the main result of this section. It follows easily from 1.25 and I.2.13 that simple rings with a minimal one-sided ideal are, up to an isomorphism, precisely the rings $\mathfrak{F}_U(V)$ for some dual pair $(U,V)$. This was established by Dieudonné [2] and rediscovered by Gluskin [4]. The equivalence of i) and iv) in the next theorem, closely related to this, is due to Jacobson [3]. The equivalence of i) and v) was established by Hotzel [1] and its proof (due to Hotzel by amplified somewhat) appears here for the first time; it represents the ring version of the Rees theorem in semigroups, see Clifford and Preston ([1], Chapter 3, Section 2). The remaining equivalences are new.

II.2.8 *THEOREM. The following conditions on a ring* $\mathfrak{R}$ *are equivalent.*

i) $\mathfrak{R}$ *is a simple ring with a minimal right ideal.*

ii) $\mathfrak{R}$ *is a prime (respectively primitive) atomic ring.*

iii) $\mathfrak{R}$ *is a simple ring satisfying d.c.c. for principal right ideals.*

iv) $\mathfrak{R}$ *is isomorphic to a dense ring of linear transformations of finite rank.*

v) $\mathfrak{M}\mathfrak{R}$ *has a completely 0-simple ideal* $\mathfrak{K}$ *and* $\mathfrak{R} = C(\mathfrak{K})$.

vi) $\mathfrak{R}$ *is isomorphic to a Rees matrix ring.*

*PROOF.* i) ⇒ v). Since $\mathfrak{R}$ satisfies condition iv) of 1.25, $\mathfrak{M}\mathfrak{R}$ has a completely 0-simple ideal $\mathfrak{K}$. But then $C(\mathfrak{K})$ is a nonzero ideal of $\mathfrak{R}$ and thus $\mathfrak{R} = C(\mathfrak{K})$ by simplicity of $\mathfrak{R}$.

v) ⇒ iv). First let $A$ be a minimal right ideal of $\mathfrak{K}$. We know by I.3.11 that $A = e\mathfrak{K}$ for some idempotent $e \in \mathfrak{K}$. If $a \in A$ and $r \in \mathfrak{R}$, then $ar = (ea)r = e(ar) \in A$ since $\mathfrak{K}$ is an ideal of $\mathfrak{M}\mathfrak{R}$, which implies that $A$ is a right ideal of $\mathfrak{M}\mathfrak{R}$. Consequently $A = e\mathfrak{K} \subseteq e(\mathfrak{M}\mathfrak{R}) \subseteq A$ and thus the equality prevails. Hence $A = e\mathfrak{R}$ which implies that $A$ is a right ideal of $\mathfrak{R}$, minimal in view of its minimality in $\mathfrak{K}$.

Next let $I$ be a nonzero ideal of $\mathfrak{M}\mathfrak{R}$ and suppose that $I \cap \mathfrak{K} = 0$. It follows that $I\mathfrak{K} = \mathfrak{K}I = 0$ and thus $I\mathfrak{R} = \mathfrak{R}I = 0$. We have just seen that every minimal right ideal of $\mathfrak{K}$ is a minimal right ideal of $\mathfrak{R}$. Hence with every element $a$ of $\mathfrak{K}$, the ideal $\mathfrak{K}$ contains its negative since $\mathfrak{K}$ is the union of its minimal right ideals. It follows that the elements of $\mathfrak{R}$ are sums of elements in $\mathfrak{K}$. Let $0 \neq t \in I$ have the property that $t = \sum_{i=1}^{n} t_i$ with $t_i \in \mathfrak{K}$ and $n$ is minimal. The hypothesis $I \cap \mathfrak{K} = 0$ implies that $n > 1$. Since $\mathfrak{K}$ is completely 0-simple, there exists $e_i \in \mathfrak{K}$ such that $t_i = t_i e_i$ for $1 \leq i \leq n$. Thus

$$\begin{aligned} 0 &= (t_1 + t_2 + \ldots + t_n)[e_1 + e_2 + \ldots + e_n - e_1(e_2 + \ldots + e_n)] \\ &= t + \sum_{i=2}^{n} t_i(e_1 + \ldots + e_{i-1} + e_{i+1} + \ldots + e_n) - \sum_{i=2}^{n} t_i e_1(e_2 + \ldots + e_n), \end{aligned}$$

so that

$$t = \sum_{i=2}^{n} t_i[e_1(e_2 + \ldots + e_n) - (e_1 + \ldots + e_{i-1} + e_{i+1} + \ldots + e_n)]$$

where all the summands are in $\mathfrak{K}$, contradicting the minimality of $n$. Therefore $I \cap \mathfrak{K} \neq 0$, and since $I \cap \mathfrak{K}$ is an ideal of $\mathfrak{K}$, we must have $I \cap \mathfrak{K} = \mathfrak{K}$ and thus $\mathfrak{K} \subseteq I$. Consequently $\mathfrak{K}$ is contained in every nonzero ideal of $\mathfrak{M}\mathfrak{R}$, which by 1.25 and I.2.13 implies that $\mathfrak{R} \cong \mathfrak{R}'$ where $\mathfrak{F}_U(V) \subseteq \mathfrak{R}' \subseteq \mathfrak{L}_U(V)$ for some dual pair $(U,V)$ of vector spaces. But $\mathfrak{F}_U(V) = C(\mathfrak{F}_{2,U}(V))$ by I.2.6 and $\mathfrak{F}_{2,U}(V)$ is the unique completely 0-simple ideal of $\mathfrak{M}\mathfrak{R}$, which by hypothesis implies that $\mathfrak{R}' = \mathfrak{F}_U(V)$. Hence $\mathfrak{R}'$ is a dense ring of linear transformations of finite rank.

iv) ⇒ iii). Again by I.2.13, we have $\mathfrak{R} \cong \mathfrak{F}_U(V)$. Further, I.3.15 says that $\mathfrak{F}_U(V)$ satisfies d.c.c. for principal right ideals.

iii) ⇒ ii). The hypothesis implies that $\mathfrak{R}$ contains a minimal principal right ideal A. If B is a right ideal of $\mathfrak{R}$ such that $0 \neq B \subset A$, then for any $0 \neq b \in B$, the principal right ideal (b) generated by b has the property $0 \neq (b) \subseteq B \subset A$ contradicting the minimality of A. As above, we conclude that $\mathfrak{R}$ is atomic. Further, 1.25 implies that $\mathfrak{R}$ is also a prime (respectively primitive) ring.

ii) ⇒ i). By 1.25, prime and primitive are equivalent in an atomic ring, and each implies that the socle of the ring is simple, hence $\mathfrak{R}$ is also simple.

iv) ⇒ vi). By I.2.13, we have $\mathfrak{R} \cong \mathfrak{F}_U(V)$ for a pair (U,V) of dual vector spaces over a division ring $\Delta$. We fix a basis $Z = \{z_\lambda\}_{\lambda\in\Lambda}$ of V and represent the elements of $\mathfrak{L}(V)$ as $\Lambda \times \Lambda$-matrices over $\Delta$ by letting $\varphi: a \to (\sigma_{\lambda\mu})$ $(a \in \mathfrak{L}(V))$ where $z_\lambda a = \sum_{\mu\in\Lambda} \sigma_{\lambda\mu} z_\mu$. We have remarked in I.6.12 that $\varphi$ is an isomorphism of $\mathfrak{L}(V)$ onto the ring $\mathfrak{M}$ of all row-finite $\Lambda \times \Lambda$-matrices over $\Delta$ under the usual addition and multiplication. For any $a \in \mathfrak{L}(V)$, the subspace of V spanned by the vectors $z_\mu$ for which $\sigma_{\lambda\mu} \neq 0$ for some $\lambda \in \Lambda$ is equal to Va, so that identifying $\mathfrak{L}(V)$ and $\mathfrak{M}$, we have

$$\mathfrak{F}(V) = \{A \in \mathfrak{M} \mid \text{only a finite number of columns of } A \text{ have nonzero elements}\}. \qquad (1)$$

Now let $W = \{w_i\}_{i\in I}$ be a basis of U, let $a \in \mathfrak{L}_U(V)$ and b be the adjoint of a in U. We set

$$P = (p_{\lambda i}) \quad \text{where} \quad p_{\lambda i} = (z_\lambda, w_i) \qquad (i \in I,\ \lambda \in \Lambda).$$

Letting $\sigma_{\lambda\mu}$ be as above and $bw_i = \sum_{j\in I} w_j \tau_{ji}$, it follows that $(z_\lambda a, w_i) = (z_\lambda, bw_i)$ implies

$$\left(\sum_\mu \sigma_{\lambda\mu} z_\mu, w_i\right) = \left(z_\lambda, \sum_j w_j \tau_{ij}\right) \qquad (i \in I,\ \lambda \in \Lambda)$$

so that $\sum_\mu \sigma_{\lambda\mu} p_{\mu i} = \sum_j p_{\lambda j} \tau_{ji}$. We let $A = (\sigma_{\lambda\mu})$, $B = (\tau_{ji})$, so in the matrix notation, we obtain $AP = PB$. Conversely, the last formula, going backwards, implies that for the linear transformation a we must have $a \in \mathfrak{L}_U(V)$. Consequently

$$\mathfrak{L}_U(V) = \{A \in \mathfrak{M} \mid \text{there exists a column finite } I \times I\text{-matrix } B \text{ over } \Delta \text{ such that } AP = PB\}. \qquad (2)$$

Recalling that $\mathfrak{F}_U(V) = \mathfrak{L}_U(V) \cap \mathfrak{F}(V)$, we have by (1) and (2) the form of matrices in $\mathfrak{F}_U(V)$.

Suppose that $\sum_{j=1}^{n}\sigma_j p_{\lambda_j i} = 0$ for every $i \in I$. Then $\sum_{j=1}^{n}(\sigma_j z_{\lambda_j}, w_i) = 0$ for all $i \in I$ and hence $(\sum_{j=1}^{n}\sigma_j z_j, u) = 0$ for all $u \in U$. By the nondegeneracy of the bilinear form, we must have $\sum_{j=1}^{n}\sigma_j z_{\lambda_j} = 0$, which by linear independence of the vectors $z_{\lambda_j}$ implies $\sigma_1 = \sigma_2 = \ldots = \sigma_n = 0$. Thus $P$ is left row-independent; one verifies analogously that $P$ is also right column-independent.

For $a \in \mathfrak{F}_U(V)$, by I.2.6, we have $a = \sum_{k=1}^{n}[i_k, \gamma_k, \lambda_k]$. Letting $u_{i_k} = \sum_j w_j \tau_{j i_k}$, $v_{\lambda_k} = \sum_\mu \sigma_{\lambda_k \mu} z_\mu$, we obtain

$$
\begin{aligned}
z_\lambda a &= \sum_{k=1}^{n}(z_\lambda, u_{i_k})\gamma_k v_{\lambda_k} \\
&= \sum_{k=1}^{n}(z_\lambda, \sum_j w_j \tau_{j i_k})\gamma_k(\sum_\mu \sigma_{\lambda_k \mu} z_\mu) = \sum_{k,j,\mu}(p_{\lambda j}\tau_{j i_k}\gamma_k \sigma_{\lambda_k \mu})z_\mu .
\end{aligned} \tag{3}
$$

As above we let $A = (\sigma_{\lambda\mu})$ where $z_\lambda a = \sum_\mu \sigma_{\lambda\mu} z_\mu$, and now form a matrix $M_A = (\sum_{k=1}^{n}\tau_{j i_k}\gamma_k \sigma_{\lambda_k \mu})$. Then $M_A$ is a $I \times \Lambda$-matrix over $\Delta$ having only a finite number of nonzero entries; in matrix notation (3) becomes $PM_A = A$. Suppose that $C = (c_{i\lambda})$ is an $I \times \Lambda$-matrix over $\Delta$ with only a finite number of nonzero entries for which $PC = A$ and let $M_A = (m_{i\lambda})$. Then $P(M_A - C) = 0$ which implies that $\sum_i p_{\lambda i}(m_{i\mu} - c_{i\mu}) = 0$ for all $\lambda, \mu \in \Lambda$. Since at most a finite number of $m_{i\mu} - c_{i\mu}$ may be nonzero, this is a finite sum, and thus the right column-independence of $P$ implies that $m_{i\mu} = c_{i\mu}$ for all $i, \mu$. Consequently $C = M_A$ which shows that the equation $PM_A = A$ uniquely determines $M_A$.

Define a function $\psi : \mathfrak{F}_U(V) \to \mathfrak{m}(I, \Delta, \Lambda; P)$ by $\psi : A \to M_A$. For $A, A' \in \mathfrak{F}_U(V)$, we then have

$$
AA' = (PM_A)(PM_{A'}) = P(M_A P M_{A'}) = P(M_A * M_{A'}),
$$

$$
AA' = PM_{AA'},
$$

which implies that $M_{AA'} = M_A * M_{A'}$. Similarly $M_{A+A'} = M_A + M_{A'}$ so that $\psi$ is a ring homomorphism. Let $M \in \mathfrak{m}(I, \Delta, \Lambda; P)$, then for $A = PM$, we see that $A$ is row-finite, only a finite number of columns of $A$ contain nonzero entries, and $AP = (PM)P = PB$ where $B = MP$ is a column-finite $I \times I$-matrix over $\Delta$. It follows from (1) and (2) above that $A \in \mathfrak{F}_U(V)$ and also $M_A = M$, which proves that $\psi$ is onto. If $M_A = 0$, then $A = PM_A = 0$ and the kernel of $\psi$ is trivial. Therefore $\psi$ is an isomorphism of $\mathfrak{F}_U(V)$ onto $\mathfrak{m}(I, \Delta, \Lambda; P)$.

vi) $\Rightarrow$ i). Let $\mathfrak{R} = \mathfrak{M}(I,\Delta,\Lambda;P)$ be a Rees matrix ring. Let $J$ be an ideal of $\mathfrak{R}$ and $A$ be a nonzero element of $J$. We may choose the notation in such a way that the nonzero rows of $A$ are labeled $1,2,\ldots,m$ and the nonzero columns are labeled $1,2,\ldots,n$. We fix $k \in I$ and $\mu \in \Lambda$ and let $b_{\mu k} = \sum_{i=1}^{m} \sum_{\lambda=1}^{n} p_{\mu i} a_{ik} p_{\lambda k}$. If $b_{\mu k} = 0$, then $\sum_{i=1}^{m} p_{\mu i}(\sum_{\lambda=1}^{n} a_{i\lambda} p_{\lambda k}) = 0$ which implies that $\sum_{\lambda=1}^{n} a_{i\lambda} p_{\lambda k} = 0$ for $1 \leq i \leq m$ since $P$ is right column-independent; but then $a_{i\lambda} = 0$ for $1 \leq i \leq m$, $1 \leq \lambda \leq n$ since $P$ is left row-independent, contradicting $A \neq 0$. Consequently $b_{\mu k} \neq 0$ for every $k \in I$, $\mu \in \Lambda$.

For any $i \in I$, $\lambda \in \Lambda$, we denote by $(c)_{i\lambda}$ the $I \times \Lambda$-matrix whose $(i,\lambda)$-entry is $c$ and all the remaining entries are zero. Continuing with the notation introduced above and letting $C = (c)_{i\mu}$ where $c \neq 0$, $D = (1)_{k\lambda}$, we obtain $C * A * D = CPAPD = (e_{j\sigma}) = (cb_{\mu k})_{i\lambda}$ since $e_{j\sigma} = \sum_{\rho,t,\theta,s} c_{j\rho} p_{\rho t} a_{t\theta} p_{\theta s} d_{s\sigma}$ is nonzero only in the $(i,\lambda)$-position, in which case $e_{i\lambda} = c \sum_{t,\theta} p_{\mu t} a_{t\theta} p_{\theta k} = cb_{\mu k}$. Now $c$ can be chosen in $\Lambda$ arbitrarily which shows that $J$ contains all matrices of the form $(g)_{i\lambda}$. But then $J$ must be equal to $\mathfrak{R}$.

The conditions on the matrix $P$ imply that there exists $p_{\lambda i} \neq 0$. Then $(p_{\lambda i}^{-1})_{i\lambda} P (p_{\lambda i}^{-1})_{i\lambda} = (p_{\lambda i}^{-1})_{i\lambda}$ showing that $(p_{\lambda i}^{-1})_{i\lambda}$ is a nonzero idempotent. Consequently $\mathfrak{R}$ is a simple ring. With the same notation, the mapping $\chi : a \to (p_{\lambda i}^{-1} a)_{i\lambda}$ $(a \in \Delta)$ is easily seen to be an isomorphism of $\Delta$ onto $\Delta_{i\lambda} = \{(a)_{i\lambda} \mid a \in \Delta\}$. It is also easy to verify that $\Delta_{i\lambda} = (p_{\lambda i}^{-1})_{i\lambda} * \mathfrak{R} * (p_{\lambda i}^{-1})_{i\lambda}$. Hence the latter is a division ring which by 1.14 implies that $(p_{\lambda i}^{-1})_{i\lambda} * \mathfrak{R}$ is a minimal right ideal of $\mathfrak{R}$ since $\mathfrak{R}$ is semiprime.

II.2.9 <u>COROLLARY</u>. <u>If $\mathfrak{K}$ is a completely 0-simple ideal of $\mathfrak{M}\mathfrak{R}$ for a ring $\mathfrak{R}$, then $\mathfrak{C}(\mathfrak{K})$ is a simple atomic ring and an ideal of $\mathfrak{R}$.</u>

II.2.10 <u>NOTATION</u>. For $\varkappa$ a cardinal number and $\Delta$ a division ring, let $\Delta_\varkappa$ denote the ring of all $\varkappa \times \varkappa$-matrices over $\Delta$ with only a finite number of nonzero entries.

This is an extension of the notation $\Delta_n$ for the ring of all $n \times n$-matrices over $\Delta$. Further, $\Delta_\varkappa \cong \mathfrak{M}(I,\Delta,I;Q)$ where $I$ has cardinality $\varkappa$ and $Q$ is the identity $\varkappa \times \varkappa$-matrix. We have seen in I.3.12-3.15 that for $\dim V < \infty$, we can identify $U$ with $V^*$ and have $\mathfrak{L}(V) = \mathfrak{F}_U(V) \cong \Delta_{\dim V}$. It is then of interest to know for which dual pairs $(U,\Delta,V)$ is $\mathfrak{F}_U(\Delta,V)$ isomorphic to a ring of the form $\Delta_\varkappa$. For convenience, we introduce a new concept.

II.2.11 DEFINITION. Let $(U,V)$ be a dual pair of vector spaces. Then the subsets $\{u_i \mid i \in I\}$ and $\{v_i \mid i \in I\}$ of $U$ and $V$, respectively, are biorthogonal if $(v_i,u_j) = \delta_{ij}$ for all $i,j \in I$.

II.2.12 NOTATION. Let ${}_{\mathfrak{R}}\mathfrak{R}$ and $\mathfrak{R}_{\mathfrak{R}}$ denote the left and the right $\mathfrak{R}$-module $\mathfrak{R}$, respectively (i.e., the action is left, respectively right, multiplication).

II.2.13 DEFINITION. If $M$ is a left (respectively right) $\mathfrak{R}$-module, then a subset $G$ of $M$ generates $M$ if every element $m$ of $M$ can be written in the form $\sum_{i=1}^{n} r_i g_i$ (respectively $\sum_{i=1}^{n} g_i r_i$) for some $r_i \in \mathfrak{R}$, $g_i \in G$.

II.2.14 DEFINITION. In a ring or a semigroup with zero, an idempotent is primitive if it is nonzero and minimal relative to the ordering

$$e \leq f \Leftrightarrow e = ef = fe.$$

The usual definition of a completely 0-simple semigroup $S$ is that $S$ has a zero, no nonzero proper ideals and a primitive idempotent, cf. 1.24. The next result is due to Hotzel [1], see also Dieudonné [2].

II.2.15 PROPOSITION. Let $(U,\Delta,V)$ be a pair of dual vector spaces with $\dim V > 1$, and let $\mathfrak{R} = \mathfrak{F}_U(\Delta,V)$, then the following statements are equivalent.

i) $\mathfrak{R} \cong \Gamma_\varkappa$ for some division ring $\Gamma$ and a cardinal number $\varkappa$.

ii) ${}_{\mathfrak{R}}\mathfrak{R}$ and $\mathfrak{R}_{\mathfrak{R}}$ have a common set $G$ of generators consisting of primitive orthogonal idempotents.

iii) $U$ and $V$ have biorthogonal bases.

Moreover, if i) occurs, then $\Delta \cong \Gamma$ and $\dim U = \dim V = \varkappa$.

PROOF. i) $\Rightarrow$ ii). The elements of $\mathfrak{R} = \Gamma_\varkappa$ are $I \times I$-matrices over $\Delta$ with only a finite number of nonzero entries, where $I$ has cardinality $\varkappa$. For every $i \in I$, let $E_i = (1)_{ii}$. It is clear that the set $E = \{E_i \mid i \in I\}$ consists of orthogonal primitive idempotents. Let $A \in \Gamma_\varkappa$ and let $i_1,i_2,\ldots,i_n$ be the nonzero columns of $A$. A simple argument shows that $A = \sum_{k=1}^{n} AE_{i_k}$ which proves that $E$ generates ${}_{\mathfrak{R}}\mathfrak{R}$. A symmetric proof shows that $E$ generates $\mathfrak{R}_{\mathfrak{R}}$.

ii) $\Rightarrow$ iii). Suppose that $G$ generates both ${}_{\mathfrak{R}}\mathfrak{R}$ and $\mathfrak{R}_{\mathfrak{R}}$, and let $Z = \{v_\lambda \mid [i,\gamma,\lambda] \in G\}$, $W = \{u_i \mid [i,\gamma,\lambda] \in G\}$. We will show that $Z$ and $W$ are biorthogonal bases of $V$ and $U$, respectively.

Let $v = \sigma v_\mu$ be a nonzero vector in $V$. Choose any $u_i \in W$ and let $x$ be such that $(x,u_i) = \sigma$. The hypothesis implies that $[i,1,\mu] = \sum_{k=1}^{n} a_k[i_k,\gamma_k,\lambda_k]$ for some $a_k \in \mathfrak{R}$, $[i_k,\gamma_k,\lambda_k] \in G$. Hence

$$v = \sigma v_\mu = (x,u_i)v_\mu = x[i,1,\mu]$$
$$= x(\sum_{k=1}^{n} a_k[i_k,\gamma_k,\lambda_k]) = \sum_{k=1}^{n} (xa_k,u_{i_k})\gamma_k v_{\lambda_k},$$

proving that $Z$ generates $V$.

Let $[i,\gamma,\lambda],[j,\delta,\mu] \in G$ be distinct. Then

$$[i,\gamma,\lambda][j,\delta,\mu] = [j,\delta,\mu][i,\gamma,\lambda] = 0$$

so that $(v_\lambda,u_j) = (v_\mu,u_i) = 0$. Since $(v_\lambda,u_i) \neq 0$, it follows that $i \neq j$ and similarly $\lambda \neq \mu$. This establishes a one-to-one correspondence between $Z$ and $W$, and writing $Z = \{z_k \mid k \in K\}$ and $W = \{w_k \mid k \in K\}$, we also have that $(w_i,z_j) = \delta_{ij}$.

Now suppose that $v_\lambda = \sum_{k=1}^{n} \sigma_k v_{\lambda_k}$, where $v_\lambda, v_{\lambda_k} \in Z$ and $v_\lambda$ is different from all $v_{\lambda_k}$. There exists $i$ such that $[i,(v_\lambda,u_i)^{-1},\lambda] \in G$; thus

$$0 \neq (v_\lambda,u_i) = \sum_{k=1}^{n} \sigma_k (v_{\lambda_k},u_i) = 0$$

by the above, a contradiction. This shows that $Z$ is linearly independent, thus proving that $Z$ is a basis of $V$. A similar proof shows that $W$ is a basis of $U$. Hence $Z$ and $W$ are biorthogonal bases.

iii) $\Rightarrow$ i). In the proof of "iv) $\Rightarrow$ vi)" in 2.8, we take the given biorthogonal bases and immediately obtain that $P = (p_{\lambda i}) = ((z_\lambda,w_i))$ is the identity matrix and the rest of the proof shows that $\mathfrak{F}_U(\Delta,V) \cong \Delta_{\dim V}$.

We finally prove the last statement of the proposition. Hence assume that $\mathfrak{F}_U(\Delta,V) \cong \Gamma_\varkappa$. Let $e = (p_{\lambda i}^{-1})_{i\lambda}$ where $p_{\lambda i} \neq 0$. We have seen in the proof of "vi) $\Rightarrow$ i)" in 2.8 that $\Delta' = e * \Gamma_\varkappa * e$ is a division ring isomorphic to $\Gamma$ and that $\Gamma_\varkappa$ is a simple atomic ring. Now 1.14, together with the proof of 1.19, shows that $\Gamma_\varkappa \cong \mathfrak{F}_{U'}(\Delta',V')$, where $U' = \Gamma_\varkappa * e$ and $V' = e * \Gamma_\varkappa$ with the induced bilinear form. Thus $\mathfrak{F}_U(\Delta,V) \cong \mathfrak{F}_{U'}(\Delta',V')$ which by I.5.15 implies that $\Delta \cong \Delta'$ and $\dim V = \dim V'$ so that $\Delta \cong \Gamma$. We have seen above that both $(U,V)$ and $(U',V')$ must have biorthogonal bases which evidently implies that $\dim U = \dim V$ and $\dim U' = \dim V'$. Hence all four vector spaces have the same dimension.

It is easy to see that $V'$ consists of all matrices in $\Gamma_\varkappa$ all of whose nonzero entries are situated in the $i^{\text{th}}$ row. Let $I$ be an index set of rows and columns of matrices in $\Gamma_\varkappa$, and let $B = \{(1)_{i\mu} \mid \mu \in I\}$. If $A \in V'$ has nonzero entries in the columns $\mu_1,\mu_2,\ldots,\mu_n$, then $A = \sum_{k=1}^{n} (a_{i\mu_k} p_{\lambda i}^{-1})_{i\lambda} * (1)_{i\mu_k}$. If $\sum_{j=1}^{m} (c_j)_{i\lambda} * (1)_{i\mu_j} = 0$, then $\sum_{j=1}^{m} (c_j p_{\lambda i})_{i\mu_j} = 0$ which evidently implies that $c_j p_{\lambda i} = 0$ for $1 \leq j \leq m$; but then $c_j = 0$ for $1 \leq j \leq m$. Consequently $B$ is a basis of $V'$. It is clear that $B$ is of cardinality $\varkappa$, which finally yields $\dim V = \dim V' = \varkappa$.

II.2.16 COROLLARY. *A ring* $\mathfrak{R}$ *is isomorphic to a ring of the form* $\Delta_{\mathfrak{n}}$ *if and only if* $\mathfrak{R}$ *is a simple atomic ring for which* ${}_{\mathfrak{R}}\mathfrak{R}$ *and* $\mathfrak{R}_{\mathfrak{R}}$ *have a common set of generators consisting of orthogonal primitive idempotents.*

II.2.17 COROLLARY. *If* $\Gamma_{\mathfrak{a}} \cong \Delta_{\mathfrak{b}}$, *then* $\Gamma \cong \Delta$ *and* $\mathfrak{a} = \mathfrak{b}$.

Note that in the case $\dim V < \infty$ and $(U,V)$ is a dual pair, I.2.3 implies that $U$ and $V$ admit biorthogonal bases, confirming what we already know, viz., that $\mathfrak{F}_U(V) = \mathfrak{L}_U(V) = \mathfrak{L}(V) \cong \Delta_n$ for $n = \dim V$ and $\Delta$ a division ring. This can be carried one step further as the next proposition, due to Mackey [2], shows.

II.2.18 PROPOSITION. *Let* $(U,V)$ *be a dual pair of vector spaces such that* $\dim U = \dim V$ *is denumerable, then* $U$ *and* $V$ *admit biorthogonal bases.*

PROOF. Let $Z = \{z_i \mid i = 1,2,3,\ldots\}$ and $W = \{w_i \mid i = 1,2,\ldots\}$ be bases of $V$ and $U$, respectively. We will construct biorthogonal bases $X = \{x_i \mid i = 1,2,\ldots\}$ and $Y = \{y_i \mid i = 1,2,\ldots\}$ of $V$ and $U$, respectively, by induction. Let $j_0 = \min\{j \mid (z_j, w_1) \neq 0\}$ and set $x_1 = (z_{j_0}, w_1)^{-1} z_{j_0}$, $y_1 = w_1$, hence $(x_1, y_1) = 1$.

Assume that $x_1,\ldots,x_k$, and $y_1,\ldots,y_k$ have been chosen in such a way that $(x_i, y_j) = \delta_{ij}$ for $1 \le i,j \le k$. Suppose first that $k$ is odd. Let

$$j_k = \min\{j \mid z_j \text{ is not contained in the subspace generated by } x_1,\ldots,x_n\}$$

and set $x_{k+1} = z_{j_k} - \sum_{i=1}^{k} (z_{j_k}, y_i) x_i$. Then for $1 \le t \le k$, we obtain

$$\begin{aligned}(x_{k+1}, y_t) &= (z_{j_k}, y_t) - \sum_{i=1}^{k} (z_{j_k}, y_i)(x_i, y_t) \\ &= (z_{j_k}, y_t) - (z_{j_k}, y_t) = 0.\end{aligned}$$

Next let $j_{k+1} = \min\{j \mid (x_{k+1}, w_j) \neq 0\}$ and set

$$y_{k+1} = \Big(w_{j_{k+1}} - \sum_{i=1}^{k} y_i (x_i, w_{j_{k+1}})\Big)(x_{k+1}, w_{j_{k+1}})^{-1}.$$

For $1 \le t \le k$, we have

$$\begin{aligned}(x_t, y_{k+1}) &= \Big[(x_t, w_{j_{k+1}}) - \sum_{i=1}^{k} (x_t, y_i)(x_i, w_{j_{k+1}})\Big](x_{k+1}, w_{j_{k+1}})^{-1} \\ &= \big[(x_t, w_{j_{k+1}}) - (x_t, y_t)(x_t, w_{j_{k+1}})\big](x_{k+1}, w_{j_{k+1}})^{-1} = 0\end{aligned}$$

and also

$$(x_{k+1}, y_{k+1}) = \Big[(x_{k+1}, w_{j_{k+1}}) - \sum_{i=1}^{k} (x_{k+1}, y_i)(x_i, w_{j_{k+1}})\Big](x_{k+1}, w_{j_{k+1}})^{-1} = 1.$$

If $k$ is even, we interchange the roles of $U$ and $V$ to define $x_{k+1}$ and $y_{k+1}$. This process defines the sequences $x_1, x_2, \ldots$ and $y_1, y_2, \ldots$. One shows that the sets $X$ and $Y$ are linearly independent as in the proof of I.2.3. That $X$ generates $V$ follows from the fact that every $z_k$ is contained in the subspace of $V$ spanned by some $x_i$'s. Similarly $Y$ generates $U$. Hence $X$ and $Y$ are biorthogonal bases of $V$ and $U$, respectively.

Recall that I.3.20 asserts that $\mathfrak{F}_U(\Delta,V)$ is a locally matrix ring over $\Delta$. We are now in a position to extend this result. This is due to Hotzel [1].

II.2.19 PROPOSITION. Let $\mathfrak{R}$ be any ring isomorphic to $\mathfrak{F}_U(\Delta,V)$ for a dual pair $(U,V)$ of vector spaces over a division ring $\Delta$, then every denumerable subset of $\mathfrak{R}$ is contained in a subring of $\mathfrak{R}$ isomorphic to $\Delta_\varkappa$ where $\varkappa \leq \aleph_0$.

PROOF. Let $B = \{b_i \mid i = 1,2,\ldots\}$ be a set consisting of nonzero elements of $\mathfrak{R}$. We construct a sequence $e_1, e_2, \ldots$ of idempotents of $\mathfrak{R}$ as follows. By the proof of I.3.20, there exists an idempotent $e_1$ such that $b_1 \in e_1\mathfrak{R}e_1$; suppose that $e_1, e_2, \ldots, e_n$ have been defined and let $e_{n+1}$ be an idempotent for which $b_{n+1}, e_n \in e_{n+1}\mathfrak{R}e_{n+1}$, this is possible again in view of the proof of I.3.20. Next let $K = \bigcup_{n=1}^{\infty} e_n\mathfrak{R}e_n$. Since $e_n\mathfrak{R}e_n \subseteq e_{n+1}\mathfrak{R}e_{n+1}$ for $n = 1,2,\ldots$, it follows that $K$ is a subring of $\mathfrak{R}$. We will show that $K$ is the desired subring.

Let $I$ be a nonzero ideal of $K$. Then $I \cap e_n\mathfrak{R}e_n \neq 0$ for some $e_n$ and thus $e_n\mathfrak{R}e_n \subseteq I$ since $e_n\mathfrak{R}e_n \cong \Delta_{\text{rank } e_n}$ by I.3.18 and the last ring is simple by I.3.6. But then $e_1\mathfrak{R}e_1 \subseteq I$ and hence $I \cap e_k\mathfrak{R}e_k \neq 0$ for $k = 1,2,\ldots$, which by the same argument implies that $e_k\mathfrak{R}e_k \subseteq I$, so that $I = K$. Since $K$ contains nonzero idempotents, it must be simple.

Next let $A$ be a minimal right ideal of $e_1\mathfrak{R}e_1$. By 1.6, $A$ is generated by an idempotent $e$ so that $A = e(e_1\mathfrak{R}e_1) = e\mathfrak{R}e_1$. Hence by 1.14, $e(e_1\mathfrak{R}e_1)e$ is a division ring, and it is clear that $e(e_1\mathfrak{R}e_1)e = e\mathfrak{R}e \cong \Delta$. Furthermore,

$$e\mathfrak{R}e = e(e_1\mathfrak{R}e_1)e \subseteq eKe \subseteq e\mathfrak{R}e$$

so that $eKe = e\mathfrak{R}e \cong \Delta$. But then 1.14 implies that $eK$ is a minimal right ideal of $K$, proving that $K$ is a simple atomic ring. The proof of 1.19 now shows that $K$ is isomorphic to $\mathfrak{F}_{Ke}(eKe,eK)$ where $(Ke,eK)$ is a dual pair of vector spaces over $eKe$ with the bilinear form induced by ring multiplication.

Note that $eK = e(\bigcup_{n=1}^{\infty} e_n\mathfrak{R}e_n) = \bigcup_{n=1}^{\infty} (e\mathfrak{R}e_n)$, where $e\mathfrak{R}e_n \subseteq e\mathfrak{R}e_{n+1}$ and $e\mathfrak{R}e_n$ is a left vector space over $e\mathfrak{R}e$ for $n = 1,2,\ldots$. Further, $\dim e\mathfrak{R}e_n = \text{rank } e_n$ since $e_n\mathfrak{R}e_n \cong \Delta_{\text{rank } e_n}$ according to I.3.18. Starting with a basis $v_1, v_2, \ldots, v_{\text{rank } e_1}$ of $e\mathfrak{R}e_1$, we may extend it to a basis $v_1, v_2, \ldots, v_{\text{rank } e_1}, \ldots, v_{\text{rank } e_2}$ of $e\mathfrak{R}e_2$, and with an obvious inductive process construct a denumerable basis of $eK$. The same construction applies to the right vector space Ke, which shows that $\varkappa = \dim Ke = \dim eK \leq \aleph_0$. If $\varkappa$ is a finite cardinal, then I.2.3 says that Ke and eK have biorthogonal bases, and if $\varkappa$ is infinite the same conclusion follows from 2.18. In either case, 2.15 implies that $\mathfrak{F}_{Ke}(eKe,eK) \cong \Delta_\varkappa$ with $\varkappa \leq \aleph_0$.

II.2.21 <u>EXERCISES</u>.

i) Show that in any semigroup Green's relations $\mathfrak{L}$ and $\mathfrak{R}$ commute and that $\mathfrak{D}$ is an equivalence relation.

ii) Let $\mathfrak{R} = \mathfrak{M}(I,\Delta,\Lambda;P)$ be a Rees matrix ring.

a) Find a Rees matrix representation of the semigroup socle of $\mathfrak{R}$.

b) Characterize the right ideals of $\mathfrak{R}$.

c) Characterize the elements in the semigroup socle of $\mathfrak{R}$.

d) Simplify the results in a), b), c) in the case that P is the identity matrix to obtain as simple characterizations as possible.

iii) Show that in a regular ring $\mathfrak{R}$, an idempotent e is primitive if and only if $e\mathfrak{R}e$ is a division ring.

iv) Let V be a vector space of countably infinite dimension. Show that $\mathfrak{L}(V)/\mathfrak{F}(V)$ is a simple regular ring with identity and without minimal one-sided ideals. Give an example of a simple ring without an identity and without minimal one sided ideals.

For further results on this subject, see Behrens ([1], Chapter II, Section 5), Dieudonné [2], Hotzel [1], Jacobson [3],([7], Chapter IV, Section 15), Pless [1], Rosenberg [1], Wolfson [1],[3].

## II.3 MAXIMAL PRIME RINGS

The theorem of this section gives a number of abstract characterizations of the ring of all linear transformations of finite rank on a vector space. The remaining results deal with further special cases of primitive rings with a nonzero socle. It will be convenient to introduce several new concepts.

II.3.1 <u>DEFINITION</u>. A semigroup S is an <u>ri-semigroup</u> (ring-ideal) if S is isomorphic to an ideal of $\mathfrak{M}\mathfrak{R}$ for some ring $\mathfrak{R}$ (equivalently, S is an ideal of $\mathfrak{M}\mathfrak{R}'$ for some ring $\mathfrak{R}'$).

II.3.2 <u>LEMMA</u>. <u>If K is a completely 0-simple ri-semigroup, then there exists a simple atomic ring $\mathfrak{R}$ such that K is the semigroup socle of $\mathfrak{R}$ and $\mathfrak{R} = C(K)$.</u>

<u>PROOF</u>. Let K be an ideal of $\mathfrak{M}\mathfrak{R}'$. Then $\mathfrak{R} = C(K)$ is a simple atomic ring by 2.8, and in view of 1.25 it is clear that K is the unique completely 0-simple ideal of $\mathfrak{M}\mathfrak{R}$ so that K must be the semigroup socle of $\mathfrak{R}$.

II.3.3 <u>DEFINITION</u>. A simple atomic ring is <u>r-maximal</u> if it is not isomorphic to a proper right ideal of a simple ring.

II.3.4 <u>DEFINITION</u>. A completely 0-simple ri-semigroup is <u>r-maximal</u> if it is not isomorphic to a proper right ideal of a completely 0-simple ri-semigroup.

We are now ready for the first main result of this section. The equivalence of i) and ii) is due to Gluskin [4], that of ii) and iv) to Wolfson [1]; the remaining part is new. Another characterization of these rings can be found in Jacobson ([7], Chapter IV, Section 16).

II.3.5 <u>THEOREM. The following conditions on a ring</u> $\mathfrak{R}$ <u>are equivalent.</u>

i) $\mathfrak{R}$ <u>is an r-maximal simple atomic ring.</u>

ii) $\mathfrak{R}$ <u>is isomorphic to the ring of all linear transformations of finite rank on a left vector space.</u>

iii) $\mathfrak{M}\mathfrak{R}$ <u>has a completely 0-simple r-maximal ideal</u> $\mathfrak{K}$ <u>and</u> $\mathfrak{R} = C(\mathfrak{K})$.

iv) $\mathfrak{R}$ <u>is a simple atomic ring in which</u> $\mathfrak{A}_{r,\mathfrak{R}}(L) = 0$ <u>for a left ideal</u> $L$ <u>implies</u> $L = \mathfrak{R}$.

<u>PROOF</u>. i) $\Rightarrow$ ii). By 2.8, we have $\mathfrak{R} \cong \mathfrak{F}_U(V)$ for some pair $(U,V)$ of dual vector spaces. By I.2.6, we have

$$\mathfrak{F}_U(V) = \{a \in \mathfrak{F}(V) \mid a^*V^* \subseteq \text{nat } U\}$$

which implies that $\mathfrak{F}_U(V)$ is a right ideal of the simple ring $\mathfrak{F}(V)$. Hence by r-maximality of $\mathfrak{R}$, we must have $\mathfrak{F}_U(V) = \mathfrak{F}(V)$. Consequently $\mathfrak{R} \cong \mathfrak{F}(V)$.

ii) $\Rightarrow$ iii). We may take $\mathfrak{R} = \mathfrak{F}(V)$, where $V$ is a left vector space, and assume that $\mathfrak{K} = \mathfrak{F}_{2,V^*}(V)$ is isomorphic to a right ideal $K$ of a completely 0-simple ri-semigroup $S$. By 3.2, we know that $S$ is the semigroup socle of a simple atomic ring $\mathfrak{R}'$. Hence there exists a dual pair $(U',V')$ of vector spaces such that $\mathfrak{R}' \cong \mathfrak{F}_{U'}(V')$ which implies that $S \cong \mathfrak{F}_{2,U'}(V')$. Thus $\mathfrak{K} \cong K \cong K'$ where $K'$ is a right ideal of $\mathfrak{F}_{2,U'}(V')$.

It is easy to verify that the nonzero right ideals of a Rees matrix semigroup $S = \mathfrak{M}^o(I,G,\Lambda;P)$ are precisely the semigroups of the form $\mathfrak{M}^o(I',G,\Lambda;P')$, where $I'$ is a nonempty subset of $I$, $P'$ is the $\Lambda \times I'$-matrix obtained from $P$ by eliminating the columns of $P$ indexed by $I \setminus I'$, where by abuse of notation, we have written $\mathfrak{M}^o(I',G,\Lambda;P')$ even though there may exist $\lambda \in \Lambda$ such that $p_{\lambda i'} = 0$ for all $i' \in I'$.

Hence $K' = \mathfrak{M}^o(I',\mathfrak{M}\Delta^-,I_{V'};P')$ where $I' \subseteq I_{V'}$ since

$$\mathfrak{F}_{2,U'}(V') = \mathfrak{M}^o(I_{V'},\mathfrak{M}\Delta^-,I_{V'};P).$$

If $\dim V = 1$, then $\mathfrak{F}_{2,U'}(V')$ is a division ring, so $K'$ is also a division ring which forces $I_{V'}$ to have only one element. But then $\dim V' = 1$ so that $K' = \mathfrak{F}_{2,U'}(V') = \mathfrak{F}_{2,V'^*}(V')$. Suppose next that $\dim V > 1$ and let $\theta$ be an isomorphism of $\mathfrak{K}$ onto $K'$. If for some $\lambda \in I_{V'}$ we have $p_{\lambda i'} = 0$ for all $i' \in I'$, then for any $j \in I'$ and $\gamma \in \Delta^-$, we obtain $[i,\gamma,\lambda]K' = 0$ with $[i,\gamma,\lambda] \neq 0$. By the isomorphism $\theta$, $\mathfrak{K}$ must have the same property, contradicting I.3.3. Consequently $\mathfrak{M}^o(I',\mathfrak{M}\Delta'^-,I_{V'};P')$ is indeed a Rees matrix semigroup. Hence I.5.6 applies and the functions $a$ and $b$ in the statement of I.5.9 can be defined. We let $U'' = \{u_i'\tau \mid i \in I',\ \tau \in \Delta\}$ and observe that in the proof of I.5.9 the implicit hypothesis that $U'$ is closed under addition is not used. We conclude that $(\omega,a)$ is a semilinear isomorphism of $(\Delta,V)$ onto $(\Delta',V')$ and $\zeta_{(\omega,a)}|_{\mathfrak{K}} = \theta$. Consequently the image of $\mathfrak{K}$ under $\theta$ is both $K'$ and $\mathfrak{F}_{2,V'^*}(V')$ which implies that $K' = \mathfrak{F}_{2,U'}(V') = \mathfrak{F}_{2,V'^*}(V')$. It then follows that $K = S$ proving that $\mathfrak{K}$ is an r-maximal completely 0-simple semigroup.

iii) ⇒ iv). We may take $\mathfrak{R} = \mathfrak{F}_U(V)$. Then $\mathfrak{K} = \mathfrak{F}_{2,U}(V)$ and since $\mathfrak{F}_{2,U}(V)$ is a right ideal of $\mathfrak{F}_{2,V^*}(V)$, the r-maximality of the former implies that $\mathfrak{F}_{2,U}(V) = \mathfrak{F}_{2,V^*}(V)$. But then $\mathfrak{F}_U(V) = \mathfrak{F}(V)$ so that $\mathfrak{R} = \mathfrak{F}(V)$.

Let $L$ be a left ideal of $\mathfrak{F}(V)$. Then the dual of I.3.4 implies that $L = \mathfrak{I}_{V^*}(S)$ for some subspace $S$ of $V$, and it is easy to see that

$$\mathfrak{A}_{r,\mathfrak{R}}(L) = \mathfrak{A}_{r,\mathfrak{R}}(\mathfrak{I}_{V^*}(S)) = \mathfrak{I}_{S^\perp}(V) \tag{1}$$

where $S^\perp = \{f \in V^* \mid sf = 0 \text{ for all } s \in S\}$. If $S \neq V$, then there exists a nonzero linear transformation $a$ on $V$ such that $Sa = 0$. Then for any $s \in S$, $f \in V^*$, we have $s(a^*f) = (sa)f = 0f = 0$. It follows that $a^*V^* \subseteq S^\perp$ and thus $0 \neq a \in \mathfrak{I}_{S^\perp}(V)$. Consequently, if $\mathfrak{A}_{r,\mathfrak{R}}(L) = 0$, then by (1) above we obtain $S = V$ which implies that $L = \mathfrak{F}_{V^*}(V) = \mathfrak{F}(V) = \mathfrak{R}$.

iv) ⇒ i). Let $\mathfrak{R}$ satisfying iv) be a right ideal of a simple ring $\mathfrak{R}'$. Further let $A$ be a minimal right ideal of $\mathfrak{R}$. Then $A = e\mathfrak{R}$ for some idempotent $e \in \mathfrak{R}$ by 1.6. Hence $A\mathfrak{R}' = (e\mathfrak{R})\mathfrak{R}' \subseteq e\mathfrak{R} = A$ so that $A'$ is a right ideal of $\mathfrak{R}'$, which is minimal in view of its minimality in $\mathfrak{R}$. Thus $\mathfrak{R}'$ is a simple atomic ring, and we may suppose that $\mathfrak{R}' = \mathfrak{F}_{U'}(V)$ where $U'$ is a t-subspace of $V^*$. By I.3.4, we have $\mathfrak{R} = \mathfrak{I}_U(V)$ for some subspace $U$ of $U'$ and by I.3.8, we have that $(U,V)$ is a dual pair.

The following argument and notation is relative to the dual pair $(U,V)$. By the dual of I.3.4 and (1) above, we obtain that if a subspace $S$ of $V$ has the property $\mathfrak{I}_{S^\perp}(V) = 0$, then $S = V$. Let $0 \neq f \in V^*$. Then $S = \{v \in V \mid vf = 0\}$ is a proper subspace of $V$, and hence by the preceding statement, we must have $\mathfrak{I}_{S^\perp}(V) \neq 0$. Consequently

$$S^{\perp} = \{u \in U \mid su = 0 \quad \text{for all} \quad s \in S\} \neq 0.$$

Let $0 \neq g \in S^{\perp}$, then for every $v \in V$ we have: if $vf = 0$, then $vg = 0$. If $f$ and $g$ are linearly independent, then there exists $v \in V$ such that $vf = 0$ and $vg \neq 0$ according to I.2.3 applied to the dual pair $(V^*, V)$, a contradiction. Thus $f = g\tau$ for some $\tau \in \Delta^-$ so that $f \in U$. Therefore $U = V^*$ and thus $U = U'$ which implies that $\mathfrak{R} = \mathfrak{R}'$.

The equivalence of ii) and iv) in the next corollary is due to Wolfson [1]; the rest is new.

II.3.6 <u>COROLLARY</u>. <u>The following conditions on a ring $\mathfrak{R}$ are equivalent</u>.

i) $\mathfrak{R}$ <u>is a primitive ring with an r-maximal socle</u>.

ii) $\mathfrak{R}$ <u>is isomorphic to a ring of linear transformations on a left vector space containing all transformations of finite rank</u>.

iii) $\mathfrak{M}\mathfrak{R}$ <u>is a dense extension of an r-maximal completely 0-simple semigroup</u>.

iv) $\mathfrak{R}$ <u>is a primitive ring with a nonzero socle</u> $\mathfrak{S}$ <u>in which</u> $\mathfrak{A}_{r,\mathfrak{R}}(L) = 0$ <u>for a left ideal</u> $L$ <u>implies</u> $L \supseteq \mathfrak{S}$.

<u>PROOF</u>. i) $\Rightarrow$ ii). We know by 1.25 that $\mathfrak{R} \cong \mathfrak{R}'$ for some ring $\mathfrak{R}'$ satisfying $\mathfrak{F}_U(V) \subseteq \mathfrak{R}' \subseteq \mathfrak{L}_U(V)$ for some dual pair $(U,V)$. Since $\mathfrak{S}$ is r-maximal, so is $\mathfrak{F}_U(V)$. But then $\mathfrak{F}_U(V) = \mathfrak{F}(V)$ and thus $\mathfrak{F}(V) \subseteq \mathfrak{R}' \subseteq \mathfrak{L}(V)$.

ii) $\Rightarrow$ iii). We may take $\mathfrak{F}(V) \subseteq \mathfrak{R} \subseteq \mathfrak{L}(V)$ for a left vector space $V$. Hence $\mathfrak{M}\mathfrak{R}$ is a dense extension of $\mathfrak{F}_{2,V^*}(V)$ by 1.25. It was shown in the proof of "ii) $\Rightarrow$ iii)" in 3.5 that $\mathfrak{F}_{2,V^*}(V)$ is an r-maximal completely 0-simple semigroup.

iii) $\Rightarrow$ iv). As before, $\mathfrak{R}$ is a primitive ring with a nonzero socle $\mathfrak{S}$. Let $L$ be a left ideal of $\mathfrak{R}$ such that $\mathfrak{A}_{r,\mathfrak{R}}(L) = 0$. Further let $0 \neq a \in \mathfrak{S}$, then $a \notin \mathfrak{A}_{r,\mathfrak{R}}(L)$ and hence $ba \neq 0$ for some $b \in L$. It follows from 1.25 and I.3.6 that $\mathfrak{S}$ is a regular ring, and thus there exists $c \in \mathfrak{S}$ such that $ba = bacba$. Hence for $x = bacb$, we have $xa \neq 0$ and $x \in L \cap \mathfrak{S}$ which shows that $a \notin \mathfrak{A}_{r,\mathfrak{S}}(L \cap \mathfrak{S})$, and thus $\mathfrak{A}_{r,\mathfrak{S}}(L \cap \mathfrak{S}) = 0$. By 3.5 we must have $L \cap \mathfrak{S} = \mathfrak{S}$ so that $L \supseteq \mathfrak{S}$.

iv) $\Rightarrow$ i). Let $L$ be a left ideal of the socle $\mathfrak{S}$ of $\mathfrak{R}$ and suppose that $\mathfrak{A}_{r,\mathfrak{S}}(L) = 0$. As above $\mathfrak{S}$ is regular, and thus $L$ is also a left ideal of $\mathfrak{R}$. If $0 \neq a \in \mathfrak{R}$, then $ab \neq 0$ for some $b \in \mathfrak{S}$ by 1.25 and I.3.3. But then $0 \neq ab \in \mathfrak{S}$ so $Lab \neq 0$, which implies that $La \neq 0$. Consequently $\mathfrak{A}_{r,\mathfrak{R}}(L) = 0$ and the hypothesis implies that $L \supseteq \mathfrak{S}$. Now 3.5 implies that $\mathfrak{S}$ is an r-maximal simple ring.

The next result characterizes rings isomorphic to $\mathfrak{L}(V)$ for some left vector space $V$; other characterizations can be found in Leptin ([2], Part I) and Wolfson [1].

II.3.7 COROLLARY. The following conditions on a ring $\mathfrak{R}$ are equivalent.

i) $\mathfrak{R}$ is a maximal primitive ring with an r-maximal socle.

ii) $\mathfrak{R}$ is isomorphic to the ring of all linear transformations on a left vector space.

iii) $\mathfrak{M}\mathfrak{R}$ is a maximal dense extension of an r-maximal completely 0-simple semigroup.

iv) $\mathfrak{R}$ is a semiprime ring having a minimal right ideal $A$ such that $\mathfrak{A}_{r,\mathfrak{R}}(A) = 0$ and is maximal among the semiprime rings $\mathfrak{R}'$ having $A$ as a right ideal with $\mathfrak{A}_{r,\mathfrak{R}'}(A) = 0$.

PROOF. The equivalence of i), ii) and iii) is an immediate consequence of 1.27 and 3.6.

i) ⇒ iv). By 1.25, $\mathfrak{R}$ is a semiprime ring and has a minimal right ideal $A$ such that $\mathfrak{A}_{r,\mathfrak{R}}(A) = 0$. Let $\mathfrak{R}'$ be a semiprime ring containing $\mathfrak{R}$ as a subring and having $A$ as a (automatically minimal) right ideal. Let $B$ be a minimal right ideal of $\mathfrak{R}$. By 1.6, we have $A = e\mathfrak{R} = e'\mathfrak{R}'$ and $B = f\mathfrak{R}$ for some idempotents $e, e' \in A$, $f \in B$. Then $A = e\mathfrak{R} \subseteq e\mathfrak{R}' = e'\mathfrak{R}' = A$, so that $A = e\mathfrak{R}'$. In view of 1.25 and 2.8, by 2.5 there exist $a,b \in \mathfrak{R}$ such that $e = ab$, $f = ba$. If $x \in B$ and $y \in \mathfrak{R}'$, then

$$xy = (fx)y = (baba)xy = b(ab)axy = fbe(axy) \in fbe\mathfrak{R}' = fbA \subseteq f\mathfrak{R} = B$$

which proves that $B$ is a right ideal of $\mathfrak{R}'$. But then $B$ must also be a minimal right ideal of $\mathfrak{R}'$. Consequently the socle $\mathfrak{S}$ of $\mathfrak{R}$ is contained in the socle $\mathfrak{S}'$ of $\mathfrak{R}'$. Since each minimal right ideal of $\mathfrak{R}$ is a minimal right ideal of $\mathfrak{R}'$, it follows that their sum $\mathfrak{S}$ is a right ideal of $\mathfrak{R}'$ and thus also of $\mathfrak{S}'$. By the r-maximality of $\mathfrak{S}$, we have $\mathfrak{S} = \mathfrak{S}'$ since $\mathfrak{R}'$ satisfies the conditions in 1.25. Now the maximality of $\mathfrak{R}$ as a primitive ring with socle $\mathfrak{S} = \mathfrak{S}'$ implies $\mathfrak{R} = \mathfrak{R}'$.

iv) ⇒ ii). Again $\mathfrak{R} \cong \mathfrak{R}'$ where $\mathfrak{F}_U(V) \subseteq \mathfrak{R}' \subseteq \mathfrak{L}_U(V)$ for some pair $(U,V)$ of dual vector spaces by 1.25 and I.2.8. We have seen in I.3.11 that the minimal right ideals of $\mathfrak{F}_U(V)$ are precisely the sets

$$R_i = \{[i,\gamma,\lambda] \mid \gamma \in \Delta,\ \lambda \in I_V\}.$$

It follows easily that the sets of all minimal right ideals of the rings $\mathfrak{F}_U(V)$, $\mathfrak{R}'$ and $\mathfrak{L}(V)$ coincide. Thus the hypothesis on $\mathfrak{R}$ implies that $\mathfrak{R}'$ is semiprime and $\mathfrak{A}_{r,\mathfrak{R}'}(R_i) = 0$ for some $R_i$. Since $\mathfrak{L}(V)$ satisfies the conditions in 1.25, it is semiprime, and we have seen that $R_i$ is a minimal right ideal of $\mathfrak{L}(V)$. If $a \in \mathfrak{A}_{r,\mathfrak{L}(V)}(R_i)$, then $[i,\gamma,\lambda]a = 0$ for all $\gamma \in \Delta$, $\lambda \in I_V$. There exists $\mu \in I_V$ such that $(v_\mu,u_i) \neq 0$, so that $(v_\mu,u_i)\gamma(v_\lambda a) = 0$ which for $\gamma \neq 0$ yields $v_\lambda a = 0$ for all $\lambda \in I_V$. But then $a = 0$ which shows that $\mathfrak{A}_{r,\mathfrak{L}(V)}(R_i) = 0$. By the maximality of $\mathfrak{R}'$, we conclude that $\mathfrak{R}' = \mathfrak{L}(V)$ which finally yields $\mathfrak{R} \cong \mathfrak{L}(V)$.

The implication "i) ⇒ ii)" in the next corollary is known as the Second Wedderburn (or Wedderburn-Artin) theorem. A multiplicative characterization of these rings was given by Gluskin [1]. For further characterizations, see Jacobson [4],([7], Chapter IV, Section 16) and Wolfson [1].

II.3.8 COROLLARY. The following conditions on a ring $\mathfrak{R}$ are equivalent.

i) $\mathfrak{R}$ is a simple ring satisfying d.c.c. for right ideals.

ii) $\mathfrak{R}$ is isomorphic to a full matrix ring $\Delta_n$ over a division ring $\Delta$.

iii) $\mathfrak{R}$ is isomorphic to the ring of all linear transformations on a finite dimensional vector space.

iv) $\mathfrak{R}$ is a prime (respectively primitive) ring satisfying d.c.c. for right ideals.

v) $\mathfrak{R}$ is a simple ring and $\mathfrak{M}\mathfrak{R}$ is a maximal dense extension of a completely 0-simple semigroup.

PROOF. i) ⇒ v). The second part of the hypothesis implies that $\mathfrak{R}$ has a minimal right ideal, which together with simplicity by 2.8 implies that $\mathfrak{R} \cong \mathfrak{F}_U(V)$ for some dual pair $(U,V)$. But then $\mathfrak{F}_U(V)$ satisfies d.c.c. for right ideals which by I.3.14 implies that $\dim V < \infty$. Consequently $\mathfrak{F}_U(V) = \mathfrak{L}(V)$ and thus $\mathfrak{R} \cong \mathfrak{L}(V)$. But then I.7.10 yields that $\mathfrak{M}\mathfrak{L}(V) = \mathfrak{S}(V)$ is a maximal dense extension of $\mathfrak{F}_{2,V^*}(V)$ which is a completely 0-simple semigroup. Consequently $\mathfrak{M}\mathfrak{R}$ is a maximal dense extension of a completely 0-simple semigroup.

v) ⇒ iv). By 1.27, the second part of the hypothesis implies that $\mathfrak{R} \cong \mathfrak{L}_U(V)$ for some dual pair $(U,V)$. Since $\mathfrak{R}$ is also simple, we must have $\mathfrak{F}_U(V) = \mathfrak{L}_U(V)$ which implies that $\dim V < \infty$. Consequently nat $U = V^*$ by I.3.13 so that $\mathfrak{L}(V)$ satisfies d.c.c. for right ideals by I.3.4. Hence $\mathfrak{R}$ satisfies d.c.c. for right ideals and is both prime and primitive by 1.25.

iv) ⇒ iii). We deduce from 1.25 and I.3.14 that $\mathfrak{R} \cong \mathfrak{L}(V)$ for a finite dimensional vector space $V$.

ii) ⇔ iii). This is a well-known fact proved in elementary texts on linear algebra.

iii) ⇒ i). First note that $\mathfrak{L}(V)$ is simple by I.3.6 since $\mathfrak{L}(V) = \mathfrak{F}(V)$. Further $\mathfrak{L}(V) = \mathfrak{F}_{V^*}(V)$ where $\dim V < \infty$ so that I.3.4 implies that $\mathfrak{L}(V)$ satisfies d.c.c. for right ideals.

Again the hypothesis on right ideals in i) and iv) can be substituted by the corresponding hypothesis on left ideals, and in iii) the linear transformations can be taken on a left or a right vector space.

II.3.9 EXERCISES.

i) Let $\mathfrak{R}$ be a primitive ring with a nonzero socle $\mathfrak{S}$. Show that a right ideal of $\mathfrak{R}$ which contains a minimal left ideal must contain $\mathfrak{S}$.

A nonempty subset $A$ of a ring $\mathfrak{R}$ is called a left annihilator if $A = \mathfrak{A}_{\ell,\mathfrak{R}}(B)$ for some nonempty subset $B$ of $\mathfrak{R}$; a right annihilator is defined dually. It is clear that a left (right) annihilator is a left (right) ideal.

ii) Let $V$ be a vector space. Show that the principal right (left) ideals of $\mathfrak{L}(V)$ coincide with right (left) annihilators of $\mathfrak{L}(V)$. Deduce that $V$ is finite dimensional if and only if every right (left) ideal of $\mathfrak{L}(V)$ is a right (left) annihilator.

iii) Let $\mathfrak{R}$ be a regular ring with identity. Show that a sum of two principal right (left) ideals of $\mathfrak{R}$ is a principal right (left) ideal of $\mathfrak{R}$. Deduce that in $\mathfrak{L}(V)$ a sum of two right (left) annihilators is a right (left) annihilator. (Hint: For the first statement, if $A$ and $B$ are principal right ideals generated by $e = e^2$ and $b$, respectively, consider $e + (1-e)bx(1-e)$ where $(1-e)b = (1-e)bx(1-e)b$.)

iv) Show that in any ring the intersection of any set of right (left) annihilators is a right (left) annihilator. Deduce that in $\mathfrak{L}(V)$ the intersection of any set of principal right (left) ideals is a principal right (left) ideal.

v) Let $V$ be a finite dimensional vector space, $U$ be a subspace of $V^*$ and $\mathfrak{F} = \mathfrak{F}(V)$. Show that

$$\mathfrak{I}_U(V) = \mathfrak{A}_{r,\mathfrak{F}}(\{c \in \mathfrak{F} \mid c^*U = 0\}).$$

vi) Give an example of a ring $\mathfrak{R}$ and two principal right ideals of $\mathfrak{R}$ whose intersection is not a principal right ideal.

vii) Give an example of a ring $\mathfrak{R}$ and two principal right ideals of $\mathfrak{R}$ whose sum is not a principal right ideal.

## II.4 SEMIPRIME RINGS

We are interested here in semiprime rings with a nonzero socle. This will give us a general result, several special cases of which will be treated in greater detail in the two succeeding sections. Some new notions will come in handy (recall 1.15).

II.4.1 DEFINITION. A family $\{A_i\}_{i\in I}$ of subrings of $\mathfrak{R}$ is independent if for every $i \in I$, we have $A_i \cap \sum_{j\neq i} A_j = 0$. A ring $\mathfrak{R}$ is a direct sum of its subrings $A_i$, $i \in I$, if $\mathfrak{R} = \sum_{i\in I} A_i$ and the family $\{A_i\}_{i\in I}$ is independent, to be denoted by $\mathfrak{R} = \bigoplus_{i\in I} A_i$.

II.4.2 DEFINITION. A semigroup $S$ with zero is an orthogonal sum of its subsemigroups $S_\alpha$, $\alpha \in A$, if $S = \bigcup_{\alpha\in A} S_\alpha$ and $S_\alpha S_\beta = S_\alpha \cap S_\beta = 0$ if $\alpha \neq \beta$, to be denoted by $S = \sum_{\alpha\in A} \oplus S_\alpha$.

Such a "sum" of semigroups may be termed "internal". Externally, we may proceed as follows. Let $\{S_\alpha\}_{\alpha\in A}$ be a family of pairwise disjoint semigroups with zeroes $0_\alpha$. Let $S = [\bigcup_{\alpha\in A} (S_\alpha \setminus 0_\alpha)] \cup 0$, where $0$ is an extra symbol, with multiplication $a * b = ab$ if $a, b \in S_\alpha$ for some $\alpha \in A$ and $ab \neq 0_\alpha$, and $a * b = 0$ in all other cases. In this case also we say that $S$ is an orthogonal sum of semigroups $S_\alpha$, $\alpha \in A$, with the same notation. For the origin of a part of the next theorem, see Dieudonné [3] and Jacobson ([7], Chapter IV, Section 3), the rest is new.

II.4.3 THEOREM. Let $\mathfrak{R}$ be a semiprime ring with the set $\{R_i\}_{i\in I}$ of all minimal right ideals assumed nonempty. Let

$$\mathfrak{K} = \bigcup_{i\in I} R_i, \qquad \mathfrak{S} = \sum_{i\in I} R_i$$

be the socles of $\mathfrak{M}\mathfrak{R}$ and $\mathfrak{R}$, respectively. On $I$ define a relation $\tau$ by: $i \,\tau\, j \Leftrightarrow R_i \cong R_j$ as $\mathfrak{R}$-modules. Then $\tau$ is an equivalence relation. Let $J = I/\tau$ and

$$\mathfrak{K}_\alpha = \bigcup_{i\in\alpha} R_i, \qquad \mathfrak{S}_\alpha = \sum_{i\in\alpha} R_i \qquad (\alpha \in J).$$

Then each $\mathfrak{K}_\alpha$ is a completely 0-simple ideal of $\mathfrak{M}\mathfrak{R}$ and $\mathfrak{K}$ is an orthogonal sum of all $\mathfrak{K}_\alpha$. Each $\mathfrak{S}_\alpha$ is a simple atomic ideal of $\mathfrak{R}$, and $\mathfrak{S}$ is a direct sum of all $\mathfrak{S}_\alpha$. Furthermore, $\mathfrak{S}_\alpha = C(\mathfrak{K}_\alpha)$, $\mathfrak{S} = C(\mathfrak{K})$, and the analogous definitions with minimal left ideals yield the same $\mathfrak{K}$, $\mathfrak{S}$, $\mathfrak{K}_\alpha$, $\mathfrak{S}_\alpha$.

PROOF. That $\tau$ is an equivalence relation follows immediately from the definition of a module isomorphism. We know by 1.16 that the left socle of $\mathfrak{R}$ agrees with the socle of $\mathfrak{R}$. If $L$ and $L'$ are minimal left ideals isomorphic as left $\mathfrak{R}$-modules, then by 1.6 we have $L = \mathfrak{R}e$, $L' = \mathfrak{R}e'$ for some idempotents $e$, $e'$. By 1.14, we have that $e\mathfrak{R}$ and $e'\mathfrak{R}$ are minimal right ideals of $\mathfrak{R}$ and by 2.5 that they are isomorphic as $\mathfrak{R}$-modules. Hence they are contained in some $\mathfrak{K}_\alpha$. By reversing each of these steps, we conclude that the corresponding definitions using minimal left ideals yield the same $\mathfrak{S}$, $\mathfrak{K}$, $\mathfrak{S}_\alpha$, $\mathfrak{K}_\alpha$. Consequently each $\mathfrak{K}_\alpha$ is also a union of minimal left ideals of $\mathfrak{M}\mathfrak{R}$ (and of $\mathfrak{R}$ since these agree in a semiprime ring by 1.13), and hence $\mathfrak{K}_\alpha$ is an ideal of $\mathfrak{M}\mathfrak{R}$. Consequently $\mathfrak{K}$ is an ideal of $\mathfrak{M}\mathfrak{R}$. By the very definition of a sum of subrings, it follows that $\mathfrak{S}_\alpha = C(\mathfrak{K}_\alpha)$, $\mathfrak{S} = C(\mathfrak{K})$, which then implies that both $\mathfrak{S}_\alpha$ and $\mathfrak{S}$ are ideals of $\mathfrak{R}$.

If $\mathfrak{K}_\alpha \cap \mathfrak{K}_\beta \neq 0$, then this intersection contains a minimal right ideal which implies that every right ideal in $\mathfrak{K}_\alpha$ is $\mathfrak{R}$-isomorphic to any minimal right ideal in

$\mathfrak{K}_\beta$, so we must have $\mathfrak{K}_\alpha = \mathfrak{K}_\beta$. It follows that $\mathfrak{K}_\alpha\mathfrak{K}_\beta \subseteq \mathfrak{K}_\alpha \cap \mathfrak{K}_\beta = 0$ if $\alpha \neq \beta$ since both $\mathfrak{K}_\alpha$ and $\mathfrak{K}_\beta$ are ideals of $\mathfrak{M}\mathfrak{R}$, so that $\mathfrak{K}$ is an orthogonal sum of all $\mathfrak{K}_\alpha$.

We prove next that $\mathfrak{S}_\alpha$ is a simple atomic ring. Let $R_i$ be a minimal right ideal of $\mathfrak{R}$ contained in $\mathfrak{S}_\alpha$ and let $0 \neq a \in R_i$. By the dual of 1.11, we have $a\mathfrak{R} = R_i$ and thus $R_i \subseteq a\mathfrak{S}_\alpha \subseteq a\mathfrak{R} = R_i$ which implies that $R_i = a\mathfrak{S}_\alpha$. Hence again by the dual of 1.11, we conclude that $R_i$ is a minimal right ideal of $\mathfrak{S}_\alpha$.

Let $I$ be a nonzero ideal of $\mathfrak{S}_\alpha$. In the proof of "v) $\Rightarrow$ iv)" in 2.8, we let $\mathfrak{K} = \mathfrak{K}_\alpha$. From the first paragraph above, we have that $\mathfrak{K}_\alpha$ is also the union of idempotent minimal left ideals, which by 1.6 are generated by idempotents. Hence for every $t \in \mathfrak{K}_\alpha$ there exists an idempotent $e$ in $\mathfrak{K}_\alpha$ such that $t = te$. The proof of "v) $\Rightarrow$ iv)" in 2.8 now goes through with obvious modifications, and we conclude that $I \cap \mathfrak{K}_\alpha \neq 0$.

Let $J$ be an ideal of $\mathfrak{K}_\alpha$ and let $0 \neq x \in J$, $y \in \mathfrak{K}_\alpha$. Then $x \in R_i$ and $y \in R_j$ for some minimal right ideals $R_i$ and $R_j$ of $\mathfrak{R}$. By 1.6, $R_i = e_i\mathfrak{R}$ and $R_j = e_j\mathfrak{R}$ for some idempotents $e_i$ and $e_j$ in $\mathfrak{K}_\alpha$. By hypothesis $R_i$ and $R_j$ are $\mathfrak{R}$-isomorphic, which by 2.5 implies the existence of $a,b \in \mathfrak{K}_\alpha$ such that $e_i = ab$, $e_j = ba$. Further, $x\mathfrak{R} = e_i\mathfrak{R}$ implies that $e_i = xz$ for some $z \in \mathfrak{R}$. It follows that $y = e_jy = b(ab)ay = be_iay = bx(zay) \in J$, since $b, zay \in \mathfrak{K}_\alpha$. Consequently $J = \mathfrak{K}_\alpha$ and $\mathfrak{K}_\alpha$ has no proper nonzero ideals.

It was shown above that $I \cap \mathfrak{K}_\alpha \neq 0$ which now yields $I \cap \mathfrak{K}_\alpha = \mathfrak{K}_\alpha$ and thus $\mathfrak{K}_\alpha \subseteq I$. Since $\mathfrak{K}_\alpha$ generates $\mathfrak{S}_\alpha$, it follows that $I = \mathfrak{S}_\alpha$. Finally, since $\mathfrak{S}_\alpha$ contains nonzero idempotents, it must be simple. Therefore $\mathfrak{S}_\alpha$ is a simple atomic ring.

We know from above that every minimal right ideal of $\mathfrak{R}$ contained in $\mathfrak{S}_\alpha$ is a minimal right ideal of $\mathfrak{S}_\alpha$. Conversely, if $A$ is a minimal right ideal of $\mathfrak{S}_\alpha$, then $A = e\mathfrak{S}_\alpha$ for some idempotent $e \in \mathfrak{S}_\alpha$ by 1.6. But then $A\mathfrak{R} = e\mathfrak{S}_\alpha\mathfrak{R} \subseteq e\mathfrak{S}_\alpha = A$ and $A$ is a right ideal of $\mathfrak{R}$, minimal in view of its minimality in $\mathfrak{S}_\alpha$. Consequently $\mathfrak{K}_\alpha$ is the union of all minimal right ideals of $\mathfrak{S}_\alpha$, that is, the semigroup socle of $\mathfrak{S}_\alpha$. But in a simple atomic ring, the semigroup socle is completely 0-simple (e.g. consider $\mathfrak{F}_U(V)$). Therefore $\mathfrak{K}_\alpha$ is completely 0-simple.

We have seen above that for $\alpha \neq \beta$, we have $\mathfrak{K}_\alpha\mathfrak{K}_\beta = 0$ so that $\mathfrak{S}_\alpha\mathfrak{S}_\beta = 0$. Hence $\mathfrak{S}_\alpha(\sum_{\beta\neq\alpha}\mathfrak{S}_\beta) = 0$ and if $a \in \mathfrak{S}_\alpha \cap \sum_{\beta\neq\alpha}\mathfrak{S}_\beta$, then $\mathfrak{S}_\alpha a = 0$ so that $a \in \mathfrak{U}_\ell(\mathfrak{S}_\alpha)$. But $\mathfrak{U}_\ell(\mathfrak{S}_\alpha)$ is equal to zero since $\mathfrak{S}_\alpha$ is a simple atomic ring. Thus $a = 0$ and we deduce $\mathfrak{S}_\alpha \cap \sum_{\beta\neq\alpha}\mathfrak{S}_\beta = 0$, that is, the family $\{\mathfrak{S}_\alpha\}_{\alpha\in A}$ is independent. Since $\mathfrak{S} = \sum_{\alpha\in J}\mathfrak{S}_\alpha$, it follows that $\mathfrak{S}$ is a direct sum of all $\mathfrak{S}_\alpha$.

II.4.4 COROLLARY. <u>A semiprime ring $\mathfrak{R}$ has a minimal right ideal if and only if $\mathfrak{M}\mathfrak{R}$ has a completely 0-simple ideal. In fact, in the notation of the theorem, if $R_i$ is a minimal right ideal of $\mathfrak{R}$ contained in $\mathfrak{S}_\alpha$, then $\mathfrak{S}_\alpha = \mathfrak{R}R_i$, the ring product, and $\mathfrak{K}_\alpha = (\mathfrak{M}\mathfrak{R})R_i$, the semigroup product.</u>

PROOF. The first statement follows from 4.3 and 1.13. Further, $\mathfrak{R}R_i$ is an ideal of $\mathfrak{R}$ contained in $\mathfrak{S}_\alpha$ and thus must coincide with $\mathfrak{S}_\alpha$ since $\mathfrak{S}_\alpha$ is simple. Also $(\mathfrak{M}\mathfrak{R})R_i$ is an ideal of $\mathfrak{M}\mathfrak{R}$ contained in $\mathfrak{K}_\alpha$ so it must coincide with $\mathfrak{K}_\alpha$ since a completely 0-simple semigroup has no proper nonzero ideals.

II.4.5 EXERCISES.

i) Let $A$ and $B$ be $\mathfrak{R}$-isomorphic minimal right ideals of a ring $\mathfrak{R}$. Show that $AB = 0$ if $B^2 = 0$ and $AB = A$ if $B^2 \neq 0$, and that an $\mathfrak{R}$-homomorphism of $A$ into $B$ is either a zero $\mathfrak{R}$-homomorphism or an $\mathfrak{R}$-isomorphism. If $A^2 \neq 0$, find the form of all $\mathfrak{R}$-homomorphisms of $A$ into $B$.

If $A$ is a minimal right ideal of a ring $\mathfrak{R}$, a sum of all minimal right ideals of $\mathfrak{R}$ which are $\mathfrak{R}$-isomorphic to $A$ is called a <u>homogeneous component</u> of the socle of $\mathfrak{R}$; a homogeneous component of the left socle is defined analogously.

ii) Show that a homogeneous component of the socle of a ring $\mathfrak{R}$ is an ideal of $\mathfrak{R}$.

iii) Show that the nonzero socle $\mathfrak{S}$ of a ring $\mathfrak{R}$ is a direct sum of its homogeneous components. Also prove that each homogeneous component of $\mathfrak{S}$ is a direct sum of pairwise $\mathfrak{R}$-isomorphic minimal right ideals of $\mathfrak{R}$. (Hint: For the second statement, consider independent families of minimal right ideals and use Zorn's lemma.)

iv) Let $F$ be a homogeneous component of the socle of a ring $\mathfrak{R}$, and let $N$ be a sum of all nilpotent minimal right ideals of $\mathfrak{R}$ contained in $F$. Show that $N$ is a nilpotent ideal of $\mathfrak{R}$ and that $N = F \cap \mathfrak{A}_{r,\mathfrak{R}}(F)$.

v) Give an example of a ring which coincides with a homogeneous component of its socle and contains both idempotent and nilpotent minimal right ideals.

vi) Prove that if a homogeneous component $F$ of the socle of a ring $\mathfrak{R}$ is semiprime, then $F$ is a simple atomic ring and a homogeneous component of the left socle of $\mathfrak{R}$.

For general references on this subject, see Behrens ([1], Chapter IV, Sections 1, 2), Jacobson ([7], Chapter IV, Sections 1,2,3).

## II.5 SEMIPRIME RINGS ESSENTIAL EXTENSIONS OF THEIR SOCLES

The principal result here is a collection of characterizations of semiprime rings which are essential extensions of their nonzero socles. As a preparation, we first study the relationship of prime and semiprime ideals in any ring. For an element $a$ of a ring $\mathfrak{R}$, let $(a)$ denote the principal ideal of $\mathfrak{R}$ generated by $a$.

II.5.1 LEMMA. *The following conditions on an ideal $I$ of $\mathfrak{R}$ are equivalent.*

i) *$I$ is a prime ideal.*

ii) *If $(a)(b) \subseteq I$, then either $a \in I$ or $b \in I$.*

iii) *If $A$ and $B$ are right ideals of $\mathfrak{R}$ such that $AB \subseteq I$, then either $A \subseteq I$ or $B \subseteq I$.*

PROOF. i) ⇒ ii). If $(a)(b) \subseteq I$, then either $(a) \subseteq I$ or $(b) \subseteq I$ so that either $a \in I$ or $b \in I$.

ii) ⇒ iii). First note that for any $a \in \mathfrak{R}$,

$$(a) = \{na + ra + as + \sum_{i=1}^{n} p_i a q_i \mid n \text{ is an integer, } r,s,p_i,q_i \in \mathfrak{R}\}. \tag{1}$$

Let $A$ and $B$ be right ideals of $\mathfrak{R}$ such that $AB \subseteq I$. Suppose that $A \not\subseteq I$; then there exists $a \in A$ such that $a \notin I$. For any $b \in B$, using (1) and the hypothesis, we obtain $(a)(b) \subseteq AB + \mathfrak{R}AB \subseteq I$ and thus $b \in I$ since $a \notin I$. Hence $B \subseteq I$.

iii) ⇒ i). Trivial.

II.5.2 LEMMA. *Every semiprime ideal of a ring $\mathfrak{R}$ is the intersection of prime ideals of $\mathfrak{R}$.*

PROOF. Let $I$ be a semiprime ideal of $\mathfrak{R}$. Let $r \in \mathfrak{R}$ and suppose that $r\mathfrak{R}r \subseteq I$; then $(\mathfrak{R}r\mathfrak{R})^2 \subseteq \mathfrak{R}(r\mathfrak{R}r)\mathfrak{R} \subseteq I$ where $\mathfrak{R}r\mathfrak{R}$ is an ideal, so $\mathfrak{R}r\mathfrak{R} \subseteq I$. But then $(r)^3 \subseteq \mathfrak{R}r\mathfrak{R} \subseteq I$ which implies that $r \in I$. We will use this in the following form: if $r \notin I$, then $rxr \notin I$ for some $x \in \mathfrak{R}$.

We let $r \in \mathfrak{R} \setminus I$ and wish to find a prime ideal $P$ such that $r \notin P$ and $P \supseteq I$, which will prove the lemma. Starting with $r_1 = r$, there exists $x_1 \in \mathfrak{R}$ such that $r_2 = r_1 x r_1 \notin I$. We can inductively define sequences $r_1, r_2, \ldots$ and $x_1, x_2, \ldots$ such that $r_{n+1} = r_n x_n r_n \notin I$. Let $M = \{r_1, r_2, \ldots\}$. For $i \leq j$, we easily obtain

$$r_i[(x_i r_i)(x_{i+1} r_{i+1}) \cdots (x_{j-1} r_{j-1}) x_j] r_j = r_{j+1} \in M,$$

$$r_j[x_j(r_{j-1} x_{j-1}) \cdots (r_{i+1} x_{i+1})(r_i x_i)] r_i = r_{j+1} \in M.$$

Consequently, if $c,d \in M$, then $cxd \in M$ for some $x \in \mathfrak{R}$.

Now let $\mathcal{P}$ be the partially ordered set (under inclusion) of all ideals $J$ of $\mathfrak{R}$ for which $J \cap M = \emptyset$ and $I \subseteq J$. Then $\mathcal{P} \neq \emptyset$ since $I \cap M = \emptyset$ so $I \in \mathcal{P}$. A standard Zorn's lemma argument shows that $\mathcal{P}$ has a maximal element $P$. In order to show that $P$ is in fact a prime ideal, we suppose that $a, b \notin P$. Then $P + (a) \supset P$ which by maximality of $P$ implies that $(P + (a)) \cap M \neq \emptyset$ and hence there exists $c \in (P + (a)) \cap M$ and similarly $d \in (P + (b)) \cap M$. We have seen above that then $cxd \in M$ for some $x \in \mathfrak{R}$. If $(a)(b) \subseteq P$, then

$$cxd \in (P + (a))(P + (b)) \subseteq P$$

and thus $cxd \in P \cap M$ contradicting $P \cap M = \emptyset$. Consequently $(a)(b) \not\subseteq P$ which by contrapositive and 5.1 implies that $P$ is a prime ideal. Since $P \in \mathcal{P}$ we have $r \notin P$ and $P \supseteq I$.

II.5.3 <u>DEFINITION</u>. Let $\{\mathfrak{R}_\alpha\}_{\alpha \in A}$ be a family of rings. On the Cartesian product $\prod_{\alpha \in A} \mathfrak{R}_\alpha$ define the addition and multiplication coordinatewise. The resulting structure is a ring called a <u>complete direct sum</u> (also <u>direct product</u>) of the rings $\mathfrak{R}_\alpha$ and is also denoted by $\prod_{\alpha \in A} \mathfrak{R}_\alpha$. The mapping which to every element of $\prod_{\alpha \in A} \mathfrak{R}_\alpha$ associates its $\alpha$-th component is the $\alpha$-th <u>projection homomorphism</u>, to be denoted by $\pi_\alpha$. Any subring $\mathfrak{R}$ of $\prod_{\alpha \in A} \mathfrak{R}_\alpha$ all of whose projection homomorphisms are onto is a <u>subdirect sum</u> (also <u>subdirect product</u>) of the rings $\mathfrak{R}_\alpha$. The subdirect sum of $\mathfrak{R}_\alpha$ consisting of all elements of $\prod_{\alpha \in A} \mathfrak{R}_\alpha$ with only a finite number of nonzero components is an <u>external</u> (also <u>discrete</u>) <u>direct sum</u> of the rings $\mathfrak{R}_\alpha$. Any ring isomorphic to any of these "sums" will be labeled the same way.

A "direct sum" of rings defined earlier is sometimes called "internal direct sum". The relationship of internal and external direct sums is very close and the distinction is rarely made.

II.5.4 <u>PROPOSITION</u>. <u>An ideal $I$ of $\mathfrak{R}$ is semiprime if and only if $I$ is the intersection of prime ideals. A ring $\mathfrak{R}$ is semiprime if and only if $\mathfrak{R}$ is a subdirect sum of prime rings</u>.

<u>PROOF</u>. Necessity of the first statement is the content of 5.2; the proof of sufficiency consists of a straightforward verification and is omitted. If $\mathfrak{R}$ is semiprime, then $0 = \bigcap_{\alpha \in A} I_\alpha$ for some prime ideals $I_\alpha$ of $\mathfrak{R}$ by the first statement of the proposition. It follows that the mapping $r \rightarrow (r + I_\alpha)_{\alpha \in A}$ is an isomorphism of $\mathfrak{R}$ into $\prod_{\alpha \in A} (\mathfrak{R}/I_\alpha)$ and the projection homomorphisms of the image are evidently onto. Thus $\mathfrak{R}$ is a subdirect sum of prime rings $\mathfrak{R}/I_\alpha$. Conversely, if $\mathfrak{R}$ is a subdirect sum of prime rings $\mathfrak{R}_\alpha$, $\alpha \in A$, we can identify $\mathfrak{R}$ with a subring of $\prod_{\alpha \in A} \mathfrak{R}_\alpha$. If now $I$ is a nilpotent ideal of $\mathfrak{R}$, then $I\pi_\alpha$ is a nilpotent ideal of $\mathfrak{R}_\alpha$, which implies that $I\pi_\alpha = 0_\alpha$. Since $\alpha \in A$ is arbitrary, we obtain $I = 0$. Consequently $\mathfrak{R}$ is semiprime.

II.5.5 DEFINITION. Let $\{S_\alpha\}_{\alpha\in A}$ be a family of semigroups. The Cartesian product $\prod_{\alpha\in A} S_\alpha$ together with coordinatewise multiplication is a direct product of semigroups $S_\alpha$, $\alpha \in A$, and is also denoted by $\prod_{\alpha\in A} S_\alpha$.

II.5.6 LEMMA. *Let* $S = \sum_{\alpha\in A} \oplus S_\alpha$, *where for each* $\alpha$, $S_\alpha^2 = S_\alpha$. *Then the mapping*

$$\psi:(\lambda,\rho) \to (\lambda|_{S_\alpha}, \rho|_{S_\alpha})_{\alpha\in A} \qquad ((\lambda,\rho) \in \Omega(S))$$

*is an isomorphism of* $\Omega(S)$ *onto* $\prod_{\alpha\in A} \Omega(S_\alpha)$.

PROOF. Let $\lambda$ be a left translation of $S$, $a \in S_\alpha$, and suppose that $\lambda a \neq 0$. Then $a = xy$ since $S_\alpha^2 = S_\alpha$ so that $\lambda a = \lambda(xy) = (\lambda x)y \neq 0$ which implies that $\lambda x \in S_\alpha$. Hence $\lambda$ maps $S_\alpha$ into itself. Consequently $\psi$ maps $\Omega(S)$ into $\prod_{\alpha\in A} \Omega(S_\alpha)$. Verification of the remaining properties of $\psi$ is left as an exercise.

II.5.7 DEFINITION. A semigroup which is an orthogonal sum of completely 0-simple semigroups is a primitive regular semigroup.

It is clear that a primitive regular semigroup is indeed regular, and furthermore that each of its nonzero idempotents is primitive; this is the usual definition of a primitive regular semigroup. The two definitions are equivalent, see Clifford and Preston ([1], Chapter 6, Section 5).

II.5.8 LEMMA. *Let* $\mathfrak{R} = \prod_{\alpha\in A} \mathfrak{R}_\alpha$, *where each* $\mathfrak{R}_\alpha$ *is a semiprime ring having minimal right ideals. A subset* $R$ *of* $\mathfrak{R}$ *is a minimal right ideal of* $\mathfrak{R}$ *if and only if for some* $\alpha \in A$ *and some minimal right ideal* $R_\alpha$ *of* $\mathfrak{R}_\alpha$, *we have*

$$R = \{a \in \mathfrak{R} \mid a\pi_\alpha \in R_\alpha,\ a\pi_\beta = 0_\beta \quad \text{if} \quad \beta \neq \alpha\}.$$

PROOF. Necessity. First note that $\mathfrak{R}$ is a semiprime ring; the proof is the same as in 5.4. Now let $R$ be a minimal right ideal of $\mathfrak{R}$, then $R\pi_\alpha \neq 0_\alpha$ for some $\alpha \in A$. Let $0_\alpha \neq a_\alpha \in R\pi_\alpha$; then there exists $0 \neq a \in R$ such that $a\pi_\alpha = a_\alpha$. By the dual of 1.11, we have $a\mathfrak{R} = R$ so that $a_\alpha\mathfrak{R}_\alpha = R\pi_\alpha$ which again by the dual of 1.11 implies that $R\pi_\alpha$ is a minimal right ideal of $\mathfrak{R}_\alpha$. Let

$$D = \{r \in R \mid r\pi_\beta = 0_\beta \quad \text{for all} \quad \beta \neq \alpha\}.$$

If $r \in D$ and $s \in R$, then $(rs)\pi_\beta = (r\pi_\beta)(s\pi_\beta) = 0_\beta$ and $rs \in R$, so that $D$ is a right ideal of $\mathfrak{R}$. Let $0_\alpha \neq a_\alpha \in R\pi_\alpha$ and $b_\alpha \in \mathfrak{R}_\alpha$ be such that $a_\alpha b_\alpha \neq 0_\alpha$ (recall that $\mathfrak{R}_\alpha$ is semiprime so it has trivial right annihilator). Then for some $a \in R$, we have $a\pi_\alpha = a_\alpha$; letting $b \in \mathfrak{R}$ be such that $b\pi_\alpha = b_\alpha$, $b\pi_\beta = 0_\beta$ if $\beta \neq \alpha$, we obtain $0 \neq ab \in D$. Consequently $D$ is a nonzero right ideal of $\mathfrak{R}$ contained in $R$ which implies that $R = D$.

Sufficiency. Let $a, r \in R$ and suppose that $a \neq 0$. Then $0_\alpha \neq a_\alpha = a\pi_\alpha \in R_\alpha$ and $r_\alpha = r\pi_\alpha \in R_\alpha$, which by the dual of 1.11 implies that $r_\alpha = a_\alpha x_\alpha$ for some

$x_\alpha \in \mathfrak{R}_\alpha$. Now letting $x \in \mathfrak{R}$ be such that $x\pi_\alpha = x_\alpha$ and $x\pi_\beta = 0_\beta$ if $\beta \neq \alpha$, we obtain $ax = r$. Consequently $a\mathfrak{R} = R$ for every $0 \neq a \in R$ which again by the dual of 1.11 implies that $R$ is a minimal right ideal of $\mathfrak{R}$.

We are now ready to prove the main result of this section. This result is new. Another characterization of the rings in the next theorem was given by Jaffard [1].

II.5.9 THEOREM. The following conditions on a ring $\mathfrak{R}$ are equivalent.

i) $\mathfrak{R}$ is a semiprime ring essential extension of its nonzero socle.

ii) Every nonzero ideal of $\mathfrak{R}$ contains an idempotent minimal right ideal.

iii) $\mathfrak{R}$ is isomorphic to a subring of $\prod_{\alpha \in A} \mathfrak{R}_\alpha$ containing the socle thereof, where each $\mathfrak{R}_\alpha$ is a prime ring with minimal right ideals.

iv) $\mathfrak{M}\mathfrak{R}$ is a dense extension of a primitive regular semigroup.

v) $\mathfrak{M}\mathfrak{R}$ has a primitive regular ideal having nonzero intersection with every nonzero ideal of $\mathfrak{M}\mathfrak{R}$.

PROOF. i) $\Rightarrow$ ii). Since $\mathfrak{R}$ is semiprime, 1.6 implies that every minimal right ideal is idempotent. Let $\mathfrak{S}$ denote the socle of $\mathfrak{R}$. For a nonzero ideal $I$, we have $\mathfrak{S} \cap I \neq 0$ since $\mathfrak{R}$ is an essential extension of $\mathfrak{S}$. Let $0 \neq a \in \mathfrak{S} \cap I$, then $a = a_1 + a_2 + \ldots + a_n$ for some $a_i \in \mathfrak{S}_{\alpha_i}$, where $\mathfrak{S}_{\alpha_i}$ are some homogeneous components of $\mathfrak{S}$. By 4.3, each $\mathfrak{S}_{\alpha_i}$ is a simple atomic ring and thus there exists $b_1 \in \mathfrak{S}_{\alpha_1}$ such that $a_1 b_1 \neq 0$; on the other hand $a_i b_1 = 0$ for $1 < i \leq n$ since distinct homogeneous components annihilate each other. Hence $0 \neq ab_1 = a_1 b_1 \in \mathfrak{S}_{\alpha_1} \cap I$ and thus $\mathfrak{S}_{\alpha_1} \subseteq I$. But then $I$ contains an idempotent minimal right ideal.

ii) $\Rightarrow$ iii). The hypothesis implies that $\mathfrak{R}$ cannot have nilpotent ideals and hence $\mathfrak{R}$ is semiprime. By 5.2, the ideal $0$ is the intersection of prime ideals of $\mathfrak{R}$. If $A$ is an index set of homogeneous components of the socle $\mathfrak{S}$ of $\mathfrak{R}$, for every $\alpha \in A$ we can choose a prime ideal $I_\alpha$ such that $I_\alpha \cap \mathfrak{S}_\alpha = 0$ since $\mathfrak{S}_\alpha$ is simple.

Assume that $0 \neq a \in \bigcap_{\alpha \in A} I_\alpha$, then $(a)$ is a nonzero ideal and hence by hypothesis contains a minimal right ideal $R$ of $\mathfrak{R}$. Hence $R \subseteq (a) \subseteq \bigcap_{\alpha \in A} I_\alpha$ and thus $R \subseteq I_\alpha$ for all $\alpha \in A$. But $R$ is contained in some homogeneous component $\mathfrak{S}_\beta$ and hence $R \subseteq \mathfrak{S}_\beta \cap I_\beta = 0$, a contradiction. Consequently $\bigcap_{\alpha \in A} I_\alpha = 0$. Hence $\mathfrak{R}_\alpha = \mathfrak{R}/I_\alpha$ is a prime ring and the mapping $\psi : r \to (r + I_\alpha)_{\alpha \in A}$ is an isomorphism of $\mathfrak{R}$ onto a subdirect sum of rings $\mathfrak{R}_\alpha$.

The canonical homomorphism $\psi_\beta : \mathfrak{R} \to \mathfrak{R}_\beta = \mathfrak{R}/I_\beta$ is one-to-one on $\mathfrak{S}_\beta$ since $\mathfrak{S}_\beta \cap I_\beta = 0_\beta$. Hence $\mathfrak{S}_\beta \cong \mathfrak{S}_\beta\psi_\beta$ so that the latter is a simple atomic ring and also an ideal of $\mathfrak{R}_\beta$. From the properties of a simple atomic ring, it follows that every minimal right ideal of $\mathfrak{S}_\beta\psi_\beta$ is a minimal right ideal of $\mathfrak{R}_\beta$ and thus $\mathfrak{S}_\beta\psi_\beta$ is contained in the socle of $\mathfrak{R}_\beta$. Since $\mathfrak{R}_\beta$ is a prime ring, its socle is contained in every nonzero ideal, and we conclude that $\mathfrak{S}_\beta\psi_\beta$ is the socle of $\mathfrak{R}_\beta$. Hence $\mathfrak{R}_\beta$ is a prime ring with minimal right ideals.

Next let $M$ be a minimal right ideal of $\prod_{\alpha\in A} \mathfrak{R}_\alpha$. Then by 5.8, there exists $\beta \in A$ and a minimal right ideal $M_\beta$ of $\mathfrak{R}_\beta$ such that

$$M = \{(a_\alpha) \in \prod_{\alpha\in A} \mathfrak{R}_\alpha \mid a_\beta \in M_\beta,\ a_\alpha = 0 \ \text{ if } \ \alpha \neq \beta\}. \tag{1}$$

Furthermore, we must have $M_\beta \subseteq \mathfrak{S}_\beta\psi_\beta$ and thus $M \subseteq \mathfrak{R}\psi$. Consequently $\mathfrak{R}\psi$ contains the socle of $\prod_{\alpha\in A} \mathfrak{R}_\alpha$.

iii) $\Rightarrow$ iv). We may suppose that $\mathfrak{R} \subseteq \prod_{\alpha\in A} \mathfrak{R}_\alpha$ with property iii), and that $\mathfrak{R}_\alpha$ are pairwise disjoint. We know by 1.25 that $\mathfrak{M}\mathfrak{R}_\alpha$ is a dense extension of the semigroup socle $\mathfrak{K}_\alpha$ of $\mathfrak{R}_\alpha$. Define

$$\mathfrak{K}'_\alpha = \{\alpha \in \mathfrak{R} \mid a\pi_\alpha \in \mathfrak{K}_\alpha,\ a\pi_\beta = 0_\beta \ \text{ if } \ \alpha \neq \beta\},$$

$$\mathfrak{K}' = \bigcup_{\alpha\in A} \mathfrak{K}'_\alpha, \qquad \mathfrak{K} = \sum_{\alpha\in A} \oplus\, \mathfrak{K}_\alpha .$$

Then $\pi_\beta|_{\mathfrak{K}'_\beta}$ is an isomorphism of $\mathfrak{K}'_\beta$ into $\mathfrak{K}_\beta$. If $M_\beta$ is a minimal right ideal of $\mathfrak{R}_\beta$, then $M$ given in (1) above is a minimal right ideal of $\prod_{\alpha\in A} \mathfrak{R}_\alpha$ by 5.8 and the hypothesis implies that $M \subseteq \mathfrak{R}$. It follows that $\pi_\beta$ maps $\mathfrak{K}'_\alpha$ onto $\mathfrak{K}_\alpha$. Consequently $\mathfrak{K}'_\alpha \cong \mathfrak{K}_\alpha$ and thus $\mathfrak{K}'_\alpha$ is completely 0-simple. If $a \in \mathfrak{K}'_\alpha$ and $b \in \mathfrak{K}'_\beta$ with $\alpha \neq \beta$, then $(ab)\pi_\gamma = (a\pi_\gamma)(b\pi_\gamma) = 0_\gamma$ for all $\gamma \in A$, and thus $ab = 0$. It follows that $\mathfrak{K}'$ is an orthogonal sum of semigroups $\mathfrak{K}'_\alpha$ and therefore $\mathfrak{K}'$ is a primitive regular semigroup. By I.7.5, the mapping

$$a \to (\lambda_a|_{\mathfrak{K}_\alpha}, \rho_a|_{\mathfrak{K}_\alpha}) \qquad (a \in \mathfrak{R}_\alpha)$$

is an isomorphism of $\mathfrak{M}\mathfrak{R}_\alpha$ into $\Omega(\mathfrak{K}_\alpha)$. Hence by 5.6, we have that

$$(a_\alpha)_{\alpha\in A} \to (\lambda_{a_\alpha}|_{\mathfrak{K}_\alpha}, \rho_{a_\alpha}|_{\mathfrak{K}_\alpha}) \to (\lambda, \rho) \qquad ((a_\alpha)_{\alpha\in A} \in \mathfrak{R}),$$

where $\lambda|_{\mathfrak{K}_\alpha} = \lambda_{a_\alpha}|_{\mathfrak{K}_\alpha}$, $\rho|_{\mathfrak{K}_\alpha} = \rho_{a_\alpha}|_{\mathfrak{K}_\alpha}$, are isomorphisms mapping

$$\mathfrak{M}\mathfrak{R} \to \prod_{\alpha\in A} \Omega(\mathfrak{K}_\alpha) \to \Omega(\mathfrak{K}).$$

For $a' \in \mathfrak{K}'_\beta$, we obtain

$$a' = (a'_\alpha)_{\alpha\in A} \to (\lambda_{a'_\alpha}|_{\mathfrak{K}_\alpha}, \rho_{a'_\alpha}|_{\mathfrak{K}_\alpha}) \to (\lambda_a, \rho_a) \in \Pi(\mathfrak{K}) \tag{2}$$

where $a = a'\pi_\beta$ and $\Pi(\mathfrak{K})$ is the inner part of $\Omega(\mathfrak{K})$. On the other hand, if $a \in \mathfrak{K}$, then $a \in \mathfrak{K}_\beta$ for some $\beta \in A$. As above, the hypothesis implies the existence of $a' \in \mathfrak{R}$ for which $a'\pi_\beta = a$ and $a'\pi_\alpha = 0_\alpha$ if $\alpha \neq \beta$; in (2), $a'$ is mapped onto $(\lambda_a, \rho_a)$. It follows that the isomorphism which maps $\mathfrak{M}\mathfrak{R}$ into $\Omega(\mathfrak{K})$ also maps $\mathfrak{K}'$ onto $\Pi(\mathfrak{K})$ which by I.7.5 implies that $\mathfrak{M}\mathfrak{R}$ is a dense extension of the primitive regular semigroup $\mathfrak{K}'$.

iv) $\Rightarrow$ v). If $I$ is an ideal of $\mathfrak{M}\mathfrak{R}$ such that $I \cap \mathfrak{K} = 0$, where $\mathfrak{M}\mathfrak{R}$ is a dense extension of the primitive regular semigroup $\mathfrak{K}$, then the congruence $\sigma$ defined on $\mathfrak{M}\mathfrak{R}$ by: $a \sigma b$ if and only if $a, b \in I$ or $a = b$, has the property that $\sigma|_{\mathfrak{K}}$ is the equality relation. But then the hypothesis implies that $\sigma$ must be the equality relation which forces $I = 0$.

v) $\Rightarrow$ i). Let $\mathfrak{K}'$ be the primitive regular ideal in v), and let $\mathfrak{K}$ be the semigroup socle of $\mathfrak{R}$. Then $\mathfrak{K}' \subseteq \mathfrak{K}$ by 4.3. Let $R$ be a minimal right ideal of $\mathfrak{R}$. If $R$ is not contained in $\mathfrak{K}'$, consider $I = (\mathfrak{M}\mathfrak{R})(\mathfrak{M}R)$. Then $I$ is an ideal of $\mathfrak{M}\mathfrak{R}$ and if $a \in \mathfrak{K}' \cap I$, then $a = sr$ for some $s \in \mathfrak{R}$, $r \in R$, and since $\mathfrak{K}'$ is regular, there exists $x \in \mathfrak{K}'$ such that $a = axa = s(rxa)$ where $rxa \in \mathfrak{K}' \cap R = 0$. Hence $a = 0$ and thus $\mathfrak{K}' \cap I = 0$. The ideal $J = I \cup \mathfrak{M}R$ of $\mathfrak{M}\mathfrak{R}$ then has the property $J \cap \mathfrak{K}' = 0$ which by hypothesis implies that $J = 0$. But then $R = 0$, a contradiction. Thus $R \subseteq \mathfrak{K}'$ which proves that $\mathfrak{K} = \mathfrak{K}'$. By 4.3, we have that $\mathfrak{S} = C(\mathfrak{K})$ is the socle of $\mathfrak{R}$. If $I$ is an ideal of $\mathfrak{R}$ such that $\mathfrak{S} \cap I = 0$, then $\mathfrak{M}I \cap \mathfrak{K} = 0$ and the hypothesis implies that $I = 0$. Consequently $\mathfrak{R}$ is an essential extension of its nonzero socle $\mathfrak{S}$. If $I$ is a nonzero ideal of $\mathfrak{R}$, then $\mathfrak{M}I \cap \mathfrak{K}$ is a nonzero ideal of $\mathfrak{K}$, and thus must contain a completely 0-simple component $\mathfrak{K}_\alpha$. But $\mathfrak{K}_\alpha$ always contains a nonzero idempotent so that $I$ cannot be nilpotent. Therefore $\mathfrak{R}$ is a semiprime ring.

Note that the rings $\mathfrak{R}$ in the theorem are subdirect sums of prime rings with minimal right ideals. For if $\mathfrak{R}$ is a subring of $\prod_{\alpha \in A} \mathfrak{R}_\alpha$ containing its socle, then according to 5.8, for each $\alpha \in A$, $\mathfrak{R}\pi_\alpha$ must contain the socle of $\mathfrak{R}_\alpha$ and is thus a prime ring by 1.25. It is clear that the rings in the theorem do not represent the general case of subdirect sums of prime rings with minimal right ideals. In order to deduce a consequence of the theorem we first need a ring analogue of 5.6.

II.5.10 <u>LEMMA</u>. <u>Let $\mathfrak{R} = \bigoplus_{\alpha \in A} \mathfrak{R}_\alpha$, where for each $\alpha \in A$, $\mathfrak{R}_\alpha^2 = \mathfrak{R}_\alpha$. Then the mapping</u>

$$\psi : (\lambda, \rho) \to (\lambda|_{\mathfrak{R}_\alpha}, \rho|_{\mathfrak{R}_\alpha})_{\alpha \in A} \qquad ((\lambda, \rho) \in \Omega(\mathfrak{R}))$$

<u>is an isomorphism of $\Omega(\mathfrak{R})$ onto $\prod_{\alpha \in A} \Omega(\mathfrak{R}_\alpha)$.</u>

PROOF. Let $\lambda$ be a left translation of $\mathfrak{R}$, $a \in \mathfrak{R}_\alpha$, and suppose that $\lambda a \neq 0$. Then $a = \sum_{i=1}^{n} x_i y_i$ for some $x_i y_i \in \mathfrak{R}_\alpha$ since $\mathfrak{R}_\alpha^2 = \mathfrak{R}_\alpha$ so that

$$0 \neq \lambda a = \lambda(\sum_{i=1}^{n} x_i y_i) = \sum_{i=1}^{n} (\lambda x_i) y_i .$$

Since $\mathfrak{R}_\alpha$ is an ideal of $\mathfrak{R}$ and $y_i \in \mathfrak{R}_\alpha$, we must have $(\lambda x_i) y_i \in \mathfrak{R}_\alpha$ and thus $\lambda a \in \mathfrak{R}_\alpha$. This shows that $\lambda$ maps $\mathfrak{R}_\alpha$ into itself. Similarly the same result is valid for right translations which implies that $\psi$ indeed maps $\Omega(\mathfrak{R})$ into $\prod_{\alpha \in A} \Omega(\mathfrak{R}_\alpha)$. That $\psi$ has the remaining properties follows easily from the definition of a direct sum and is left as an exercise.

II.5.11 <u>COROLLARY</u>. <u>The following conditions on a ring</u> $\mathfrak{R}$ <u>are equivalent</u>.

i) $\mathfrak{R}$ <u>is a semiprime ring maximal essential extension of its nonzero socle</u>.

ii) $\mathfrak{R}$ <u>is a complete direct sum of maximal primitive rings</u>.

iii) $\mathfrak{M}\mathfrak{R}$ <u>is a maximal dense extension of a primitive regular semigroup</u>.

PROOF. i) $\Rightarrow$ ii). Since $\mathfrak{R}$ is semiprime, it has a trivial annihilator and hence the ring analogue of I.7.5 implies that $\mathfrak{R} \cong \Omega(\mathfrak{S})$, where $\mathfrak{S}$ is the socle of $\mathfrak{R}$. By 4.3, we have $\mathfrak{S} = \oplus_{\alpha \in A} \mathfrak{S}_\alpha$, where each $\mathfrak{S}_\alpha$ is isomorphic to a ring of the form $\mathfrak{F}_{U_\alpha}(\Delta_\alpha, V_\alpha)$ by 2.8 so that $\mathfrak{S}_\alpha^2 = \mathfrak{S}_\alpha$. Hence 5.10 yields $\Omega(\mathfrak{S}) \cong \prod_{\alpha \in A} \Omega(\mathfrak{S}_\alpha)$. By I.7.13, we have $\Omega(\mathfrak{S}_\alpha) \cong \mathfrak{L}_{U_\alpha}(\Delta_\alpha, V_\alpha)$, where $\mathfrak{L}_{U_\alpha}(\Delta_\alpha, V_\alpha)$ is a maximal primitive ring.

ii) $\Rightarrow$ iii). By 1.27 and I.7.13, we have $\mathfrak{M}\mathfrak{R} \cong \prod_{\alpha \in A} \mathfrak{S}_{U_\alpha}(\Delta_\alpha, V_\alpha)$ for some dual pairs $(U_\alpha, \Delta_\alpha, V_\alpha)$. But by I.7.10, $\mathfrak{S}_{U_\alpha}(\Delta_\alpha, V_\alpha)$ is a maximal dense extension of the completely 0-simple semigroup $\mathfrak{F}_{2,U_\alpha}(\Delta_\alpha, V_\alpha)$. Now using I.7.5 and 5.6, we obtain

$$\mathfrak{M}\mathfrak{R} \cong \prod_{\alpha \in A} \mathfrak{S}_{U_\alpha}(\Delta_\alpha, V_\alpha) \cong \prod_{\alpha \in A} \Omega(\mathfrak{F}_{2,U_\alpha}(\Delta_\alpha, V_\alpha)) \cong \Omega(\sum_{\alpha \in A} \oplus \mathfrak{F}_{2,U_\alpha}(\Delta_\alpha, V_\alpha))$$

where $\sum_{\alpha \in A} \oplus \mathfrak{F}_{2,U_\alpha}(\Delta_\alpha, V_\alpha)$ is a primitive regular semigroup which again by I.7.5 implies that iii) holds.

iii) $\Rightarrow$ i). By 5.9, $\mathfrak{R}$ is a semiprime ring essential extension of its nonzero socle. By the ring analogue of I.7.5, we may identify $\mathfrak{R}$ with a subring of $\Omega(\mathfrak{S})$ containing $\Pi(\mathfrak{S})$, where $\mathfrak{S}$ is the socle of $\mathfrak{R}$. It follows from the proof of 5.9 that $\mathfrak{M}\mathfrak{R}$ is a dense extension of the semigroup socle $\mathfrak{K}$ of $\mathfrak{R}$ which is a primitive regular semigroup. On the other hand $\mathfrak{S} = \oplus_{\alpha \in A} \mathfrak{S}_\alpha$, $\mathfrak{S}_\alpha \cong \mathfrak{F}_{U_\alpha}(\Delta_\alpha, V_\alpha)$, $\mathfrak{K} = \sum_{\alpha \in A} \oplus \mathfrak{K}_\alpha$, $\mathfrak{K}_\alpha \cong \mathfrak{F}_{2,U_\alpha}(\Delta_\alpha, V_\alpha)$ as before. In the proof of I.7.13, we have seen that the correspondence which to each bitranslation of $\mathfrak{F}_U(\Delta, V)$ associates its restriction to $\mathfrak{F}_{2,U}(\Delta, V)$ is an isomorphism of $\mathfrak{M}\Omega(\mathfrak{F}_U(\Delta, V))$ onto $\Omega(\mathfrak{F}_{2,U}(\Delta, V))$ so the same must hold for $\mathfrak{S}_\alpha$ and $\mathfrak{K}_\alpha$. Hence $\mathfrak{M}\Omega(\mathfrak{S}_\alpha) \cong \Omega(\mathfrak{K}_\alpha)$ which together with 5.10 and 5.6, by taking direct products, yields $\mathfrak{M}\Omega(\mathfrak{S}) \cong \Omega(\mathfrak{K})$. By I.7.5, $\Omega(\mathfrak{K})$ is a maximal dense extension of $\Pi(\mathfrak{K})$, identified above with $\mathfrak{K}$, so $\mathfrak{M}\Omega(\mathfrak{S})$ is a maximal dense extension of $\mathfrak{K}$. We have already seen that $\mathfrak{M}\mathfrak{R}$ is a dense extension of $\mathfrak{K}$ which by hypothesis of maximality implies that $\mathfrak{M}\mathfrak{R} = \mathfrak{M}\Omega(\mathfrak{S})$ and thus $\mathfrak{R} = \Omega(\mathfrak{S})$. The ring analogue of I.7.5 yields that $\mathfrak{R}$ is a maximal essential extension of its non-zero socle $\mathfrak{S}$.

II.5.12 <u>EXERCISES</u>.

i) Establish the relationship between external and internal direct sums.

ii) Show that in 5.9, part v), "ideal of $\mathfrak{M}\mathfrak{R}$" may be replaced by "ideal of $\mathfrak{R}$".

iii) Show that an ideal I of a ring $\mathfrak{R}$ is prime if and only if for any $a,b \in \mathfrak{R}$, $a\mathfrak{R}b \subseteq I$ implies that either $a \in I$ or $b \in I$. Establish an analogue of this for semiprime ideals.

For more information on prime and semiprime ideals, see McCoy ([1], Sections 18-20), and on subdirect sums, Behrens ([1], Chapter III, Section 2).

## II.6 SEMIPRIME ATOMIC RINGS

We will first establish several characterizations of semiprime atomic rings and then will deduce some consequences thereof. For an element $a$ of a ring $\mathfrak{R}$, we denote by $(a)_r$ the principal right ideal of $\mathfrak{R}$ generated by $a$. The equivalence of ii) and iii) in the next theorem is due to Faith [1] and also to F. Szász [1],[2]; the last two references contain further information on rings satisfying d.c.c for principal right ideals. The rest of the theorem is new.

II.6.1 <u>THEOREM</u>. <u>The following conditions on a ring $\mathfrak{R}$ are equivalent.</u>

i) $\mathfrak{R}$ <u>is a semiprime atomic ring</u>.

ii) $\mathfrak{R}$ <u>is a semiprime ring satisfying d.c.c for principal right ideals</u>.

iii) $\mathfrak{R}$ <u>is a direct sum of simple atomic rings</u>.

iv) $\mathfrak{M}\mathfrak{R}$ <u>has a primitive regular ideal $\mathfrak{K}$ such that $\mathfrak{R} = C(\mathfrak{K})$</u>.

<u>PROOF</u>. i) ⇒ ii). By 4.3, we have $\mathfrak{R} = \bigoplus_{\alpha\in A} \mathfrak{S}_\alpha$, where $\mathfrak{S}_\alpha$ are simple atomic rings. Let $(a_1)_r \supseteq (a_2)_r \supseteq \dots$ be a descending chain of principal right ideals of $\mathfrak{R}$. Then $a_1 = x_{11} + x_{12} + \dots + x_{1n}$ for some $x_{1i} \in \mathfrak{S}_{\alpha_i}$, $1 \le i \le n$. We know from 2.8 and I.3.6 that each $\mathfrak{S}_\alpha$ is a regular ring. Hence there exists $e_i \in \mathfrak{S}_{\alpha_i}$ such that $x_{1i} = x_{1i}e_i$, and thus letting $e = e_1 + e_2 + \dots + e_n$, we obtain $a_1 = a_1 e$. It follows that

$$\begin{aligned}(a_1)_r &= a_1\mathfrak{R} = (x_{11}+x_{12}+\dots+x_{1n})\mathfrak{R}\\ &= x_{11}\mathfrak{S}_{\alpha_1}+\dots+x_{1n}\mathfrak{S}_{\alpha_n} = x_{11}\mathfrak{R}+\dots+x_{1n}\mathfrak{R}\\ &= (x_{11})_r+\dots+(x_{1n})_r.\end{aligned}$$

Since for any $i$ we have $(a_i)_r \subseteq (a_1)_r$, this discussion is valid for any $a_i = x_{i1}+\dots+x_{in}$, and we obtain

$$(x_{11})_r+(x_{12})_r+\dots+(x_{1n})_r \supseteq (x_{21})_r+(x_{22})_r+\dots+(x_{2n})_r \supseteq \dots .$$

Since the sum $\bigoplus_{i=1}^{n}\mathfrak{S}_{\alpha_i}$ is direct and $(x_{ji})_r \subseteq \mathfrak{S}_{\alpha_i}$, we conclude that

$$(x_{1i})_r \supseteq (x_{2i})_r \supseteq \dots \quad \text{for } 1 \le i \le n,$$

where each of these is a principal right ideal of $\mathfrak{S}_{\alpha_i}$. By 2.8, each $\mathfrak{S}_\alpha$ satisfies d.c.c. for principals right ideals. Hence we can find $m$ such that

$$(x_{mi})_r = (x_{m+1,i})_r = \dots \quad \text{for } 1 \le i \le n.$$

But then $(a_m)_r = (a_{m+1})_r = \dots$ and thus $\mathfrak{R}$ satisfies d.c.c. for principal right ideals.

ii) ⇒ iii). We have seen in the proof of "iii) ⇒ ii)" in 2.8 that every minimal principal right ideal is a minimal right ideal. Since $\mathfrak{R}$ is semiprime, by 1.6, every minimal right ideal is of the form $e\mathfrak{R}$ for some idempotent $e$. Hence every principal right ideal of $\mathfrak{R}$ contains a minimal right ideal of the form $e\mathfrak{R}$. Let $a \in \mathfrak{R}$ and consider $(a)_r$. If $(a)_r$ is not a minimal right ideal, then $(a)_r \supset e_1\mathfrak{R}$ for some idempotent $e_1$ such that $e_1\mathfrak{R}$ is a minimal right ideal. Since

$$a = e_1a+(a-e_1a) \in e_1\mathfrak{R}+(a-e_1a)_r,$$

we obtain $(a)_r \subseteq e_1\mathfrak{R}+(a-e_1a)_r$. Conversely, for any integer n, we have

$$n(a-e_1a) = na-e_1(na) \subseteq (a)_r+e_1\mathfrak{R} \subseteq (a)_r$$

and for $s \in \mathfrak{R}$, we have $(a-e_1a)s = as-e_1as \in (a)_r+e_1\mathfrak{R} \subseteq (a)_r$ which implies $e_1\mathfrak{R}+(a-e_1a)_r \subseteq (a)_r$. Hence letting $a_1 = a-e_1a$, we obtain $(a)_r = e_1\mathfrak{R}+(a_1)_r$. If $(a_1)_r$ is not a minimal right ideal, we can find a minimal right ideal of $\mathfrak{R}$ of the form $e_2\mathfrak{R}$ contained in $(a_1)_r$, where $e_2$ is an idempotent. We deduce that for $a_2 = a_1-e_2a_1$, we have $(a_1)_r = e_2\mathfrak{R}+(a_2)_r$. Hence

$$(a)_r = e_1\mathfrak{R}+e_2\mathfrak{R}+(a_2)_r \supset e_2\mathfrak{R}+(a_2)_r \supset (a_2)_r.$$

This process must end after a finite number of steps because of the d.c.c. for principal right ideals. Thus $(a)_r = e_1\mathfrak{R}+e_2\mathfrak{R}+\dots+e_n\mathfrak{R}$ so that $a$ can be written as a sum of elements of minimal right ideals $e_i\mathfrak{R}$. Therefore $\mathfrak{R}$ is an atomic ring which by 4.3 implies that $\mathfrak{R}$ is a direct sum of simple atomic rings.

iii) ⇒ iv). Let $\mathfrak{R} = \bigoplus_{\alpha \in A} \mathfrak{S}_\alpha$, where each $\mathfrak{S}_\alpha$ is a simple atomic ring. By 2.8, we have $\mathfrak{S}_\alpha = C(\mathfrak{K}_\alpha)$, where $\mathfrak{K}_\alpha$ is the semigroup socle of $\mathfrak{S}_\alpha$. For $\alpha \neq \beta$, we obtain $\mathfrak{S}_\alpha \mathfrak{S}_\beta \subseteq \mathfrak{S}_\alpha \cap \mathfrak{S}_\beta = 0$ since the sum is direct. Hence in particular, $\mathfrak{K}_\alpha \mathfrak{K}_\beta \subseteq \mathfrak{K}_\alpha \cap \mathfrak{K}_\beta = 0$ and $\mathfrak{K} = \bigcup_{\alpha \in A} \mathfrak{K}_\alpha$ is actually an orthogonal sum of $\mathfrak{K}_\alpha$. Thus $\mathfrak{K}$ is a primitive regular semigroup. Since for each $\alpha \in A$, $\mathfrak{K}_\alpha$ is an ideal of $\mathfrak{M}\mathfrak{S}_\alpha$, and $\mathfrak{K}$ is a regular semigroup, it follows that $\mathfrak{K}$ is an ideal of $\mathfrak{M}\mathfrak{R}$. Further, each element of $\mathfrak{R}$ is a sum of elements of some $\mathfrak{S}_\alpha$, and for each $\alpha \in A$, $\mathfrak{S}_\alpha = C(\mathfrak{K}_\alpha)$ by 2.8; it follows that each element of $\mathfrak{R}$ is a sum of elements of some $\mathfrak{K}_\alpha$. Consequently $\mathfrak{R} = C(\mathfrak{K})$.

iv) ⇒ i). Let $\mathfrak{R} = C(\mathfrak{K})$, where $\mathfrak{K}$ is a primitive regular ideal of $\mathfrak{M}\mathfrak{R}$. Let I be a nonzero ideal of $\mathfrak{R}$. A part of the proof of "v) ⇒ iv)" in 2.8 can be applied to the present situation showing that $I \cap \mathfrak{K} \neq 0$. Hence $\mathfrak{K}_\alpha \subseteq I$ for some $\alpha \in A$, where $\mathfrak{K} = \sum_{\alpha \in A} \oplus \mathfrak{K}_\alpha$ and $\mathfrak{K}_\alpha$ is completely 0-simple, so that I cannot be nilpotent. Consequently $\mathfrak{R}$ is semiprime. By 1.13, the minimal right ideals of $\mathfrak{R}$ and $\mathfrak{M}\mathfrak{R}$ coincide which together with the hypothesis implies that $\mathfrak{R}$ is atomic.

II.6.2 <u>COROLLARY</u>. <u>Let $\mathfrak{R}$ be a semiprime atomic ring and $\mathfrak{K}$ be its semigroup socle. Then $\mathfrak{M}\mathfrak{K}$ is a dense extension of $\mathfrak{K}$ and $\mathfrak{R}$ is a regular ring.</u>

<u>PROOF</u>. We have seen in the proof of "iv) ⇒ i)" in 6.1 that every nonzero ideal of $\mathfrak{R}$ has a nonzero intersection with $\mathfrak{K}$. Hence by 5.9, we conclude that $\mathfrak{M}\mathfrak{K}$ is a dense extension of $\mathfrak{K}$. By 6.1, we have $\mathfrak{R} = \bigoplus_{\alpha \in A} \mathfrak{S}_\alpha$, where each $\mathfrak{S}_\alpha$ is simple and atomic hence regular by 2.8 and I.3.6, and we know that the $\mathfrak{S}_\alpha$ annihilate each other. Hence for $a = a_1 + a_2 + \ldots + a_n$, where $a_i \in \mathfrak{S}_{\alpha_i}$ and $\mathfrak{S}_{\alpha_i} \neq \mathfrak{S}_{\alpha_j}$ if $i \neq j$, there exists $b_i \in \mathfrak{S}_{\alpha_i}$ such that $a_i = a_i b_i a_i$ and thus

$$\begin{aligned} a = a_1 + a_2 + \ldots + a_n &= a_1 b_1 a_1 + a_2 b_2 a_2 + \ldots + a_n b_n a_n \\ &= (a_1 + a_2 + \ldots + a_n)(b_1 + b_2 + \ldots + b_n)(a_1 + a_2 + \ldots + a_n) \end{aligned}$$

and hence $\mathfrak{R}$ is a regular ring.

II.6.3 <u>COROLLARY</u>. <u>If $\mathfrak{K}$ is a primitive regular ideal of $\mathfrak{M}\mathfrak{R}$ for a ring $\mathfrak{R}$, then $\mathfrak{M}C(\mathfrak{K})$ is a dense extension of $\mathfrak{K}$.</u>

<u>PROOF</u>. This follows immediately from 6.1 and 6.2.

The next corollary is known as the First Wedderburn (or Wedderburn-Artin) Theorem with "semisimple" instead of "semiprime" which are in this case equivalent. For other characterizations of these rings, see Fuchs and Szele [1], Goldman [1], Jans ([1], Chapter 2). Also consult Behrens ([1], Chapter III, Section 3).

II.6.4 <u>COROLLARY</u>. <u>A ring $\mathfrak{R}$ is semiprime and satisfies d.c.c. for right ideals if and only if it is a direct sum of a finite number of matrix rings over division rings.</u>

PROOF. Necessity. By 6.1, we have $\mathfrak{R} = \bigoplus_{\alpha\in A} \mathfrak{S}_\alpha$, where each $\mathfrak{S}_\alpha$ is simple atomic. Since the sum is direct, every right ideal of $\mathfrak{S}_\alpha$ is also a right ideal of $\mathfrak{R}$. Hence $\mathfrak{S}_\alpha$ satisfies d.c.c. for right ideals, which by 2.8 and I.3.14 implies that $\mathfrak{S}_\alpha \cong \Delta^\alpha_{n_\alpha}$, $n_\alpha \times n_\alpha$ matrices over the division ring $\Delta^\alpha$. If $A$ is infinite, then for any infinite sequence $\alpha_1, \alpha_2, \ldots \in A$, letting $J_i = \bigoplus_{k\geq i} \mathfrak{S}_{\alpha_k}$, we obtain an infinite descending chain $J_1 \supset J_2 \supset \ldots$, contradicting the hypothesis. Therefore $A$ must be finite

The proof of the sufficiency is left as an exercise.

II.6.5 EXERCISES.

i) Prove that each of the following conditions on a ring $\mathfrak{R}$ is sufficient for $\mathfrak{R}$ to be a division ring.

a) $\mathfrak{M}\mathfrak{R}$ is a primitive regular semigroup.

b) $\mathfrak{R}$ is a regular ring with only one nonzero idempotent.

c) For every $0 \neq a \in \mathfrak{R}$, there exists a unique $x \in \mathfrak{R}$ such that $a = axa$.

(These conditions are trivially necessary.)

ii) Let $\mathfrak{S} = \bigoplus_{\alpha\in A} \mathfrak{S}_\alpha$, where each $\mathfrak{S}_\alpha$ is a simple atomic ring. Show that every ideal of $\mathfrak{S}$ is a direct sum of some $\mathfrak{S}_\alpha$ and conversely. Deduce that

a) $A$ is finite if and only if $\mathfrak{S}$ satisfies d.c.c. for two-sided ideals,

b) each $\mathfrak{S}_\alpha$ is isomorphic to a matrix ring over a division ring if and only if every minimal two-sided ideal of $\mathfrak{S}$ satisfies d.c.c. for right ideals.

## II.7 ISOMORPHISMS

We consider first isomorphisms of some of the rings we have encountered in the two preceding sections. After that we give a construction of all isomorphisms of two Rees matrix rings.

II.7.1 LEMMA. Let $\mathfrak{R}$ and $\mathfrak{R}'$ be semiprime atomic rings, $\mathfrak{K}$ and $\mathfrak{K}'$ be the semigroup socles of $\mathfrak{R}$ and $\mathfrak{R}'$, respectively, and assume that $\mathfrak{R}$ has no ideal which is a division ring. Then every isomorphism of $\mathfrak{K}$ onto $\mathfrak{K}'$ can be uniquely extended to a ring isomorphism of $\mathfrak{R}$ onto $\mathfrak{R}'$.

PROOF. Let $\psi$ be an isomorphism of $\mathfrak{K}$ onto $\mathfrak{K}'$. By 6.1, we have $\mathfrak{R} = \bigoplus_{\alpha\in A} \mathfrak{S}_\alpha$ and $\mathfrak{R}' = \bigoplus_{\alpha'\in A'} \mathfrak{S}'_{\alpha'}$, where each $\mathfrak{S}_\alpha$ and $\mathfrak{S}'_{\alpha'}$ is a simple atomic ring. Hence $\mathfrak{K} = \sum_{\alpha\in A} \oplus \mathfrak{K}_\alpha$ and $\mathfrak{K}' = \sum_{\alpha'\in A'} \oplus \mathfrak{K}'_{\alpha'}$, where $\mathfrak{K}_\alpha$ and $\mathfrak{K}'_{\alpha'}$ are the semigroup socles of $\mathfrak{S}_\alpha$ and $\mathfrak{S}'_{\alpha'}$, respectively. It is easy to see that $\psi$ induces a one-to-one function $\xi$ of $A$ onto $A'$ such that $\mathfrak{K}_\alpha\psi = \mathfrak{K}'_{\alpha\xi}$ for all $\alpha \in A$. Letting $\psi_\alpha = \psi|_{\mathfrak{K}_\alpha}$, we have an isomorphism $\psi_\alpha$ of $\mathfrak{K}_\alpha$ onto $\mathfrak{K}'_{\alpha\xi}$.

By 2.8, we have $\mathfrak{S}_\alpha \cong \mathfrak{J}_{U_\alpha}(\Delta_\alpha, V_\alpha)$ and $\mathfrak{S}'_{\alpha\xi} \cong \mathfrak{J}_{U'_\alpha}(\Delta'_\alpha, V'_\alpha)$, where $\dim V_\alpha > 1$ in view of the hypothesis on $\mathfrak{R}$. We know from I.5.12 that every isomorphism $\theta$ of $\mathfrak{J}_{2,U}(\Delta, V)$ onto $\mathfrak{J}_{2,U'}(\Delta', V')$, where $\dim V > 1$, can be uniquely extended to an isomorphism $\bar{\theta}$ of $\mathfrak{L}_U(\Delta, V)$ onto $\mathfrak{L}_{U'}(\Delta', V')$. Using I.5.12 again, we see that the restriction of $\bar{\theta}$ to $\mathfrak{J}_U(\Delta, V)$ is the unique extension of $\theta$ to an isomorphism of $\mathfrak{J}_U(\Delta, V)$ onto $\mathfrak{J}_{U'}(\Delta', V')$.

We deduce that $\psi_\alpha$ admits a unique extension $\bar{\psi}_\alpha$ to an isomorphism of $\mathfrak{S}_\alpha$ onto $\mathfrak{S}'_{\alpha\xi}$. Since $\mathfrak{R} = \bigoplus_{\alpha \in A} \mathfrak{S}_\alpha$, the function $\psi$ defined by

$$\psi : r \to r_1 \bar{\psi}_{\alpha_1} + \ldots + r_n \bar{\psi}_{\alpha_n},$$

where $r = r_1 + \ldots + r_n$, $r_i \in \mathfrak{S}_{\alpha_i}$, is a unique extension of each $\bar{\psi}_\alpha$ to an isomorphism of $\mathfrak{R}$ onto $\mathfrak{R}'$. At each step, the extension isomorphisms are unique, hence $\bar{\psi}$ is the unique extension of $\psi$ to an isomorphism of $\mathfrak{R}$ onto $\mathfrak{R}'$.

II.7.2 <u>LEMMA</u>. <u>Let</u> $\theta$ <u>be an isomorphism of a ring</u> $\mathfrak{R}$ <u>onto a ring</u> $\mathfrak{R}'$. <u>Then the function</u>

$$\bar{\theta} : (\lambda, \rho) \to (\bar{\lambda}, \bar{\rho}) \qquad ((\lambda, \rho) \in \Omega(\mathfrak{R})),$$

<u>where</u> $\bar{\lambda}x = [\lambda(x\theta^{-1})]\theta$, $x\bar{\rho} = [(x\theta^{-1})\rho]\theta$ $(x \in \mathfrak{R}')$, <u>is an isomorphism of</u> $\Omega(\mathfrak{R})$ <u>onto</u> $\Omega(\mathfrak{R}')$ <u>such that</u> $\pi_r \bar{\theta} = \pi_{r\theta}$ $(r \in \mathfrak{R})$.

<u>PROOF</u>. For $(\lambda, \rho) \in \Omega(\mathfrak{R})$ and $x, y \in \mathfrak{R}'$, we obtain

$$\bar{\lambda}(xy) = \{\lambda[(xy)\theta^{-1}]\}\theta = \{\lambda[(x\theta^{-1})(y\theta^{-1})]\}\theta$$
$$= \{[\lambda(x\theta^{-1})](y\theta^{-1})\}\theta = [\lambda(x\theta^{-1})]\theta y = (\bar{\lambda}x)y,$$

$$\bar{\lambda}(x+y) = \{\lambda[(x+y)\theta^{-1}]\}\theta = [\lambda(x\theta^{-1} + y\theta^{-1})]\theta$$
$$= [\lambda(x\theta^{-1})]\theta + [\lambda(y\theta^{-1})]\theta = \bar{\lambda}x + \bar{\lambda}y,$$

so that $\bar{\lambda}$ is a left translation, analogously $\bar{\rho}$ is a right translation; further

$$x(\bar{\lambda}y) = x\{\lambda(y\theta^{-1})]\theta = \{(x\theta^{-1})[\lambda(y\theta^{-1})]\}\theta$$
$$= \{[(x\theta^{-1})\rho](y\theta^{-1})\}\theta = [(x\theta^{-1})\rho]\theta y = (x\bar{\rho})y,$$

which proves that $(\bar{\lambda}, \bar{\rho}) \in \Omega(\mathfrak{R}')$. If also $(\varphi, \psi) \in \Omega(\mathfrak{R})$, then

$$(\bar{\lambda}\bar{\varphi})x = \bar{\lambda}\{[\varphi(x\theta^{-1})]\theta\} = \{\lambda[\varphi(x\theta^{-1})]\}\theta = \overline{\lambda\varphi}x$$

and similarly $x(\bar{\varphi}\bar{\psi}) = x\overline{\varphi\psi}$, showing that $\bar{\theta}$ is a homomorphism. That $\bar{\theta}$ is also one-to-one and onto follows easily from the fact that $\theta$ shares these properties.

From $r \in \mathfrak{R}$ and $x \in \mathfrak{R}'$, we obtain

$$\overline{\lambda_r}x = [\lambda_r(x\theta^{-1})]\theta = [r(x\theta^{-1})]\theta = (r\theta)x = \lambda_{r\theta}x$$

so that $\overline{\lambda_r} = \lambda_{r\theta}$, and analogously $\overline{\rho_r} = \rho_{r\theta}$ which proves the formula $\pi_r\overline{\theta} = \pi_{r\theta}$.

II.7.3 <u>LEMMA</u>. <u>Let</u> $\mathfrak{R}$ <u>be an essential extension of a ring</u> $I$, $\mathfrak{A}(I) = 0$, $\mathfrak{R}'$ <u>be a maximal essential extension of a ring</u> $I'$. <u>Then every isomorphism</u> $\theta$ <u>of</u> $I$ <u>onto</u> $I'$ <u>can be uniquely extended to an isomorphism</u> $\chi$ <u>of</u> $\mathfrak{R}$ <u>into</u> $\mathfrak{R}'$. <u>Furthermore</u>, $\chi$ <u>maps</u> $\mathfrak{R}$ <u>onto</u> $\mathfrak{R}'$ <u>if and only if</u> $\mathfrak{R}$ <u>is a maximal essential extension of</u> $I$.

<u>PROOF</u>. Let $\theta$ be an isomorphism of $I$ onto $I'$. Then $\mathfrak{A}(I') = 0$ and hence the ring analogue of I.7.5 implies that $\tau = \tau(\mathfrak{R}:I)$ is an isomorphism of $\mathfrak{R}$ into $\Omega(I)$, and $\tau' = \tau(\mathfrak{R}':I')$ is an isomorphism of $\mathfrak{R}'$ onto $\Omega(I')$. Now 7.2 asserts that $\overline{\theta}$ is an isomorphism of $\Omega(I)$ onto $\Omega(I')$ with the property $\pi_x\overline{\theta} = \pi_{x\theta}$ for all $x \in I$. Hence $\chi = \tau\overline{\theta}\tau'^{-1}$ is an isomorphism of $\mathfrak{R}$ into $\mathfrak{R}'$ such that for all $x \in I$, we have

$$x\chi = (x\tau)\overline{\theta}\tau'^{-1} = \pi_x\overline{\theta}\tau'^{-1} = \pi_{x\theta}\tau'^{-1} = x\theta$$

so that $\chi$ extends $\theta$.

Let also $\varphi$ be an extension of $\theta$ to an isomorphism of $\mathfrak{R}$ into $\mathfrak{R}'$. Then for any $r \in \mathfrak{R}$ and $x \in I$, we obtain

$$(r\varphi)(x\theta) = (r\varphi)(x\varphi) = (rx)\varphi = (rx)\chi = (r\chi)(x\chi) = (r\chi)(x\theta)$$

and similarly $(x\theta)(r\varphi) = (x\theta)(r\chi)$. Since $x \in I$ is arbitrary, it follows that $r\varphi - r\chi \in \mathfrak{A}_{\mathfrak{R}'}(I')$. But $\mathfrak{A}_{\mathfrak{R}'}(I')$ is an ideal of $\mathfrak{R}'$ satisfying

$$\mathfrak{A}_{\mathfrak{R}'}(I') \cap I' = \mathfrak{A}(I') = 0$$

which together with the hypothesis that $\mathfrak{R}'$ is an essential extension of $I'$ implies $\mathfrak{A}_{\mathfrak{R}'}(I') = 0$. Consequently $r\varphi = r\chi$ and thus $\varphi = \chi$ since $r \in \mathfrak{R}$ is arbitrary.

The necessity in the last statement of the lemma is obvious; sufficiency follows from the fact that $\mathfrak{R}\chi$ is an essential extension if $I'$.

The next result is new.

II.7.4 <u>THEOREM</u>. <u>Let</u> $\mathfrak{R}$ <u>be a semiprime ring essential extension of its nonzero socle</u> $\mathfrak{S}$ <u>and assume that</u> $\mathfrak{R}$ <u>has no ideal which is a division ring</u>. <u>Let</u> $\mathfrak{R}'$ <u>be a ring which is a maximal essential extension of its socle</u> $\mathfrak{S}'$. <u>Then every isomorphism</u> $\psi$ <u>of the semigroup socle</u> $\mathfrak{K}$ <u>of</u> $\mathfrak{R}$ <u>onto the semigroup socle</u> $\mathfrak{K}'$ <u>of</u> $\mathfrak{R}'$ <u>can be uniquely extended to a ring isomorphism</u> $\chi$ <u>of</u> $\mathfrak{R}$ <u>into</u> $\mathfrak{R}'$. <u>Furthermore</u>, $\chi$ <u>maps</u> $\mathfrak{R}$ <u>onto</u> $\mathfrak{R}'$ <u>if and only if</u> $\mathfrak{R}$ <u>is a maximal essential extension of</u> $\mathfrak{S}$.

<u>PROOF</u>. Let $\psi$ be an isomorphism of $\mathfrak{K}$ onto $\mathfrak{K}'$. It follows from 5.9 that $\mathfrak{K}$ is a primitive regular semigroup. Hence $\mathfrak{K}'$ is a primitive regular ideal of $\mathfrak{M}\mathfrak{R}'$. Since $\mathfrak{S}' = \mathfrak{C}(\mathfrak{K}')$, by 6.1 we conclude that $\mathfrak{S}'$ is a semiprime atomic ring. According to 7.1, $\psi$ admits a unique extension to a ring isomorphism $\theta$ of $\mathfrak{S}$ onto $\mathfrak{S}'$. By 7.3, $\theta$ can be uniquely extended to an isomorphism $\chi$ of $\mathfrak{R}$ into $\mathfrak{R}'$.

Consequently $\chi$ provides an extension of $\psi$ to an isomorphism of $\mathfrak{R}$ into $\mathfrak{R}'$. If $\varphi$ is another such extension of $\psi$, then $\mathfrak{S} = C(\mathfrak{K})$ and $\mathfrak{S}' = C(\mathfrak{K}')$ imply that $\mathfrak{S}\varphi = \mathfrak{S}'$. But then $\varphi|_{\mathfrak{S}} = \theta$ and hence $\varphi = \chi$ by the uniqueness of both $\theta$ and $\chi$. The last statement follows immediately from 7.3.

Recall that by I.5.22, exercise i), a ring $\mathfrak{R}$ has unique addition if and only if every multiplicative isomorphism onto another ring is additive (and is thus a ring isomorphism). The next corollary is due to Rickart [1], see also Gluskin [4].

II.7.5 COROLLARY. *A semiprime ring $\mathfrak{R}$ which has no ideal which is a division ring, and is an essential extension of its nonzero socle, has unique addition.*

PROOF. Let $\varphi$ be an isomorphism of $\mathfrak{M}\mathfrak{R}$ onto $\mathfrak{M}\mathfrak{R}'$ for some ring $\mathfrak{R}'$. Since a ring with the properties that $\mathfrak{R}$ has can be characterized multiplicatively according to 5.9, it follows that $\mathfrak{R}'$ is a semiprime ring essential extension of its nonzero socle. By the ring analogue of I.7.5, we may suppose that $\mathfrak{R}'$ is embedded in a maximal essential extension $\mathfrak{R}''$ of the socle $\mathfrak{S}'$ of $\mathfrak{R}'$. Hence $\psi = \varphi|_{\mathfrak{K}}$, where $\mathfrak{K}$ is the semigroup socle of $\mathfrak{R}$, satisfies the conditions of 7.4 and hence admits a unique extension to an isomorphism $\chi$ of $\mathfrak{R}$ into $\mathfrak{R}''$. Similarly as in the proof of uniqueness in 7.3, it follows that

$$(r\varphi)(x\psi) = (r\chi)(x\psi), \quad (x\psi)(r\varphi) = (x\psi)(r\chi) \qquad (x \in \mathfrak{K},\ r \in \mathfrak{R}).$$

Letting $\mathfrak{K}'$ be the semigroup socle of $\mathfrak{R}'$, we have that $\tau^{r\varphi} = \tau^{r\chi}$, where $\tau = \tau(\mathfrak{M}\mathfrak{R}'':\mathfrak{K}')$. But $\mathfrak{M}\mathfrak{R}''$ is a dense extension of $\mathfrak{K}'$ by 5.9 which by I.7.5 yields $r\varphi = r\chi$. Consequently $\varphi = \chi$ which proves that $\varphi$ is additive.

We now consider isomorphisms of Rees matrix rings. An $I \times I'$-matrix $B$ over a division ring $\Delta$ is said to be *invertible* if there exists an $I' \times I$-matrix $A$ over $\Delta$ such that $AB$ and $BA$ are the identity matrices of the corresponding sizes. The following proposition is due to Hotzel [1], its proof appears here for the first time.

II.7.6 PROPOSITION. *Let $\mathfrak{R} = \mathfrak{M}(I,\Delta,\Lambda;P)$ and $\mathfrak{R}' = \mathfrak{M}(I',\Delta',\Lambda';P')$ be Rees matrix rings. Let $\omega$ be an isomorphism of $\Delta$ onto $\Delta'$, and for any matrix $X = (x_{ij})$ over $\Delta$ we write $X\hat{\omega} = (x_{ij}\omega)$. Next let $A$ be an invertible row finite $\Lambda \times \Lambda'$-matrix over $\Delta'$, $B$ be an invertible column finite $I' \times I$-matrix over $\Delta'$, and suppose that $P\hat{\omega} = AP'B$. Then the function $\chi$ defined by*

$$\chi: X \to B(X\hat{\omega})A \qquad (X \in \mathfrak{R})$$

*is an isomorphism of $\mathfrak{R}$ onto $\mathfrak{R}'$. Conversely every isomorphism of $\mathfrak{R}$ onto $\mathfrak{R}'$ can be so constructed.*

PROOF. To prove the direct part, we let $X,Y \in \mathfrak{R}$ and obtain

$$\begin{aligned}(X * Y)\chi &= B(X * Y)\hat{\omega}A = B(XPY)\hat{\omega}A \\ &= B(X\hat{\omega})(P\hat{\omega})(Y\hat{\omega})A = B(X\hat{\omega})(AP'B)(Y\hat{\omega})A \\ &= (X\chi)P'(Y\chi) = (X\lambda) * (Y\chi).\end{aligned}$$

Since $\chi$ is obviously additive, it is a homomorphism. The invertibility of $A$ and $B$ easily implies that $\chi$ maps $\mathfrak{R}$ onto $\mathfrak{R}'$ and is one-to-one.

For the converse, we let $\chi$ be an isomorphism of $\mathfrak{R}$ onto $\mathfrak{R}'$. If $\mathfrak{R}$ is a division ring, the statement in the proposition is trivially satisfied. We thus assume that $\mathfrak{R}$ is not a division ring and consider the commutative diagram

$$\begin{array}{ccc} \mathfrak{R} & \xrightarrow{\quad\chi\quad} & \mathfrak{R}' \\ \uparrow & & \uparrow \\ \mathfrak{F}_U(\Delta,V) & \xrightarrow{\ \zeta_{(\omega,a)}\ } & \mathfrak{F}_{U'}(\Delta',V') \end{array}$$

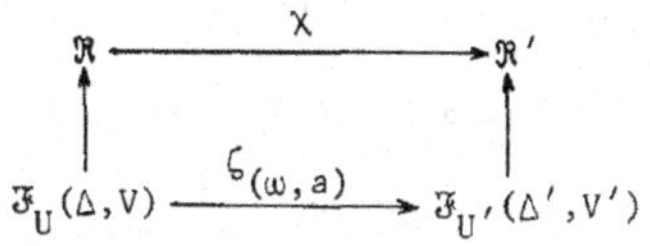

where the vertical arrows denote the isomorphisms in the proof of "iv) $\Rightarrow$ vi)" in 2.8, and $\zeta_{(\omega,a)}$ is the isomorphism induced by the semilinear isomorphism $(\omega,a)$ of $(U,\Delta,V)$ onto $(U',\Delta',V')$ provided by I.5.13. The idea of the proof consists of describing the isomorphism $\chi$ as in the statement of the proposition by exploiting the commutativity of the above diagram.

Let

$$Z = \{z_\lambda \mid \lambda \in \Lambda\}, \qquad W = \{w_i \mid i \in I\},$$
$$Z' = \{z'_{\lambda'} \mid \lambda' \in \Lambda'\}, \qquad W' = \{w'_{i'} \mid i' \in I'\}$$

be bases of $V$, $U$, $V'$ and $U'$, respectively. Let $b$ be the adjoint of the semilinear isomorphism $(\omega,a)$ and let

$$A = (\alpha_{\lambda\lambda'})_{\lambda\in\Lambda,\lambda'\in\Lambda'}, \qquad B = (\beta_{i'i})_{i'\in I',i\in I}$$

where

$$z_\lambda a = \sum_{\lambda'\in\Lambda'} \alpha_{\lambda\lambda'} z'_{\lambda'}, \qquad b^{-1}w_i = \sum_{i'\in I'} w'_{i'}\beta_{i'i}. \tag{1}$$

Then $A$ is row finite, $B$ is column finite and both $A$ and $B$ have the required size and are invertible since both $a$ and $b^{-1}$ are one-to-one and onto. Further, we obtain

$$\begin{aligned} p_{\lambda i}\omega &= (z_\lambda, w_i)\omega = (z_\lambda, b(b^{-1}w_i))\omega = (z_\lambda a, b^{-1}w_i)' \\ &= \Big(\sum_{\lambda'\in\Lambda'} \alpha_{\lambda\lambda'} z'_{\lambda'}, \sum_{i'\in I'} w'_{i'}\beta_{i'i}\Big)' \\ &= \sum_{\lambda'\in\Lambda'} \sum_{i'\in I'} \alpha_{\lambda\lambda'}(z'_{\lambda'}, w'_{i'})'\beta_{i'i} \\ &= \sum_{\lambda'\in\Lambda'} \sum_{i'\in I'} \alpha_{\lambda\lambda'} p'_{\lambda'i'}\beta_{i'i} \end{aligned} \tag{2}$$

which in terms of matrices has the form $P\hat{\omega} = AP'B$ as required.

We also have $z'_{\lambda'}a^{-1} = \sum_{\lambda\in\Lambda} \sigma_{\lambda'\lambda} z_\lambda$ so that by (1), we obtain

$$z'_{\lambda'} = z'_{\lambda'}a^{-1}a = (\sum_{\lambda\in\Lambda} \sigma_{\lambda'\lambda}z_\lambda)a = \sum_{\lambda\in\Lambda}(\sigma_{\lambda'\lambda}\omega)(z_\lambda a)$$

$$= \sum_{\lambda\in\Lambda}(\sigma_{\lambda'\lambda}\omega)\sum_{\mu'\in\Lambda'}\alpha_{\lambda\mu'}z'_{\mu'} = \sum_{\mu'\in\Lambda'}(\sum_{\lambda\in\Lambda}(\sigma_{\lambda'\lambda}\omega)\alpha_{\lambda\mu'})z'_{\mu'}$$

which implies

$$\sum_{\lambda\in\Lambda}(\sigma_{\lambda'\lambda}\omega)\alpha_{\lambda\mu'} = \delta_{\lambda'\mu'} \qquad (\lambda',\mu' \in \Lambda'), \tag{3}$$

where $\delta_{\lambda'\mu'}$ is the Kronecker delta function. It follows from (2) and (3) that

$$\sum_{\lambda\in\Lambda}(\sigma_{\lambda'\lambda}p_{\lambda i})\omega = \sum_{\lambda\in\Lambda}\sum_{\mu'\in\Lambda'}\sum_{i'\in I'}(\sigma_{\lambda'\lambda}\omega)\alpha_{\lambda\mu'}p'_{\mu'i'}\beta_{i'i}$$

$$= \sum_{\mu'\in\Lambda'}\sum_{i'\in I'}(\sum_{\lambda\in\Lambda}(\sigma_{\lambda'\lambda}\omega)\alpha_{\lambda\mu'})p'_{\mu'i'}\beta_{i'i}$$

$$= \sum_{i'\in I'}p'_{\lambda'i'}\beta_{i'i}. \tag{4}$$

Finally let $X = (x_{i\lambda}) \in \mathfrak{R}$. Following the above diagram, we have

$$X \to PX = (\sum_{i\in I}p_{\lambda i}x_{i\mu})_{\lambda,\mu\in\Lambda} \to c \to a^{-1}ca \tag{5}$$

and hence we must express $a^{-1}ca$ in matrix form. Using (1) and (4), we obtain

$$(z'_{\lambda'}a^{-1})ca = (\sum_{\lambda\in\Lambda}\sigma_{\lambda'\lambda}(z_\lambda c))a$$

$$= (\sum_{\lambda\in\Lambda}\sigma_{\lambda'\lambda}(\sum_{i\in I}p_{\lambda i}x_{i\mu})z_\mu)a$$

$$= \sum_{\lambda\in\Lambda}\sum_{i\in I}(\sigma_{\lambda'\lambda}p_{\lambda i}x_{i\mu})\omega(z_\mu a)$$

$$= \sum_{i\in I}(\sum_{\lambda\in\Lambda}(\sigma_{\lambda'\lambda}p_{\lambda i})\omega)(x_{i\mu}\omega)(z_\mu a)$$

$$= \sum_{i\in I}(\sum_{i'\in I'}p'_{\lambda'i'}\beta_{i'i})(x_{i\mu}\omega)(\sum_{\lambda'\in\Lambda'}\alpha_{\mu\lambda'}z'_{\lambda'})$$

$$= \sum_{\lambda'\in\Lambda'}\sum_{i'\in I'}\sum_{i\in I}(p'_{\lambda'i'}\beta_{i'i}(x_{i\mu}\omega)\alpha_{\mu\lambda'})z'_{\lambda'}$$

which in matrix notation becomes $P'B(X\hat{\omega})A$. Hence continuing the sequence of isomorphisms in (5), we have

$$a^{-1}ca \to P'B(X\hat{\omega})A \to B(X\hat{\omega})A$$

proving that $X\chi = B(X\hat{\omega})A$.

II.7.7 <u>COROLLARY</u>. <u>If</u> $\mathfrak{M}(I,\Delta,\Lambda;P) \cong \mathfrak{M}(I',\Delta',\Lambda';P')$, <u>then</u> <u>card</u> $I$ = <u>card</u> $I'$, $\Delta \cong \Delta'$, <u>card</u> $\Lambda$ = <u>card</u> $\Lambda'$. <u>Further</u>, $\Delta_I \cong \Delta'_{I'}$ <u>if and only if</u> $\Delta \cong \Delta'$, <u>card</u> $I$ = <u>card</u> $I'$.

II.7.8 <u>EXERCISES</u>.

i) Show that an essential extension of a semiprime ring is semiprime. Deduce that every semiprime ring can be embedded into a semiprime ring with identity. Do the same for "prime" instead of "semiprime".

ii) Let $A$ be a minimal right ideal of a semiprime ring $\mathfrak{R}$. Show that if $A$ has no nonzero nilpotent elements, then $A$ must be a division ring and an ideal of $\mathfrak{R}$.

iii) Let $\{\mathfrak{R}_\alpha\}_{\alpha\in A}$ and $\{\mathfrak{R}'_{\alpha'}\}_{\alpha'\in A'}$ be two families of rings and assume that each $\mathfrak{R}_\alpha$ is a simple atomic ring. For an isomorphism $\psi$ of $\prod_{\alpha\in A}\mathfrak{R}_\alpha$ onto $\prod_{\alpha'\in A'}\mathfrak{R}_{\alpha'}$, show that there exists a one-to-one function $\xi$ of $A$ onto $A'$ and for each $\alpha\in A$, an isomorphism $\psi_\alpha$ of $\mathfrak{R}_\alpha$ onto $\mathfrak{R}'_{\alpha\xi}$ such that for every $(r_\alpha)\in\prod_{\alpha\in A}\mathfrak{R}_\alpha$, we have $(r_\alpha)\psi\pi_{\alpha\xi} = r_\alpha\psi_\alpha$. Also consider the following variants:

a) substitute a complete direct sum by a direct sum,

b) assume that each $\mathfrak{R}_\alpha$ is a maximal primitive ring,

c) combine items a) and b).

For a discussion of isomorphisms of the translational hull of a semigroup, see Petrich ([5], Chapter V, Section 2), and for a survey consult Petrich [3].

# PART III

## LINEARLY TOPOLOGIZED VECTOR SPACES AND RINGS

This is essentially a topological treatment of some of the rings studied previously provided with a topology induced either by the vector space or by multiplication. For a pair of dual vector spaces $(U,V)$ over a division ring $\Delta$, the U-topology on $V$ is discussed here in considerable detail. Several properties of subspaces and of semilinear transformations relative to this topology are established. It is proved that $U^*$ with the finite topology is a completion of $V$ with the U-topology. Linearly compact modules are shown to be complete, and linearly compact vector spaces are characterized in several ways. The finite and the uniform topologies for the ring of all continuous linear transformations are discussed. For the ring of all continuous linear transformations of finite rank various properties of one-sided ideals relative to the relativized finite topology are established. A completion of a semiprime ring which is an essential extension of its nonzero socle, with a topology and thus a uniformity induced by its socle, is constructed.

### III.1 A TOPOLOGY FOR A VECTOR SPACE

For a pair of dual vector spaces $(U,V)$ we will introduce a topology on $V$ induced in a natural way by $U$ and the bilinear form. A rudimentary knowledge of general topology will be assumed; for the terminology and basic results we will follow Kelley [1].

The Cartesian product of a family of sets $\{X_\alpha\}_{\alpha\in A}$ is denoted by $\prod_{\alpha\in A} X_\alpha$. In the particular case when $X = X_\alpha$ for all $\alpha \in A$, we write $X^A$ instead of $\prod_{\alpha\in A} X_\alpha$, since in this case we are actually dealing with the set of all functions from $A$ into $X$.

We will denote a topological space by $(X,\tau)$ where $X$ is a nonempty set and $\tau$ is a topology on $X$ given as a collection of (open) sets; sometimes we will simply write $X$ if there is no danger of confusion.

For a nonempty family $\{X_\alpha\}_{\alpha\in A}$ of topological spaces, their _product_ is the topological space on the set $\prod_{\alpha\in A} X_\alpha$ with the topology whose subbase consists of the sets $\hat{U} = \{(x_\alpha) \mid x_\beta \in U_\beta\}$ where $\beta \in A$ and $U_\beta$ is an open set in $X_\beta$ for fixed $\beta$ and $U_\beta$ (it suffices to take $U_\beta$ ranging over either a base or a subbase of $X_\beta$). This topology is called the <u>product</u> (or <u>Tychonoff</u>) <u>topology</u> and the topological space is denoted by $\prod_{\alpha\in A} X_\alpha$.

We now take each $X_\alpha$ to have the discrete topology. Then the product $\prod_{\alpha\in A} X_\alpha$ has a subbase consisting of the sets

$$S_{(\beta;y)} = \{(x_\alpha) \mid x_\beta = y\},$$

where $\beta \in A$ and $y \in X_\beta$. If $X_\alpha = X$ for all $\alpha \in A$, we obtain a subbase for the product topology by taking all sets of the form $S_{(\beta;y)} = \{f \in X^A \mid \beta f = y\}$ where $\beta \in A$ and $y \in X$. Specializing further by letting $A = X$, we obtain $X^X$, the set of all functions from $X$ into itself, and a subbase for it is given by the sets $S_{(x;y)} = \{f \in X^X \mid xf = y\}$.

The following concepts are due to Jacobson [5].

III.1.1 <u>DEFINITION</u>. For $A$ and $X$ nonempty sets, the product topology on $X^A$, where $A$ has the discrete topology, is the <u>finite topology</u> of $X^A$.

Hence a base for the finite topology consists of the sets

$$B_{(\alpha_1,\ldots,\alpha_n;x_1,\ldots,x_n)} = \{f \in X^A \mid \alpha_i f = x_i, \quad i = 1,2,\ldots,n\}$$
$$= \bigcap_{i=1}^{n} B_{(\alpha_i;x_i)}.$$

III.1.2 <u>DEFINITION</u>. Let $(\Delta,V)$ be a vector space and give $\Delta$ the discrete topology. Then the finite topology of $\Delta^V$ relativized to $V^*$ is the <u>finite topology</u> of $V^*$.

We now fix a pair of dual vector spaces $(U,\Delta,V)$ and recall that nat $U$ is the image of $U$ in $V^*$ under the isomorphism $f: u \to f_u$, where $vf_u = (v,u)$ $(v \in V)$, see I.1.1.

III.1.3 <u>DEFINITION</u>. For $(U,\Delta,V)$ a dual pair, the V-<u>topology</u> of $U$ is the topology on $U$ for which the isomorphism $f: u \to f_u$ $(u \in U)$ is a homeomorphism of $U$ into $V^*$, where the latter is endowed with the finite topology.

Equivalently, we may require that $f$ be a homeomorphism of $U$ onto nat $U$, where the latter is given the relativized finite topology of $V^*$. This amounts to declaring as open sets in $U$ just the sets of the form $Kf^{-1}$, where $K$ ranges over all open sets in nat $U$ (or $K = L \cap$ nat $U$ where $L$ runs over all open sets in $V^*$). One defines symmetrically the U-<u>topology</u> of $V$ by considering nat $V$ in $U^*$ with the finite topology.

III.1.4 <u>NOTATION</u>. <u>For</u> $(U,\Delta,V)$ <u>a dual pair</u>, $\tau_U(V)$ <u>denotes the</u> U-<u>topology of</u> $V$. <u>If needed for emphasis, we write</u> $\tau_U(\Delta,V)$ <u>instead of</u> $\tau_U(V)$.

III.1.5 LEMMA. A base for the finite topology of $V^*$ consists of all sets of the form

$$B_{(v_1,v_2,\ldots,v_n;\delta_1,\delta_2,\ldots,\delta_n)} = \{f \in V^* \mid v_i f = \delta_i, \quad i = 1,2,\ldots,n\}$$

where each set $\{v_1,v_2,\ldots,v_n\}$ is linearly independent.

PROOF. From above, a base is given by all such sets without the requirement of linear independence. The proof that for $\{v_1,v_2,\ldots,v_n\}$ linearly dependent, the set $B_{(v_1,v_2,\ldots,v_n;\delta_1,\delta_2,\ldots,\delta_n)}$ is either empty or equal to some $B_{(z_1,z_2,\ldots,z_k;\gamma_1,\gamma_2,\ldots,\gamma_k)}$ with $\{z_1,z_2,\ldots,z_k\}$ linearly independent, is left as an exercise.

III.1.6 COROLLARY. A base for $\tau_U(V)$ consists of all sets

$$C_{(u_1,u_2,\ldots,u_n;\delta_1,\delta_2,\ldots,\delta_n)} = \{v \in V \mid (v,u_i) = \delta_i, \quad i = 1,2,\ldots,n\}$$

where $\{u_1,u_2,\ldots,u_n\}$ is a linearly independent subset of $U$.

PROOF. Exercise.

III.1.7 NOTATION. For a dual pair $(U,V)$, $S$ a subspace of $V$ and $T$ a subspace of $U$, let

$$S^\perp = \{u \in U \mid (v,u) = 0 \quad \text{for all} \quad v \in S\},$$

$$T^\perp = \{v \in V \mid (v,u) = 0 \quad \text{for all} \quad u \in T\}.$$

If $A$ is a nonempty subset of a vector space $W$, then $[A]$ denotes the subspace of $W$ generated by $A$; we also write $[w_1,w_2,\ldots,w_n]$ instead of $[\{w_1,w_2,\ldots,w_n\}]$. For $v \in V$ and $u \in U$, we write $v^\perp = [v]^\perp$, $u^\perp = [u]^\perp$.

III.1.8 COROLLARY. The neighborhood system $\tau_U(V)$ of $0$ has

i) $\{T^\perp \mid T$ subspace of $U$, $\dim T < \infty\}$ as an open base,

ii) $\{u^\perp \mid u \in U\}$ as an open subbase.

PROOF. Exercise.

A topological group is a triple $(G, \ ,\tau)$ consisting of a set $G$, a binary operation $\cdot$ on $G$ making it a group, and a topology $\tau$ on $G$ for which the binary operation $(x,y) \to x \cdot y$ is jointly continuous and the operation of inversion $x \to x^{-1}$ is continuous. In such a case, $\tau$ is said to be compatible with the group structure of $G$. We write $(G,\tau)$ or simply $G$ if there is no need for emphasizing the group operation or the topology. Since for any two elements $x,y \in G$, there exists a left and a right translation of $G$ sending $x$ onto $y$, and in a topological group these are homeomorphisms, in order to give a topology on a group, it suffices to give the neighborhood system (or a base or subbase thereof) of the identity of $G$. We will use these facts without express mention. For this reason, the preceding corollary will be particularly useful.

III.1.9 COROLLARY. *The neighborhood system of* $v \in V$ *for the topology* $\tau_U(V)$ *has*

i) $\{v + T^{\perp} \mid T$ *subspace* $U$, $\dim T < \infty\}$ *as an open base,*

ii) $\{v + u^{\perp} \mid u \in U\}$ *as an open subbase.*

III.1.10 LEMMA. *In a topological group* $(G, \tau)$ *the following statements are equivalent.*

i) $\tau$ *is a Hausdorff topology.*

ii) *The intersection of all neighborhoods of the identity of* $G$ *equals the identity.*

iii) *The identity of* $G$ *has a neighborhood base (subbase) whose intersection is the identity.*

PROOF. Exercise (see the remarks above).

III.1.11 LEMMA. *Every left and every right translation in a topological group* $G$ *is a homeomorphism of* $G$ *onto itself.*

PROOF. Exercise.

III.1.12 COROLLARY. *Every open subgroup of a topological group is closed.*

PROOF. If $H$ is an open subgroup of $G$, then the complement of $H$ in $G$ is the union of cosets $xH$ each of which is open by 1.11.

A *topological vector space* is a triple $(\Delta, V, \tau)$ where $(\Delta, V)$ is a vector space, $\Delta$ is a topological division ring (with the obvious definition), $\tau$ is a topology on $V$ making it a topological group under addition and such that the scalar multiplication $(\delta, v) \to \delta v$ is jointly continuous.

We are interested only in the following special case, which will be assumed throughout, viz. $\Delta$ *has the discrete topology.*

III.1.13 DEFINITION. A subset $H$ of a vector space $V$ is a *hyperplane* if $H$ is under inclusion a maximal subspace of $V$.

III.1.14 LEMMA. *The following statements concerning a subspace* $S$ *of a vector space* $V$ *are equivalent.*

i) $S$ *is a hyperplane.*

ii) $\dim(V/S) = 1$.

iii) $S = f^{\perp}$ *for some* $0 \neq f \in V^*$, *where* $f^{\perp}$ *is taken relative to the dual pair* $(V^*, V)$.

PROOF. The equivalence of i) and ii) follows from the well-known correspondence between subspaces of $V$ containing $S$ and subspaces of $V/S$.

i) $\Rightarrow$ iii). Let $v$ be a vector in $V$ not contained in $S$. Then $V = S \oplus [v]$, where the sum is direct in the group sense, since $S$ is a hyperplane. Hence for any $x \in V$, we can write $x = \bar{x} + (xf)v$ for unique $\bar{x} \in S$ and $xf \in \Delta$. It follows easily that $f$ is linear, so $f \in V^*$ and $f^\perp = S$.

iii) $\Rightarrow$ i). Let $v$ be a vector in $V$ such that $vf = \sigma \neq 0$. Then for every $x \in V$, we can write

$$x = (x - (xf)\sigma^{-1}v) + (xf)\sigma^{-1}v \in f^\perp + [v].$$

Thus $v \notin f^\perp$ implies $f^\perp + [v] = V$, where $f^\perp + [v]$ is the subspace of $V$ generated by $f^\perp$ and $v$. Hence $f^\perp$ is a hyperplane.

III.1.15 LEMMA. For $(U,V)$ a dual pair and $S$ a finite dimensional subspace of $V$, we have $S^{\perp\perp} = S$.

PROOF. It is easy to see that $S \subseteq S^{\perp\perp}$ for any subspace $S$ of $V$. If $\{v_1, v_2, \ldots, v_n\}$ is a basis of $S$ and $v \notin S$, then the set $\{v, v_1, v_2, \ldots, v_n\}$ is linearly independent which by I.2.3 implies the existence of $u \in U$ such that $(v,u) \neq 0$ and $(v_i,u) = 0$ for $i = 1,2,\ldots,n$. But then $(v,u) \neq 0$ and $u \in S^\perp$ which implies that $v \notin S^{\perp\perp}$. Hence $S^{\perp\perp} \subseteq S$ and the equality holds.

III.1.16 THEOREM. $(\Delta, V, \tau_U(V))$ is a Hausdorff topological vector space, $\tau_U(V)$ is the weakest topology compatible with the additive group structure of $V$ and in which all functions $f_u$ are continuous $(f_u : v \to (v,u) \quad (v \in V))$. Moreover, nat $U = \{f \in V^* \mid f$ is continuous$\}$.

PROOF. For any $x,y \in V$ and finite dimensional subspace $T$ of $U$, we have $(x + T^\perp) + (y + T^\perp) \subseteq (x + y) + T^\perp$, $-(x + T^\perp) = -x + T^\perp$, proving that $\tau_U(V)$ is compatible with the additive group structure of $V$ (i.e., addition and inversion are continuous). If $\delta \in \Delta$, then $\delta(x + T^\perp) = \delta x + T^\perp$, where $\delta$ is its own neighborhood. Consequently the scalar multiplication is continuous, so $(\Delta, V, \tau_U(V))$ is a topological vector space. Since the bilinear form associated with $(U,V)$ is nondegenerate, it follows that $\bigcap_{u \in U} u^\perp = 0$ which by 1.10 implies that $\tau_U(V)$ is Hausdorff.

In order to prove that $f_u$ is continuous, it suffices to show that $\delta f_u^{-1}$ is open for any $\delta \in \Delta$. First note that $\delta f_u^{-1} = \{v \in V \mid (v,u) = \delta\}$. For $z \in \delta f_u^{-1}$ and $x \in u^\perp$, we obtain $(z,u) = \delta$ and $(x,u) = 0$ so that $(z + x, u) = \delta$. Hence $z + x \in \delta f_u^{-1}$ and thus $z + u^\perp \subseteq \delta f_u^{-1}$, proving that $\delta f_u^{-1}$ is open and hence that $f_u$ is continuous.

Next let $\tau$ be a topology on $V$ compatible with the additive group structure of $V$ and in which all $f_u$ are continuous. For $u \in U$, the set $u^\perp = 0 f_u^{-1}$ is $\tau$-open since $f_u$ is continuous. But then $\tau_U(V) \subseteq \tau$ since $(V,\tau)$ is a topological group.

If $f \in \text{nat } U$, then $f = f_u$ for some $u \in U$ and we have seen above that $f_u$ is continuous. To prove the converse, we first let $f$ be a nonzero element of $V^*$ for which $f^\perp = 0f^{-1}$ is a closed subspace of $V$. Since $f \neq 0$, there exists $v \in V$ such that $v \notin f^\perp$. Since $V \setminus f^\perp$ is open, there exists a finite dimensional subspace $T$ of $U$ such that $(v + T^\perp) \cap f^\perp = \emptyset$. It follows easily that $v \notin T^\perp + f^\perp$ so that $T^\perp + f^\perp \neq V$. By 1.14 we know that $f^\perp$ is a hyperplane so we must have $T^\perp \subseteq f^\perp$. Considering the dual pair $(V^*, V)$, we obtain $(\text{nat } T)^\perp \subseteq [f]^\perp$ which implies $[f] = [f]^{\perp\perp} \subseteq (\text{nat } T)^{\perp\perp} = \text{nat } T$ by 1.15. Consequently $f \in \text{nat } T \subseteq \text{nat } U$. By contrapositive, we deduce that $f \notin \text{nat } U$ implies that $0f^{-1}$ is not closed, which implies that $f$ cannot be continuous. This proves the last assertion of the theorem.

We consider next a converse of the foregoing theorem.

III.1.17 <u>LEMMA</u>. <u>In a topological vector space</u> $(\Delta, V, \tau)$, <u>every linear combination of continuous linear forms is continuous</u>.

<u>PROOF</u>. It suffices to show that for $f, g \in V^*$ which are continuous and $\sigma \in \Delta$, both $f + g$ and $f\sigma$ are continuous. Assuming that $f, g \in V^*$ are continuous, we let $\delta \in \Delta$ and $v \in \delta(f+g)^{-1}$. Then $vf + vg = \delta$, so that $vf = \delta - vg \in \Delta$. Since $f$ is continuous, there exists a neighborhood $N$ of $v$ such that $N \subseteq (\delta - vg)f^{-1}$. Hence $Nf = \delta - vg$ and thus $Nf + vg = \delta$. Similarly, there exists a neighborhood $M$ of $v$ such that $vf + Mg = \delta$. For any $x \in N \cap M$, we now have

$$x(f+g) = (xf + vg) + (vf + xg) - \delta = 2\delta - \delta = \delta.$$

Consequently $v \in N \cap M \subseteq \delta(f+g)^{-1}$, proving that $\delta(f+g)^{-1}$ is open and thus $f + g$ is continuous.

Next let $\sigma \in \Delta$. If $\sigma = 0$, then $f\sigma$ is the zero function and hence continuous. If $\sigma \neq 0$, then for any $\delta \in \Delta$, we have $\delta(f\sigma)^{-1} = (\delta\sigma^{-1})f^{-1}$ which is open since $f$ is continuous, proving that $f\sigma$ is continuous.

III.1.18 <u>THEOREM</u>. <u>Let</u> $(\Delta, V, \tau)$ <u>be a topological vector space</u>. <u>Assume that</u> $\tau$ <u>has an open subbase</u> $\mathcal{H}$ <u>for the neighborhood system of</u> 0 <u>consisting of hyperplanes satisfying</u> $\bigcap_{H \in \mathcal{H}} H = 0$ <u>and let</u>

$$U = \{f \in V^* \mid f \text{ is continuous}\}.$$

<u>Then</u> $U$ <u>is a</u> t-<u>subspace of</u> $V^*$, $\tau = \tau_U(V)$ <u>and</u>

$$U = \{f \in V^* \mid f^\perp \supseteq \bigcap_{i=1}^n H_i \text{ for some } H_i \in \mathcal{H}\}.$$

<u>PROOF</u>. First let $H \in \mathcal{H}$. By 1.14, we have $H = f^\perp$ for some $f \in V^*$ and thus $V = f^\perp \oplus [v]$ for any vector $v \notin f^\perp$. Then for any $\delta \in \Delta$, we obtain

$$\delta f^{-1} = \{x \in V \mid xf = \delta\} = \{y + \sigma v \mid y \in H, (\sigma v)f = \delta\} = \bigcup_{\sigma \in \Delta} (H + \sigma v).$$

Since $H$ is open so is its translate $H + \sigma v$ for any $\sigma \in \Delta$. But the union of open sets is again open which implies that $\delta f^{-1}$ is open. Hence $f$ is continuous.

In view of 1.17, $U$ is a subspace of $V^*$. Let $0 \neq v \in V$. The hypothesis implies the existence of $H \in \mathcal{H}$ such that $v \notin H$. By the above, $H = f^{\perp}$ for some $f \in U$ so that $vf \neq 0$ which proves that $U$ is a t-subspace of $V^*$.

For any $H \in \mathcal{H}$, we also have $H = f^{\perp} = 0f^{-1}$ for some $f \in U$ which shows that $H$ is $\tau_U(V)$-open. Since $\mathcal{H}$ is a subbase of the neighborhood system of $0$ for the topology $\tau$, it follows that $\tau \subseteq \tau_U(V)$. Conversely, if $u \in U$, then $u^{\perp} = 0u^{-1}$ is $\tau$-open. But the sets $u^{\perp}$ form a subbase of the neighborhood system of $0$ in the topology $\tau_U(V)$ so that $\tau_U(V) \subseteq \tau$. Consequently $\tau = \tau_U(V)$.

If $f \in U$, then $f^{\perp} = 0f^{-1}$ is an open neighborhood of $0$ and hence must contain a basic open set $\bigcap_{i=1}^{n} H_i$ with $H_i \in \mathcal{H}$. Conversely, let $f \in V^*$ and assume that $f^{\perp} \supseteq \bigcap_{i=1}^{n} H_i$ for some $H_i \in \mathcal{H}$. Then $H_i = f_i^{\perp}$ for some $f_i \in U$ as above. Hence $f^{\perp} \supseteq \bigcap_{i=1}^{n} f_i^{\perp} = [f_1, f_2, \ldots, f_n]^{\perp}$. But then

$$f \in [f]^{\perp\perp} = f^{\perp\perp} \subseteq [f_1, f_2, \ldots, f_n]^{\perp\perp} = [f_1, f_2, \ldots, f_n] \subseteq U$$

in view of 1.15.

III.1.19 <u>LEMMA</u>. <u>Let $S$ be a proper subspace of a vector space $V$. Let $A$ be a basis of $S$ and extend $A$ to a basis $B$ of $V$. Then $C = \{b + S \mid b \in B \setminus A\}$ is a basis of $V/S$. Conversely, if $A$ is a basis of $S$, $C$ is basis of $V/S$ and $D$ is a system of representatives of the cosets of $S$ making up $C$, then $B = A \cup D$ is a basis of $V$.</u>

<u>PROOF</u>. Let $A$, $B$ and $C$ be as in the first part of the lemma. If $v \in V$, then $v = \sum_{b_i \in B} \sigma_i b_i$ so that

$$v + S = \sum_{b_i \in B} \sigma_i b_i + S = \sum_{b_i \in B} \sigma_i (b_i + S),$$

showing that $C$ generates $V/S$. Assume that $\sum_{b_i \in B\setminus A} \sigma_i (b_i + S) = S$. Then $\sum_{b_i \in B\setminus A} \sigma_i b_i \in S$ and hence $\sum_{b_i \in B\setminus A} \sigma_i b_i$ must be a linear combination of elements of $A$. Since $B$ is linearly independent, this is possible only if all $\sigma_i = 0$. Hence $C$ is linearly independent and thus a basis of $V/S$.

Now let $A$, $B$, $C$ and $D$ be as in the second part of the lemma. For $v \in V$, we obtain $v + S = \sum_{d_i \in D} \sigma_i (d_i + S) = \sum_{d_i \in D} \sigma_i d_i + S$ so that $v = \sum_{d_i \in D} \sigma_i d_i + s$ for some $d_i \in D$ and $s \in S$. Hence $s = \sum_{a_i \in A} \tau_i a_i$ and thus $v = \sum_{d_i \in D} \sigma_i d_i + \sum_{a_i \in A} \tau_i a_i$ proving that $B$ generates $V$. Suppose that $\sum_{d_i \in D} \sigma_i d_i + \sum_{a_i \in A} \tau_i a_i = 0$. Then

$$\sum_{d_i \in D} \sigma_i (d_i + S) = \sum_{d_i \in D} \sigma_i d_i + S = - \sum_{a_i \in A} \tau_i a_i + S = S,$$

where $d_i + S \in C$ and $d_i + S \neq d_j + S$ if $i \neq j$. Hence all $\sigma_i = 0$ by linear independence of $C$ so that $\sum_{a_i \in A} \tau_i a_i = 0$ which yields that all $\tau_i = 0$ by linear independence of $A$. Therefore $B$ is linearly independent and hence a basis of $V$.

III.1.20 DEFINITION. For a subspace $S$ of a vector space $V$, the codimension of $S$ is defined by: $\operatorname{codim} S = \dim V/S$.

III.1.21 THEOREM. The following conditions on a topological vector space $(\Delta, V, \tau)$ are equivalent.

i) $\tau = \tau_U(V)$ for some t-subspace $U$ of $V$.

ii) $\tau$ has an open subbase $\mathcal{H}$ for the neighborhood system of $0$ consisting of hyperplanes for which $\bigcap_{H \in \mathcal{H}} H = 0$.

iii) $\tau$ has an open base $\mathcal{B}$ for the neighborhood system of $0$ consisting of subspaces of finite codimension for which $\bigcap_{B \in \mathcal{B}} B = 0$.

PROOF. i) $\Rightarrow$ ii). This follows from 1.18 and 1.14.

ii) $\Rightarrow$ iii). It suffices to show that the codimension of the intersection of a finite number of hyperplanes is finite. The statement is trivial for $n = 1$. Let $G = H_1 \cap H_2 \cap \ldots \cap H_n$ where $H_i$ are hyperplanes, suppose that $\operatorname{codim} G < \infty$ and let $H$ be a hyperplane. Then $G + H = H$ or $G + H = V$ since $H$ is a hyperplane. In the former case, $G \subseteq H$ so that $\operatorname{codim}(G \cap H) = \operatorname{codim} G < \infty$. Thus suppose that $G + H = V$. Then $V/H = (G+H)/H \cong G/(G \cap H)$ so that $\dim(G/(G \cap H)) = \dim V/H = 1$. Further, we have $V/G \cong (V/(G \cap H))/(G/(G \cap H))$, where $\dim V/G < \infty$ and $\dim G/(G \cap H) = 1$. It then follows from 1.19 that $\dim V/(G \cap H) = \dim V/G + 1 < \infty$.

iii) $\Rightarrow$ ii). Let $\mathcal{H}$ be the set of all hyperplanes which contain at least one member of $\mathcal{B}$. Let $H \in \mathcal{H}$ and $v \in H$. Then for some $B \in \mathcal{B}$, we have $B \subseteq H$ and thus $v + B$ is an open neighborhood of $v$ contained in $H$ which shows that $H$ is open. Let $\tau'$ be the topology on $V$ having $\mathcal{H}$ as an open subbase of the neighborhood system of $0$. Hence $\tau' \subseteq \tau$. If $\operatorname{codim} B = n$ for $B \in \mathcal{B}$, then it follows from 1.19 that any basis $C$ of $B$ can be extended to a basis $D = C \cup \{x_1, x_2, \ldots, x_n\}$ of $V$. It also follows from 1.19 that for $1 \leq i \leq n$, the subspace $H_i$ of $V$ generated by $C \cup \{x_1, x_2, \ldots, x_{i-1}, x_{i+1}, \ldots, x_n\}$ is a hyperplane. Since $B = \bigcap_{i=1}^{n} H_i$, it follows that each $B \in \mathcal{B}$ is the intersection of a finite number of members of $\mathcal{H}$. Hence each $B \in \mathcal{B}$ is $\tau'$-open implying $\tau \subseteq \tau'$. Therefore $\tau = \tau'$. Assume that $0 \neq v \in \bigcap_{H \in \mathcal{H}} H$. Then there exists $B \in \mathcal{B}$ such that $v \notin B$ since $\bigcap_{B \in \mathcal{B}} B = 0$. Let $A$ be a basis of $B$ and $C = A \cup \{v, x_2, \ldots, x_n\}$ be a basis of $V$. Then the subspace $H$ of $V$ generated by $A \cup \{x_2, \ldots, x_n\}$ is a hyperplane containing $B$ so that

$H \in \mathcal{H}$. But $v \notin H$, contradicting the assumption. Therefore $\bigcap_{H \in \mathcal{H}} H = 0$.

ii) $\Rightarrow$ i). This is a part of 1.18.

III.1.22 EXERCISES.

i) Show that $\tau_U(V)$ is the discrete topology if and only if $V$ is finite dimensional.

ii) Show that for every finite dimensional subspace $S$ of $V$, $\tau_U(V)$ relativized to $S$ is discrete.

iii) Show that for any $f_1, f_2, \ldots, f_n, f \in V^*$ satisfying: for any $v \in V$, if $vf_i = 0$ for $i = 1,2,\ldots,n$, then $vf = 0$, we must have that $f$ is a linear combination of $f_1, f_2, \ldots, f_n$.

The origin of most of the concepts introduced and statements proved in this section is to be found in Dieudonné [1], [2], [3] and Jacobson [5]. For more information on this subject, see Behrens ([1], Chapter II, Section 2), Bourbaki ([1], § 1), Jacobson ([6], Chapter IX, Section 6), ([7], Chapter II, Section 3 and Chapter IV, Section 18).

## III.2 TOPOLOGICAL PROPERTIES OF SUBSPACES

We consider here some simple properties of subspaces in relation to the topology. We will tacitly assume in the whole section that a dual pair $(U,V)$ is given relative to which the topology on $V$ as well as $S^\perp$, $S^{\perp\perp}$ etc. are defined.

III.2.1 PROPOSITION. A subspace $S$ of $V^*$, where $V^*$ is endowed with the finite topology, is dense if and only if it is a t-subspace.

PROOF. Assume that $S$ is dense in $V^*$ and let $0 \neq v \in V$. There exists $f \in V^*$ such that $vf = \delta \neq 0$. Since $S$ is dense we have $S \cap B_{(v;\delta)} \neq \emptyset$. Letting $g \in S \cap B_{(v;\delta)}$, we obtain $vg = \delta \neq 0$ showing that $S$ is a t-subspace of $V^*$.

Conversely, let $S$ be a t-subspace of $V^*$. Further let $f \in V^*$ and $B_{(v_1,v_2,\ldots,v_n;\delta_1,\delta_2,\ldots,\delta_n)}$ be a basic neighborhood of $0$, where $v_1, v_2, \ldots, v_n$ are linearly independent. By I.2.3 there exist $g_1, g_2, \ldots, g_n \in S$ such that $v_i g_j = \delta_{ij}$ for $1 \leq i,j \leq n$. Set $g = \sum_{j=1}^{n} g_j(v_j f + \delta_j)$; then for any $1 \leq i \leq n$, we have

$$v_i g = \sum_{j=1}^{n} (v_i g_j)(v_j f + \delta_j) = v_i f + \delta_i$$

so that $g - f \in B_{(v_1,v_2,\ldots,v_n;\delta_1,\delta_2,\ldots,\delta_n)}$. Consequently

$$g \in (f + B_{(v_1,v_2,\ldots,v_n;\delta_1,\delta_2,\ldots,\delta_n)}) \cap S$$

proving that $S$ is dense in $V^*$.

The following lemma is due to Dieudonné [1].

III.2.2 LEMMA. *For* $S$ *a subspace of* $V$, $S^{\perp\perp}$ *is the closure of* $S$.

PROOF. We denote the closure of $S$ by $\bar{S}$. By 1.19, we have

$$\bar{S} = \{v \in V \mid (v+T^{\perp}) \cap S \neq \emptyset \text{ for every subspace } T \text{ of } U \text{ with } \dim T < \infty\}, \tag{1}$$

also note that

$$S^{\perp\perp} = \{v \in V \mid \text{if } (s,u) = 0 \text{ for all } s \in S, \text{ then } (v,u) = 0\}. \tag{2}$$

Now let $v \in \bar{S}$ and suppose that for some $u \in U$, we have $(s,u) = 0$ for all $s \in S$. Then $(v+u^{\perp}) \cap S \neq \emptyset$ by (1), say $s \in (v+u^{\perp}) \cap S$. Hence $s = v+x$ where $(x,u) = 0$ and thus $(v,u) = (s,u) = 0$ by the assumption on $u$. It follows from (2) that $v \in S^{\perp\perp}$.

To prove the converse, we identify $U$ with nat $U$ and let $v \in V\backslash\bar{S}$. Hence by (1) there exists a finite dimensional subspace $T$ of $U$ for which $(v+T^{\perp}) \cap S = \emptyset$ which implies that $v \notin S+T^{\perp}$. Let $B$ be a basis of $S+T^{\perp}$ and extend $B \cup \{v\}$ to a basis $C$ of $V$. Define a function $f$ on $C$ by letting $xf = 0$ for all $x \in B$, $vf \neq 0$, and otherwise arbitrary. Also denote by $f$ its linear extension to all of $V$. Then $f \in V^*$ and $T^{\perp} \subseteq f^{\perp}$, where $\perp$ is taken relative to the dual pair $(V^*,V)$. Using 1.15, we obtain $f \in f^{\perp\perp} \subseteq T^{\perp\perp} = T$ since $\dim T < \infty$. But then $f \in U$, $vf \neq 0$, and $xf = 0$ for all $x \in S$ which by (2) shows that $v \notin S^{\perp\perp}$.

III.2.3 COROLLARY. *Every finite dimensional subspace of* $V$ *is closed.*

PROOF. Exercise.

III.2.4 LEMMA. *For any subspace* $T$ *of* $U$, *we have* $T^{\perp\perp\perp} = T^{\perp}$.

PROOF. Exercise.

III.2.5 COROLLARY. *A subspace* $S$ *of* $V$ *is closed if and only if* $S = T^{\perp}$ *for some subspace* $T$ *of* $U$.

PROOF. If $S$ is closed, then $(S^{\perp})^{\perp} = S$ by 2.2 with $S^{\perp}$ a subspace of $U$. Conversely, if $S = T^{\perp}$ for some subspace $T$ of $U$, then by 2.4, we have

$$S^{\perp\perp} = T^{\perp\perp\perp} = T^{\perp} = S$$

and thus $S$ is closed by 2.2.

III.2.6 LEMMA. *If* $T$ *is an* $n$*-dimensional subspace of* $U$, *then* codim $T^{\perp} = n$.

PROOF. Let $\{t_1,t_2,\ldots,t_n\}$ be a basis of $T$. By 1.2.3, there exist $x_1,x_2,\ldots,x_n \in V$ such that $(x_i,t_j) = \delta_{ij}$. If $\sum_{i=1}^{n} \alpha_i(x_i+T^{\perp}) = T^{\perp}$, then $\sum_{i=1}^{n} \alpha_i x_i \in T^{\perp}$. In particular, we have

$$0 = (\sum_{k=1}^{n} \alpha_k x_k, t_i) = \sum_{k=1}^{n} \alpha_k (x_k,t_i) = \alpha_i$$

so that $\{x_i + T^{\perp}\}_{i=1}^{n}$ is a linearly independent set of vectors in $V/T^{\perp}$. For a fixed $v \in V$, let $\alpha_i = (v, t_i)$ for $1 \leq i \leq n$. Then

$$(v - \sum_{i=1}^{n} \alpha_i x_i, t_j) = (v, t_j) - \sum_{i=1}^{n} (v, t_i)(x_i, t_j) = (v, t_j) - (v, t_j) = 0$$

and thus $v - \sum_{i=1}^{n} \alpha_i x_i \in T^{\perp}$. Consequently $v + T^{\perp} = \sum_{i=1}^{n} \alpha_i (x_i + T^{\perp})$ which proves that the vectors $x_i + T^{\perp}$ generate $V/T^{\perp}$ and thus form a basis. Hence $\operatorname{codim} T^{\perp} = \dim V/T^{\perp} = n$.

III.2.7 PROPOSITION. *The following conditions on a subspace* $S$ *of* $V$ *are equivalent.*

i) $S$ *is open.*

ii) $S$ *is closed and* codim $S < \infty$.

iii) $S = T^{\perp}$ *for some finite dimensional subspace* $T$ *of* $U$.

PROOF. i) ⇒ ii). In any topological group, every open subgroup is closed. Since $S$ is open, it contains a basic open neighborhood of $0$, say $T^{\perp}$, where $\dim T < \infty$. Hence $T^{\perp} \subseteq S$ which implies $T = T^{\perp\perp} \supseteq S^{\perp}$ by 1.15, so $\dim S^{\perp} < \infty$. By 2.2 and 2.6, we have

$$\operatorname{codim} S = \dim V/S = \dim V/S^{\perp\perp} = \dim S^{\perp} < \infty.$$

ii) ⇒ iii). It follows from 2.5 that $S = T^{\perp}$ for some subspace $T$ of $U$. The beginning of the proof of 2.6 shows that if $T$ is infinite dimensional, then $V/T^{\perp}$ is also which by contrapositive implies that $T$ is finite dimensional.

iii) ⇒ i). In such a case, $S$ is a basic open neighborhood of $0$.

III.2.8 COROLLARY. *The basic open neighborhoods of* $0$ *are the only open subspaces.*

III.2.9 COROLLARY. *The following conditions on a subspace* $S$ *of* $V$ *are equivalent.*

i) $S$ *is a closed hyperplane.*

ii) $S$ *is an open hyperplane.*

iii) $S = u^{\perp}$ *for some* $0 \neq u \in U$.

Also note that a hyperplane $S$ is either closed or dense since $S^{\perp\perp}$ is a subspace containing $S$ and thus must be either $S$ or $V$.

III.2.10 COROLLARY. *Every subspace of* $V$ *is closed if and only if* nat $U = V^*$.

PROOF. Necessity. In particular, for any $0 \neq f \in V^*$, $F = \{v \in V \mid vf = 0\}$ is a closed hyperplane so that by 2.9 we have that $F = u^{\perp}$ for some $0 \neq u \in U$. But then $f = f_u \in \text{nat } U$ and thus nat $U = V^*$.

Sufficiency. Let $S$ be a subspace of $V$ and $v \in V\setminus S$. Extend $v$ to a basis $B$ of $V$. Let $f$ be the linear form on $V$ for which $vf = 1$ and $xf = 0$ if $x \in B\setminus\{v\}$. Then $f \in V^* = \text{nat } U$ and hence for some $u \in U$, we have $f = f_u$ and obtain $S \subseteq u^\perp$ and $v \notin u^\perp$. Hence $S$ is the intersection of some closed hyperplanes $u^\perp$ and thus must be closed.

III.2.11 <u>LEMMA</u>. <u>Every closed subspace of</u> $V$ <u>is the intersection of (closed and open) hyperplanes of the form</u> $u^\perp$ <u>with</u> $0 \neq u \in U$.

<u>PROOF</u>. Let $S$ be a closed subspace of $V$ and let $v \in V\setminus S$. Then for some finite dimensional subspace $T$ of $U$, we have $(v + T^\perp) \cap S = \emptyset$ so that $v \notin T^\perp + S$. Let $\{u_1, u_2, \ldots, u_n\}$ be a basis of $T$. Then

$$v \notin T^\perp + S = \bigcap_{i=1}^{n} u_i^\perp + S = \bigcap_{i=1}^{n} (u_i^\perp + S)$$

and thus $v \notin u_i^\perp + S$ for some $i$. Hence $u_i^\perp + S \neq V$ and since $u_i^\perp$ is a hyperplane, it follows that $u_i^\perp + S = u_i^\perp$, i.e., $S \subseteq u_i^\perp$. Consequently $v \notin u_i^\perp$ while $S \subseteq u_i^\perp$, as required.

The following proposition is due to Dieudonné [1].

III.2.12 <u>PROPOSITION</u>. <u>If</u> $S$ <u>is a closed subspace of</u> $V$, <u>then the quotient topology of</u> $V/S$ <u>coincides with</u> $\tau_{S^\perp}(V/S)$, <u>where</u> $(S^\perp, V/S)$ <u>is a dual pair with the bilinear form</u> $(v + S, u) = (v, u)$ $(v \in V,\ u \in S^\perp)$.

<u>PROOF</u>. If $v + S = v' + S$, then $v - v' \in S$ and thus $(v - v', u) = 0$ for all $u \in S^\perp$ so that $(v,u) = (v',u)$ which shows that the function which to each pair $(v + S, u)$ associates the value $(v,u)$, is single valued. This function is clearly bilinear. If $v + S \neq S$, then $v \notin S$ so by 2.11 there exists $u \in S^\perp$ such that $(v,u) \neq 0$ and thus $(v + S, u) \neq 0$. For every $u \in S^\perp$, there exists $v \in V$ such that $(v,u) \neq 0$ so that $(v + S, u) \neq 0$. Consequently the new bilinear form is non-degenerate.

Let $\pi: V \to V/S$ be the canonical homomorphism and let $\gamma$ denote the quotient topology of $V/S$. A subbase for $\gamma$ is given by

$$\mathcal{S} = \{A \subseteq V/S \mid A\pi^{-1} = u^\perp,\ u \in U\},$$

and a subbase for the topology $\delta = \tau_{S^\perp}(V/S)$ is given by

$$\mathfrak{J} = \{u^o \mid u \in S^\perp\},$$

where

$$u^o = \{v + S \mid (v + S, u) = 0\} = \{v + S \mid (v,u) = 0\} = \{v + S \mid v \in u^\perp\}.$$

Then for $u \in S^\perp$, we have

$$u^o\pi^{-1} = \{v \in V \mid v + S \in u^o\} = \{v \in V \mid v \in u^\perp\} = u^\perp,$$

so $u^o \in \mathcal{S}$ which proves that $\mathcal{J} \subseteq \mathcal{S}$. Conversely, let $A \in \mathcal{S}$. Then $A\pi^{-1} = u^\perp$ for some $u \in U$. Since $0 \in u^\perp$, it follows that $0 \in A$, so $S \subseteq A\pi^{-1} = u^\perp$ and hence $u \in u^{\perp\perp} \subseteq S^\perp$. Consequently

$$A = \{v\pi \mid v \in u^\perp\} = u^o \in \mathcal{J}$$

so that $\mathcal{S} \subseteq \mathcal{J}$. Therefore $\gamma = \delta$.

The next result is due to Mackey [2].

III.2.13 PROPOSITION. Let $A$ be a closed and $B$ be a finite dimensional subspace of $V$. Then $A+B$ is closed.

PROOF. Using induction on the dimension of $B$, it clearly suffices to show that if $v \notin A$, then $A+[v]$ is closed. We suppose that $v \notin A$. Then $v \notin (A^\perp)^\perp = A$ implies that $(v,u_o) \neq 0$ for some $u_o \in A^\perp$ and we may suppose that $(v,u_o) = 1$. For every $u \in A^\perp$, we have $u - u_o(v,u) \in A^\perp$ and also

$$(v, u - u_o(v,u)) = (v,u) - (v,u_o)(v,u) = (v,u) - (v,u) = 0$$

so that $u - u_o(v,u) \in (A+[v])^\perp$. Now let $x \in (A+[v])^{\perp\perp}$. Then

$$0 = (x, u - u_o(v,u)) = (x,u) - (x,u_o)(v,u) = (x - (x,u_o)v, u)$$

for all $u \in A^\perp$ so that $x - (x,u_o)v \in A^{\perp\perp} = A$. But then

$$x \in A + (x,u_o)v \subseteq A + [v]$$

and thus $(A+[v])^{\perp\perp} \subseteq A+[v]$; the opposite inclusion always holds.

III.2.14 EXERCISES.

i) Let $(U,V)$ be a dual pair. Prove that $(U,V')$ is a dual pair for no proper subspace $V'$ of $V$ if and only if $\operatorname{nat} U = V^*$.

ii) Let $(U,V)$ be a dual pair and write $\mathcal{F} = \mathcal{F}_U(V)$. Prove the following statements.

a) The mapping

$$S \to S^\perp \quad (S \text{ is a subspace of } V)$$

is an inclusion inverting function mapping the lattice of all subspaces of $V$ into the lattice of all subspaces of $U$.

b) $S \subseteq S^{\perp\perp}$ for every subspace $S$ of $V$.

c) $\mathfrak{A}_{\ell,\mathcal{F}}(\mathfrak{I}_T(V)) = \mathfrak{I}_U(T^\perp)$ for every subspace $T$ of $U$.

d) $\mathfrak{A}_{r,\mathcal{F}}(\mathfrak{I}_U(S)) = \mathfrak{I}_{S^\perp}(V)$ for every subspace $S$ of $V$.

e) $\mathfrak{I}_T(V) = \{a \in \mathcal{F} \mid T^\perp \subseteq N_a\}$ for every finite dimensional subspace $T$ of $U$.

iii) For the dual pair $(V^*,V)$, prove that $T^{\perp\perp} = T$ for every subspace $T$ of $V^*$ if and only if $V$ is finite dimensional.

iv) Let $(U,V)$ be a dual pair. Show that closed subspaces of $V$ form a complete lattice (a lattice $L$ is <u>complete</u> if every nonempty subset of $L$ has a l.u.b. and a g.l.b. in $L$).

v) Let $(U,V)$ be a dual pair and $S$ be a subspace of $V$. Show that $(U/S^{\perp},S)$ can be made into a dual pair in such a way that $\tau_{U/S^{\perp}}(S)$ is weaker than the topology $\tau_U(V)$ relativized to $S$.

The references for this section include Bourbaki ([1], §1), Dieudonné [1],[2], Jacobson ([6], Chapter IX, Sections 7,8),([7], Chapter II, Section 4 and Chapter IV, Sections 5,6), Köthe ([1], §10), Mackey [2], Ornstein [1], Ribenboim ([1], Chapter III, Section 2).

## III.3 TOPOLOGICAL PROPERTIES OF SEMILINEAR TRANSFORMATIONS

For dual pairs $(U,V)$ and $(U',V')$ and a semilinear transformation $a:V \to V'$, we examine here the conditions under which $a$ is continuous and if so, when it is also one-to-one or onto. The next theorem is due to Jacobson [5], see also Dieudonné [1].

III.3.1 <u>THEOREM</u>. <u>Let</u> $(U,\Delta,V)$ <u>and</u> $(U',\Delta',V')$ <u>be dual pairs and</u> $(\omega,a)$ <u>be a semilinear transformation of</u> $(\Delta,V)$ <u>into</u> $(\Delta',V')$. <u>Then</u> $a$ <u>is a continuous function from</u> $(V,\tau_U(V))$ <u>into</u> $(V',\tau_{U'}(V'))$ <u>if and only if</u> $a$ <u>has an adjoint</u>.

<u>PROOF</u>. Suppose that $a$ is continuous, and for every $u' \in U'$ define a function $\underline{u}'$ by: $v\underline{u}' = (va,u')'\omega^{-1}$ $(v \in V)$. Then $\underline{u}'$ is a function from $V$ into $\Delta$, and for any $\sigma,\tau \in \Delta$ and $x,y \in V$, we obtain

$$\begin{aligned}(\sigma x + \tau y)\underline{u}' &= ((\sigma x + \tau y)a,u')'\omega^{-1}\\ &= ((\sigma\omega)(xa) + (\tau\omega)(ya),u')'\omega^{-1}\\ &= (\sigma\omega\omega^{-1})[(xa,u')'\omega^{-1}] + (\tau\omega\omega^{-1})[(ya,u')'\omega^{-1}]\\ &= \sigma(x\underline{u}') + \tau(y\underline{u}')\end{aligned}$$

proving that $\underline{u}' \in V^*$. Further, $\underline{u}'$ is a composition of the following functions: $a$ which is continuous by hypothesis, $f_{u'}$ which is $\tau_{U'}(V')$-continuous by 1.16, and $\omega^{-1}$ which is continuous since both $\Delta$ and $\Delta'$ are discrete. Consequently $\underline{u}'$ is a continuous linear form which by 1.16 implies that $\underline{u}' \in$ nat $U$. Now define a function $b$ by

$$bu' = u \quad \text{if} \quad \underline{u}' = f_u \qquad (u \in U).$$

For any $v \in V$, $u' \in U'$, letting $bu' = u$, we have $\underline{u}' = f_u$ so that

$$(v,bu')\omega = (v,u)\omega = (vf_u)\omega = (v\underline{u}')\omega = (va,u')'\omega^{-1}\omega = (va,u')$$

showing that $b$ is the adjoint of $a$.

Conversely, let $b$ be the adjoint of $a$. It suffices to prove continuity of $a$ at $0$, and since by 1.8 the family $\{u'^{\perp} \mid u' \in U'\}$ constitutes a subbase for the neighborhood system of $0$ in $V'$, the proof reduces to showing that $u'^{\perp}a^{-1}$ is open in $V$. For $v \in u'^{\perp}a^{-1}$, we will show that $v + (bu')^{\perp} \subseteq u'^{\perp}a^{-1}$; here $bu' \in U$ so that $(bu')^{\perp}$ is a neighborhood of $0$ and thus $v + (bu')^{\perp}$ is a neighborhood of $v$. Let $x \in (bu')^{\perp}$; then $(x,bu') = 0$ and hence $(xa,u')' = 0$. On the other hand, $v \in u'^{\perp}a^{-1}$ implies that $(va,u')' = 0$ and thus $(v+xa,u')' = 0$. But then $v + x \in u'^{\perp}a^{-1}$ showing that $v + (bu')^{\perp} \subseteq u'^{\perp}a^{-1}$ which completes the proof.

III.3.2 <u>COROLLARY</u>. <u>For a dual pair</u> $(U,V)$, <u>we have</u>

$$\mathcal{L}_U(V) = \{a \in \mathcal{L}(V) \mid a \text{ is continuous}\}.$$

A function $f$ from a topological space $A$ into a topological space $B$ is <u>open</u> if $Cf$ is open for every open set $C$ in $A$. The next result is due to Dieudonné [1].

III.3.3 <u>THEOREM</u>. <u>Let</u> $(U,\Delta,V)$ <u>and</u> $(U',\Delta',V')$ <u>be dual pairs and</u> $(\omega,a)$ <u>be a semilinear transformation of</u> $(\Delta,V)$ <u>into</u> $(\Delta',V')$ <u>with adjoint</u> $b$. <u>Relative to the topological spaces</u> $(U,\tau_V(U))$ <u>and</u> $(U',\tau_{V'}(U'))$, <u>the following statements hold</u>.

i) $a$ <u>is one-to-one if and only if</u> $bU'$ <u>is dense in</u> $U$.

ii) $Va = V'$ <u>if and only if</u> $b$ <u>is one-to-one and open</u>.

<u>PROOF</u>. i) First note that for any $u \in U$, we have

$$\begin{aligned} v = 0a^{-1} &\Leftrightarrow va = 0 \Leftrightarrow (va,u')' = 0 \text{ for all } u' \in U' \\ &\Leftrightarrow (v,bu') = 0 \text{ for all } u' \in U' \Leftrightarrow v \in (bU')^{\perp}. \end{aligned} \tag{1}$$

Consequently, using 2.2, we obtain

$$\begin{aligned} &a \text{ is one-to-one} \Leftrightarrow 0a^{-1} = 0 \Leftrightarrow (bU')^{\perp} = 0 \\ &\Leftrightarrow (bU')^{\perp\perp} = U \Leftrightarrow bU' \text{ is dense in } U. \end{aligned}$$

ii) Suppose that $Va = V'$. Then I.5.4 asserts that $b$ is one-to-one. In order to show that $b$ is open, we first let $S'$ be a finite dimensional subspace of $V'$. Let $\{v_1',v_2',\ldots,v_n'\}$ be a basis of $S'$ and for each $1 \leq i \leq n$, choose a vector $v_i \in V$ for which $v_ia = v_i'$, and let $S = [v_1,v_2,\ldots,v_n]$. Then

$$\begin{aligned} u \in b(S'^{\perp}) &\Leftrightarrow u = bu' \text{ for some } u' \in S'^{\perp} \\ &\Leftrightarrow u = bu' \text{ for some } u' \in U' \text{ such that } (v_i',u')' = 0 \\ &\qquad \text{for } i = 1,2,\ldots,n, \end{aligned}$$

and since

$$(v_i',u') = 0 \Leftrightarrow (v_ia,u')' = 0 \Leftrightarrow (v_i,bu') = 0 \Leftrightarrow (v_i,u) = 0$$

we obtain $b(S'^{\perp}) = S^{\perp}$. Here $S$ is a finite dimensional subspace of $V$ so that $S^{\perp}$ is an open subspace of $U$. Hence $b(S'^{\perp})$ is open, and thus for any $u' \in U'$,

$b(u' + S'^{\perp}) = bu' + b(S'^{\perp})$ is also open. Since the sets $u' + S'^{\perp}$ form a base of the neighborhood system of $u'$ in $U'$, it follows that $b$ is open.

Suppose that $b$ is open; we will show that $Va$ is closed. Interchanging the roles of $a$ and $b$ in (1), we duduce that $(Va)^{\perp} = b^{-1}0$ so that

$$Va \subseteq (Va)^{\perp\perp} = (b^{-1}0)^{\perp}. \tag{2}$$

Let $v' \in (b^{-1}0)^{\perp}$. Then $v'^{\perp}$ is an open subspace of $U'$ and the hypothesis implies that $b(v'^{\perp})$ is an open subspace of $U$. Consequently there exists a finite dimensional subspace $S$ of $V$ with the property $S^{\perp} \subseteq b(v'^{\perp})$.

Let $u' \in U'$ and assume that $(va, u')' = 0$ for all $v \in S$. Then $(v, bu') = 0$ and thus $v \in (bu')^{\perp}$ for all $v \in S$. Hence $S \subseteq (bu')^{\perp}$ which implies

$$bu' \in (bu')^{\perp\perp} \subseteq S^{\perp} \subseteq b(v')^{\perp}.$$

This entails the existence of $s' \in v'^{\perp}$ for which $bu' = bs'$. Hence $b(u' - s') = 0$ which together with the hypothesis that $v' \in (b^{-1}0)^{\perp}$ implies $(v', u' - s')' = 0$. But then $s' \in v'^{\perp}$ implies $(v', u')' = (v', s')' = 0$. Now (2) in 2.2 shows that $v' \in (Sa)^{\perp\perp}$ and hence by 1.15, $v' \in Sa$ since $Sa$ is finite dimensional. Therefore $(b^{-1}0)^{\perp} \subseteq Va$ which together with (2) shows that $Va = (Va)^{\perp\perp}$. By 2.2, $(Va)^{\perp\perp}$ is the closure of $Va$ which proves that $Va$ is closed.

If $b$ is one-to-one, then interchanging the roles of $a$ and $b$ in part i), we duduce that $Va$ is dense in $V'$. Finally, if $b$ is both one-to-one and open, then $Va$ must be both dense and closed and thus $Va = V'$.

III.3.4 <u>COROLLARY</u>. <u>With the notation of</u> 3.1, $a$ <u>is a homeomorphism of</u> $(V, \tau_U(V))$ <u>onto</u> $(V', \tau_{U'}(V'))$ <u>if and only if</u> $a$ <u>is a semilinear isomorphism of</u> $(U, \Delta, V)$ <u>onto</u> $(U', \Delta', V')$.

<u>PROOF</u>. Exercise.

III.3.5 <u>EXERCISES</u>.

i) Let $a$ be a semilinear transformation of a vector space $V$ into a vector space $V'$. Construct the adjoint $b$ of $a$ relative to the dual pairs $(V^*, V)$ and $(V'^*, V')$. Then $b$ is called the <u>conjugate</u> of $a$ and is denoted by $a^*$.

ii) Let $(U, V)$ and $(U', V')$ be dual pairs, let $a$ be a semilinear transformation of $V$ into $V'$, and let $V$ (respectively $V'$) have the $U$-(respectively $U'$-) topology. Show that $a$ is continuous if and only if $a^*(\text{nat } U') \subseteq \text{nat } U$.

For more information on this subject, see Dieudonné [1],[2], Jacobson [5],([7], Chapter IV, Section 7), Ribenboim ([1], Chapter III, Section 2); on continuous linear transformations, see Rosenberg and Zelinsky [1].

## III.4 COMPLETION OF A VECTOR SPACE

Before aboarding the main subject of this section, we briefly summarize the notions related to completeness of uniform spaces, see Kelley ([1], Chapter 6).

Let $X$ be a nonempty set; a (binary) relation on $X$ is a subset of the Cartesian product $X \times X$. The set of all binary relations on $X$ forms a semigroup under the operation

$$\lambda \circ \rho = \{(x,z) \in X \times X \mid (x,y) \in \lambda \text{ and } (y,z) \in \rho \text{ for some } y \in X\}.$$

Further, $\lambda^{-1}$ is defined by $\lambda^{-1} = \{(x,y) \in X \times X \mid (y,x) \in \lambda\}$ so that $(\lambda^{-1})^{-1} = \lambda$ and $(\lambda \circ \rho)^{-1} = \rho^{-1} \circ \lambda^{-1}$. The identity relation or the diagonal is defined by $\Delta(X) = \{(x,x) \mid x \in X\}$, and is usually denoted by $\Delta$. For $\lambda \subseteq X \times X$ and $A \subseteq X$, we let

$$\lambda[A] = \{y \in X \mid (x,y) \in \lambda \text{ for some } x \in A\},$$

and for $A = \{x\}$, we simply write $\lambda[x]$. It follows that $(\lambda \circ \rho)[A] = \lambda[\rho[A]]$.

A uniformity $\mathcal{U}$ for a set $X$ is a nonempty family of binary relations on $X$ satisfying the following requirements:

($\alpha$) if $\lambda \in \mathcal{U}$, then $\lambda \supseteq \Delta$;

($\beta$) if $\lambda \in \mathcal{U}$, then $\rho \circ \rho \subseteq \lambda$ for some $\rho \in \mathcal{U}$;

($\gamma$) if $\lambda \in \mathcal{U}$, then $\lambda^{-1} \in \mathcal{U}$;

($\delta$) if $\lambda, \rho \in \mathcal{U}$, then $\lambda \cap \rho \in \mathcal{U}$;

($\varepsilon$) if $\lambda \in \mathcal{U}$ and $\lambda \subseteq \rho \subseteq X \times X$, then $\rho \in \mathcal{U}$.

The pair $(X,\mathcal{U})$ is a uniform space. A subfamily $\mathcal{B}$ of $\mathcal{U}$ is a base for $\mathcal{U}$ if each member of $\mathcal{U}$ contains a member of $\mathcal{B}$. Note that in view of ($\varepsilon$), a base of $\mathcal{U}$ completely determines $\mathcal{U}$. One easily verifies the following assertion.

A nonempty family $\mathcal{B}$ of binary relations on a set $X$ is a base for some uniformity if and only if $\mathcal{B}$ satisfies ($\alpha$), ($\beta$) and

($\gamma'$) if $\lambda \in \mathcal{B}$, then $\lambda^{-1}$ contains a member of $\mathcal{B}$,

($\delta'$) the intersection of any two members of $\mathcal{B}$ contains a member of $\mathcal{B}$.

For a uniform space $(X,\mathcal{U})$, the topology $\mathcal{T}$ of the uniformity $\mathcal{U}$, briefly the uniform topology of $\mathcal{U}$, is the family

$$\{T \subseteq X \mid \text{for every } x \in T \text{ there exists } \lambda \in \mathcal{U} \text{ such that } \lambda[x] \subseteq T\}.$$

One shows easily that $\mathcal{T}$ is indeed a topology, and thus $(X,\mathcal{T})$ is a topological space.

A function mapping a uniform space $(X,\mathcal{U})$ into a uniform space $(Y,\mathcal{V})$ is uniformly continuous if for each $\rho \in \mathcal{V}$, we have

$$\{(x,y) \in X \times X \mid (f(x),f(y)) \in \rho\} \in \mathcal{U}.$$

Further, in the case that $f$ is one-to-one, onto and both $f$ and $f^{-1}$ are uniformly continuous, then $f$ is a uniform isomorphism and the spaces $(X,\mathcal{U})$ and $(Y,\mathcal{V})$ are uniformly equivalent.

As in the case of topological spaces, if $Y$ is a nonempty subset of $X$ and $(X,\mathcal{U})$ is a uniform space, the set $\mathcal{V} = \{\lambda \cap (Y \times Y) \mid \lambda \in \mathcal{U}\}$ is the relativization of $\mathcal{U}$ to $Y$, or the relative uniformity for $Y$, and $(Y,\mathcal{V})$ is a uniform subspace of $(X,\mathcal{U})$.

A binary relation, to be denoted by $\geq$, on a nonempty set $D$, directs $D$ if $\geq$ is reflexive, transitive, and if for any $m,n \in D$, there exists $p \in D$ such that $p \geq m$ and $p \geq n$. In such a case $D$ is a directed set. A net is a pair $(S,\geq)$, where $S$ is a function on a set $D$ directed by $\geq$, and is also denoted by $\{S_n \mid n \in D\}$. A net $\{S_n \mid n \in D\}$ is in a set $A$ if $S_n \in A$ for all $n \in D$; it is eventually in $A$ if there exists $m \in D$ such that $S_n \in A$ whenever $n \geq m$. A net $(S,\geq)$ in a topological space $(X,\mathfrak{J})$ converges to a point $x$ relative to $\mathfrak{J}$ if it is eventually in every $\mathfrak{J}$-neighborhood of $x$.

A net $\{S_n \mid n \in D\}$ in a uniform space $(X,\mathcal{U})$ is a Cauchy net if for each $\lambda \in \mathcal{U}$ there exists $k \in D$ such that $(S_m,S_n) \in \lambda$ whenever $m,n \geq k$. Further, $(X,\mathcal{U})$ is complete if every Cauchy net in the space converges to a point in the space. A uniform subspace of a uniform space $(X,\mathcal{U})$ is dense if it is dense in the uniform topology of $\mathcal{U}$. A completion of $(X,\mathcal{U})$ is a pair $(f,(X^*,\mathcal{U}^*))$, where $(X^*,\mathcal{U}^*)$ is a complete uniform space and $f$ is a uniform isomorphism onto a dense subspace of $X^*$; a completion is Hausdorff if $(X^*,\mathcal{U}^*)$ is a Hausdorff uniform space (i.e., the uniform topology is Hausdorff). A Hausdorff completion of a Hausdorff uniform space is unique up to a uniform isomorphism and can be referred to as the completion.

If $(G,\mathfrak{J})$ is an abelian topological group, for every neighborhood $N$ of the identity $0$ of $G$, let $\bar{N} = \{(x,y) \in G \times G \mid x - y \in N\}$. Then the family

$$\{\bar{N} \mid N \text{ is a neighborhood of } 0\}$$

is a base for a uniformity $\mathcal{U}$ on $G$ and the uniform topology for $\mathcal{U}$ coincides with $\mathfrak{J}$ (verify!). It follows that the topology of any topological group (abelian can be omitted) is a uniform topology. The concept of a complete topological group refers to this uniformity; in addition, specializing to a complete module, vector space or a ring, we have in mind the uniformity of their additive structure.

III.4.1 LEMMA. *If $V$ is a vector space, then $V^*$ is complete in the finite topology.*

PROOF. Since $\Delta$ is discrete, it is complete. Hence by Kelley ([1], p. 194), $\Delta^V$ is complete being the Cartesian product of complete spaces. In view of Kelley ([1], p. 192), in order to show that $V^*$ is complete, it suffices to show that $V^*$ is a closed subspace of $\Delta^V$. By Kelley ([1], p. 66), it suffices to show that every convergent net in $V^*$ converges to some point in $V^*$. Hence let $\{f_n \mid n \in D\}$ be a net in $V^*$ which converges to some $f \in \Delta^V$. Since the topology of $V^*$ is the relativized product topology, by Kelley ([1], p. 217), the convergence of $\{f_n \mid n \in D\}$ to $f$ is equivalent to the convergence of $\{vf_n \mid n \in D\}$ to the point $vf$ for every $v \in V$ (pointwise convergence). Since $\Delta$ is discrete, it follows that for every $v \in V$, there exists $m \in D$ such that $vf_n = vf$ whenever $n \geq m$. For any $\sigma, \tau \in \Delta$ and $x, y \in V$, there exists $m_1$ such that $(\sigma x + \tau y)f_n = (\sigma x + \tau y)f$ whenever $n \geq m_1$; there exists $m_2$ such that $xf_n = xf$ whenever $n \geq m_2$; there exists $m_3$ such that $yf_n = yf$ whenever $n \geq m_3$. Finally there exists $m \geq m_1, m_2, m_3$ with the property

$$(\sigma x + \tau y)f = (\sigma x + \tau y)f_n = \sigma(xf_n) + \tau(yf_n) = \sigma(xf) + \tau(yf)$$

whenever $n \geq m$. But then $f$ is linear so that $f \in V^*$. Hence $V^*$ is closed and thus complete.

III.4.2 THEOREM. *Let* $(U,V)$ *be a dual pair and* $f: v \to f_v$ *be the natural isomorphism of* $V$ *into* $U^*$. *Then* $(f,(U^*,\phi))$ *is the completion of* $(V,\tau_U(V))$, *where* $\phi$ *is the finite topology.*

PROOF. By 2.1, nat $V$ is a dense subspace of $U^*$, and $U^*$ is complete in the finite topology by 4.1. The function $f$ is an algebraic isomorphism of $V$ onto nat $V$. By definition, $\tau_U(V)$ is the topology on $V$ which makes $f$ a homeomorphism of $V$ onto nat $V$, the latter having the relativized finite topology of $U^*$. For any finite dimensional subspace $T$ of $U$ and $x, y \in V$, we obtain

$$x - y \in T^{\perp} \Leftrightarrow f_{x-y}t = (x - y, t) = 0 \quad \text{for all} \quad t \in T$$
$$\Leftrightarrow (f_x - f_y)t = 0 \quad \text{for all} \quad t \in T.$$

The set $\{(x,y) \mid x - y \in T^{\perp}\}$ is a uniform neighborhood in $V$ while

$$\{(f_x, f_y) \mid (f_x - f_y)t = 0 \quad \text{for all} \quad t \in T\}$$

is a uniform neighborhood in $U^*$. Since $f$ is one-to-one, we conclude that both $f$ and $f^{-1}$ carry uniform neighborhoods onto uniform neighborhoods, proving that both are uniformly continuous. But then $f$ is a uniform isomorphism of $V$ onto a dense subspace of $U^*$, where the latter is complete. Thus $(f,(U^*,\phi))$ is a completion of $(V,\tau_U(V))$. Since both $(V,\tau_U(V))$ and $(U^*,\phi)$ are Hausdorff, the completion is unique up to a uniform isomorphism by Kelley ([1], p. 197).

III.4.3 COROLLARY. *For* $(U,V)$ *a dual pair,* $V$ *is complete in the* $U$-*topology if and only if* nat $V = U^*$.

## III.5 LINEARLY COMPACT VECTOR SPACES

We consider here linearly and weakly topologized and linearly compact vector spaces and establish relationships of these notions with those already encountered.

III.5.1 DEFINITION. If $M$ is a module (left or right) over a ring $\mathfrak{R}$, then $M$ is a topological module if $\mathfrak{R}$ is a topological ring, $M$ is a topological group under addition and the action of $\mathfrak{R}$ upon $M$ is jointly continuous. Further, $M$ is a linearly topologized module if the topology of $M$ admits an open base for the neighborhood system of $0$ consisting of submodules. A topological vector space is weakly topologized if it has an open base for the neighborhood system of $0$ consisting of subspaces of finite codimension.

We are interested here in Hausdorff topological vector spaces over discrete division rings.

As a supplement to and partly a consequence of 1.21, we have the following result.

III.5.2 PROPOSITION. *The following conditions on a topological vector space* $(V,\tau)$ *are equivalent.*

i) $V$ *is weakly topologized.*

ii) $\tau = \tau_U(V)$ *for some* t-*subspace of* $V^*$.

iii) $V$ *is linearly topologized and every open subspace has finite codimension.*

PROOF. Items i) and ii) are equivalent by 1.21.

i) ⇒ iii). Obviously a weak topology is linear; the second part of iii) follows immediately from 2.7.

iii) ⇒ i). Each subspace in an open base of the neighborhood system of $0$ by hypothesis must have finite codimension, so $V$ is indeed weakly topologized.

The next result further elucidates the relationship between linearly and weakly topologized spaces (cf. 1.18).

III.5.3 PROPOSITION. *If* $(\Delta,V,\tau)$ *is linearly topologized, then the set* $U$ *of all continuous linear forms on* $V$ *is a* t-*subspace of* $V^*$. *Furthermore,* $\tau_U(V) \subseteq \tau$ *and the equality holds if and only if* $\tau$ *is a weak topology.*

PROOF. By 1.17, $U$ is a subspace of $V^*$. We show next that $U$ is a t-subspace of $V^*$. Let $\mathfrak{B}$ be an open base of the neighborhood system of $0$ consisting of subspaces of $V$. Since $\tau$ is Hausdorff, we infer that $\bigcap_{B\in\mathfrak{B}} B = 0$. Hence for any $0 \neq x \in V$, there exists $B \in \mathfrak{B}$ such that $x \notin B$. Let $D$ be a basis of $V$ containing $x$ and whose intersection with $B$ forms a basis of $B$. Define a function $f$ on $D$ by letting: $vf = 1$ if $v \in D\backslash B$, $vf = 0$ if $v \in D \cap B$, and extend $f$ linearly to all of $V$. For any $\sigma \in \Delta$, we obtain

$$\begin{aligned}\sigma f^{-1} &= \{v \in V \mid vf = \sigma\} \\ &= \{y+z \mid y \in B,\ z \in [D\backslash B],\ zf = \sigma\} \\ &= B+\{z \in [D\backslash B] \mid zf = \sigma\} = \bigcup_{\substack{z \in [D\backslash B] \\ zf=\sigma}} (B+z)\end{aligned}$$

which is a union of open sets and is thus open. Consequently $f \in U$ with $xf = 1$ and thus $U$ is a t-subspace of $V^*$. With this notation, $f_u = u$ for any $u \in U$, so $f_u$ is continuous and 1.16 yields $\tau_U(V) \subseteq \tau$. If $\tau = \tau_U(V)$, then 5.2 implies that $\tau$ is a weak topology. Conversely, let $\tau$ be a weak topology. By 5.2, 1.21 and 1.18, we have $\tau = \tau_{U'}(V)$ where $U'$ is the vector space of all continuous linear forms on $V$. Since $U$ also has this property, we have $U = U'$ and thus $\tau = \tau_{U'}(V) = \tau_U(V)$ as required.

Note that 2.12 can be rephrased thus: if $V$ is weakly topologized and $S$ is a closed subspace of $V$, then $V/S$ is also weakly topologized. The next result asserts that the corresponding statement also holds for linearly topologized modules.

III.5.4 PROPOSITION. *If $N$ is a closed submodule of a linearly topologized module $M$, then $M/N$ is linearly topologized in the quotient topology.*

PROOF. Let $\{K_\alpha\}_{\alpha\in A}$ be an open base for the neighborhood system of $0$ consisting of submodules and let $\pi: M \to M/N$ be the canonical homomorphism. Then for every $\alpha \in A$, the set $K_\alpha\pi = K_\alpha + N$ is a union of open sets $\bigcup_{n\in N} (K_\alpha + n)$ and is thus open which makes $K_\alpha\pi$ an open submodule of $M/N$. Let $m \in M$ be such that $m+N \neq N$. Then $m \notin N$ and since $N$ is closed, there exists $K_\beta$ such that $(m+K_\beta) \cap N = \emptyset$ so that $m \notin K_\beta + N$. But then $m\pi \notin K_\beta\pi$ which proves that $\bigcap_{\alpha\in A} K_\alpha\pi = 0$ in $M/N$.

Let $P$ be an open neighborhood of $0$ in $M/N$. Then $P\pi^{-1}$ is an open neighborhood of $0$ in $M$ and thus $K_\beta \subseteq P\pi^{-1}$ for some $\beta \in A$. Since $N \subseteq P\pi^{-1}$, it follows that $K_\beta + N \subseteq P\pi^{-1}$ and hence $K_\beta\pi \subseteq P$ which proves that $\{K_\alpha\pi\}_{\alpha\in A}$ is an open base for the neighborhood system of $0$ in $M/N$ consisting of submodules.

III.5.5 DEFINITION. If $N$ is a submodule and $m$ is an element of a module $M$, then the set $m+N$ is a *linear variety*, or simply a *subvariety*, of $M$.

A family $\mathfrak{F}$ of sets has the *finite intersection property* if the intersection of any finite number of members in $\mathfrak{F}$ is nonempty. We now come to a fundamental concept, due to Lefschetz [1]; it represents a weakening of the notion of compactness.

III.5.6 DEFINITION. A linearly topologized module $M$ is *linearly compact* if every family of closed subvarieties which has the finite intersection property has nonempty intersection.

III.5.7 PROPOSITION. *If N is a closed submodule of a linearly compact module M, then M/N is linearly compact in the quotient topology.*

PROOF. By 5.4, M/N is linearly topologized. Let $\{C_\alpha\}_{\alpha\in A}$ be a family of closed subvarieties of M/N having the finite intersection property and let $\pi: M \to M/N$ be the canonical homomorphism. Then $\{C_\alpha\pi^{-1}\}_{\alpha\in A}$ is a family of closed subvarieties of M having the finite intersection property and thus has a nonempty intersection, say C. Thus

$$\emptyset \neq C\pi = \left(\bigcap_{\alpha\in A} C_\alpha\pi^{-1}\right)\pi \subseteq \bigcap_{\alpha\in A} C_\alpha\pi^{-1}\pi = \bigcap_{\alpha\in A} C_\alpha$$

and hence M/N is linearly compact.

III.5.8 PROPOSITION. *A discrete linearly compact vector space is finite dimensional.*

PROOF. Let V be a discrete and infinite dimensional vector space. Let B be a basis of V and for every $x \in B$, let $S_x$ be the subspace of V generated by $B\setminus\{x\}$. For any $x_1, x_2, \ldots, x_n \in B$, we obtain

$$(x_1 + x_2 + \ldots + x_{i-1}) + x_i + (x_{i+1} + \ldots + x_n) \in x_i + S_{x_i}$$

so $\bigcap_{i=1}^{n} (x_i + S_{x_i}) \neq \emptyset$. Since V is discrete, the family $\{x + S_x\}_{x\in B}$ consists of closed subvarieties and has the finite intersection property. Let $y = \sum_{i=1}^{n} \alpha_i x_i \in x + S_x$ where $x_i, x \in B$. Then $\sum_{i=1}^{n} \alpha_i x_i = x + \sum_{j=1}^{m} \beta_j z_j$ so that $x = \sum_{i=1}^{n} \alpha_i x_i - \sum_{j=1}^{m} \beta_j z_j$ where $z_j \in B$ and $x \neq z_j$ for $1 \leq j \leq m$. By linear independence, we must have $x = x_i$ for some i. It follows that $y \notin x + S_x$ for all $x \neq x_i$ for $1 \leq i \leq n$ and thus $\bigcap_{x\in B} (x + S_x) = \emptyset$. Consequently V is not linearly compact.

A family $\mathfrak{F}$ of subsets of a uniform space $(X, \mathcal{U})$ is said to *contain small sets* if for each $\lambda \in \mathcal{U}$ there exists $F \in \mathfrak{F}$ such that $F \subseteq \lambda[x]$ for some $x \in X$. The next result is due to Dieudonné [5].

III.5.9 THEOREM. *Every linearly compact module is complete.*

PROOF. In view of Kelley ([1], p. 193), it suffices to show that every family of closed sets which has the finite intersection property and contains small sets has nonempty intersection. Hence let $\mathfrak{F}$ be a family of subsets of a linearly compact module M satisfying these requirements and let $\mathcal{K}$ be an open base for the neighborhood system of 0 consisting of submodules of M. For every $K \in \mathcal{K}$, there exists $F \in \mathfrak{F}$ such that $F \subseteq \overline{K}[x]$ for some $x \in M$, where $\overline{K} = \{(y,z) \mid y - z \in K\}$. It follows that $F \subseteq \{y \in M \mid x - y \in K\}$ which, we claim, is equivalent to $x \in \bigcap_{f\in F} (f + K)$. For if $F \subseteq \{y \in M \mid x - y \in K\}$, then for any $f \in F$, $x - f \in K$ so $x \in f + K$ and thus $x \in \bigcap_{f\in F} (f + K)$. Conversely, if $x \in \bigcap_{f\in F} (f + K)$, then $x \in f + K$

for every $f \in F$, so $x - f \in K$ and thus $f \in \{y \in M \mid x - y \in K\}$ for every $f \in F$. Let

$$\mathcal{G} = \{F + K \mid F \in \mathfrak{F},\ K \in \mathcal{K},\ \bigcap_{f \in F} (f + K) \neq \emptyset\}.$$

We have just seen that for every $K \in \mathcal{K}$, there exists $F \in \mathfrak{F}$ such that $F + K \in \mathcal{G}$. Let $F + K \in \mathcal{G}$ and $x \in \bigcap_{f \in F} (f + K)$. Then for any $f \in F$, we have $x \in f + K$ and thus $x + K \subseteq F + K$. If $f \in F$, then the hypothesis implies that $x = f + k'$ for some $k' \in K$. For any $k \in K$, we then have

$$f + k = (x - k') + k = x + (k - k') \in x + K$$

so that $F + K \subseteq x + K$. Therefore $F + K = x + K$. By hypothesis $K$ is open and since it is a submodule it must also be closed. It follow that $F + K = x + K$ is a closed subvariety. If $F_i + K_i \in \mathcal{G}$ for $i = 1, 2, \ldots, n$, then

$$\bigcap_{i=1}^{n} (F_i + K_i) \supseteq \bigcap_{i=1}^{n} F_i \neq \emptyset$$

since $\mathfrak{F}$ has the finite intersection property. Therefore $\mathcal{G}$ is a family of closed subvarieties having the finite intersection property which by linear compactness implies that the family $\mathcal{G}$ has nonempty intersection; let $x \in \bigcap_{G \in \mathcal{G}} G$.

Suppose $x \notin \bigcap_{F \in \mathfrak{F}} F$. Then there exists $F \in \mathfrak{F}$ such that $x \notin F$. Since $F$ is closed, there exists $K \in \mathcal{K}$ such that $(x + K) \cap F = \emptyset$. For this $K$ there exists $H \in \mathfrak{F}$ such that $H + K \in \mathcal{G}$, as we have seen above. Since $x \in H + K$, we have $x = h + k$ for some $h \in H$, $k \in K$. Further $H + K \in \mathcal{G}$ implies the existence of $z \in M$ such that $z - m \in K$ for all $m \in H$. Now let $g \in H$. In particular, we have $z - h, z - g \in K$ and thus $(z - h) - (z - g) \in K$ since $K$ is a submodule. Thus $g - h \in K$ which implies $g \in h + K = h + k + K = x + K$ proving that $H \subseteq x + K$. Consequently $H \cap F \subseteq (x + K) \cap F = \emptyset$, contradicting the hypothesis that $\mathfrak{F}$ has the finite intersection property. Therefore $x \in \bigcap_{F \in \mathfrak{F}} F$.

III.5.10 <u>LEMMA</u>. <u>If</u> $V$ <u>is a vector space, then</u> $V^*$ <u>is linearly compact in the finite topology</u>.

<u>PROOF</u>. An open base for the neighborhood system of $0$ consists of the sets $\{B_{(v_1, v_2, \ldots, v_n; 0, 0, \ldots, 0)} \mid v_i \in V\}$ each of which is a subspace of $V^*$ and thus $V^*$ is linearly topologized.

Let $\mathfrak{B}$ be a family of closed subvarieties of $V^*$ having the finite intersection property. Then each element of $\mathfrak{B}$ is of the form $f + T$, where $T$ is a closed subspace of $V^*$. Letting $S = T^{\perp}$, we obtain $f + T = f + T^{\perp\perp} = f + S^{\perp}$. Hence $\mathfrak{B} = \{f_\alpha + S_\alpha^{\perp}\}_{\alpha \in A}$, where each $S_\alpha$ is a subspace of $V$. Let $W$ be the subspace of $V$ generated by all the $S_\alpha$. Let $x_{\alpha_i} \in S_{\alpha_i}$, $i = 1, 2, \ldots, m$ and $x_{\beta_j} \in S_{\beta_j}$, $j = 1, 2, \ldots, n$ with $x_{\alpha_1} + x_{\alpha_2} + \ldots + x_{\alpha_m} = x_{\beta_1} + x_{\beta_2} + \ldots + x_{\beta_n}$. Let

$$g \in [\bigcap_{i=1}^{m} (f_{\alpha_i} + S_{\alpha_i}^{\perp})] \cap [\bigcap_{j=1}^{n} (f_{\beta_j} + S_{\beta_j}^{\perp})].$$

It follows that $g \in f_{\alpha_i} + S_{\alpha_i}^{\perp}$, whence $g - f_{\alpha_i} \in S_{\alpha_i}^{\perp}$ so that $g|_{S_{\alpha_i}} = f_{\alpha_i}|_{S_{\alpha_i}}$, and similarly $g|_{S_{\beta_j}} = f_{\beta_j}|_{S_{\beta_j}}$. Consequently

$$x = x_{\alpha_1} f_{\alpha_1} + \dots + x_{\alpha_m} f_{\alpha_m} = x_{\alpha_1} g + \dots + x_{\alpha_m} g$$

$$= (x_{\alpha_1} + \dots + x_{\alpha_m}) g = (x_{\beta_1} + \dots + x_{\beta_m}) g = x_{\beta_1} f_{\beta_1} + \dots + x_{\beta_m} f_{\beta_m}.$$

It follows that the function $f$ defined on $W$ by

$$(x_{\alpha_1} + \dots + x_{\alpha_m}) f = x_{\alpha_1} f_{\alpha_1} + \dots + x_{\alpha_m} f_{\alpha_m} \quad \text{if} \quad x_{\alpha_i} \in S_{\alpha_i}$$

is single valued. Further, we obtain

$$[\gamma(x_{\alpha_1} + \dots + x_{\alpha_m}) + \delta(x_{\beta_1} + \dots + x_{\beta_n})] f$$

$$= (\gamma x_{\alpha_1}) f_{\alpha_1} + \dots + (\gamma x_{\alpha_m}) f_{\alpha_m} + (\delta x_{\beta_1}) f_{\beta_1} + \dots + (\delta x_{\beta_n}) f_{\beta_n}$$

$$= \gamma(x_{\alpha_1} f_{\alpha_1} + \dots + x_{\alpha_m} f_{\alpha_m}) + \delta(x_{\beta_1} f_{\beta_1} + \dots + x_{\beta_n} f_{\beta_n})$$

$$= \gamma[(x_{\alpha_1} + \dots + x_{\alpha_m}) f] + \delta[(x_{\beta_1} + \dots + x_{\beta_n}) f]$$

and hence $f$ is linear. Let $f$ also denote any linear extension of $f$ to all of $V$. Then $f \in V^*$ and for every $\alpha \in A$, $f|_{S_\alpha} = f_\alpha|_{S_\alpha}$ so $f - f_\alpha \in S_\alpha^{\perp}$, that is $f \in f_\alpha + S_\alpha^{\perp}$. Consequently $f \in \bigcap_{\alpha \in A} (f_\alpha + S_\alpha^{\perp})$.

We may now put together some of the results on linearly compact vector spaces as follows.

III.5.11 THEOREM. The following conditions on a topological vector space $(\Delta, V, \tau)$ are equivalent.

i) $V$ is linearly compact.

ii) $V$ is weakly topologized and complete.

iii) $V$ is algebraically and topologically isomorphic to $U^*$ with the finite topology, where $U$ is some right vector space over $\Delta$.

PROOF. i) ⇒ ii). Let $\mathcal{K}$ be an open base for the neighborhood system of 0 consisting of subspaces. For any $K \in \mathcal{K}$, in the quotient topology of $V/K$, we have that $V/K$ is discrete since $K$ is open, and that $V/K$ is linearly topologized since $K$ is also closed. But then 5.8 asserts that $V/K$ must be finite dimensional, so every $K \in \mathcal{K}$ has finite codimension and thus $V$ is weakly topologized The completeness of $V$ follows from 5.9.

ii) $\Rightarrow$ iii). By 5.2, $\tau = \tau_U(V)$ for some t-subspace $U$ of $V^*$. Now the completion of $(V,\tau)$ is $(f,(U^*,\phi))$ as in 4.2 which by our hypothesis yields that $f$ maps $V$ onto $U^*$ and provides the needed algebraic and topological isomorphism.

iii) $\Rightarrow$ i). This follows immediately from 5.10.

III.5.12 EXERCISES.

i) Let $V$ and $V'$ be weakly topologized vector spaces. Show that every continuous semilinear transformation of $V$ into $V'$ is uniformly continuous.

ii) Show that in any vector space $V$, the finite topology of $V^*$ is a weak topology. Also show that the space of all continuous linear forms on $V^*$ coincides with the canonical image of $V$ in $V^{**}$ and is dense in $V^{**}$, where $V^{**}$ is endowed with the finite topology.

For information on linearly compact vector spaces, consult Dieudonné [5], Köthe ([1], §10); see also Fischer and Gross [1] and the Appendix.

## III.6 A TOPOLOGY FOR $\mathcal{L}_U(V)$

The topology we have in mind here is the finite topology of $V^V$ relativized to $\mathcal{L}_U(V)$. We consider a few simple properties of this topology particularly relative to the ring structure of $\mathcal{L}_U(V)$.

III.6.1 DEFINITION. For a pair of dual vector spaces $(U,V)$, the finite topology of $V^V$ relativized to $\mathcal{L}_U(V)$ is the finite topology $\phi_U(V)$ of $\mathcal{L}_U(V)$.

From the observations at the beginning of Section 1, we see that a base for the finite topology of $\mathcal{L}_U(V)$ is given by the sets

$$B_{(x_1,x_2,\ldots,x_n;y_1,y_2,\ldots,y_n)} = \{a \in \mathcal{L}_U(V) \mid x_i a = y_i,\ i = 1,2,\ldots,n\},$$

where as in 1.5, the set $\{x_1,x_2,\ldots,x_n\}$ can be taken to be linearly independent. In particular the sets $B_{(x;y)}$, with $x \neq 0$, form a subbase and

$$B_{(x_1,x_2,\ldots,x_n;y_1,\ldots,y_n)} = \bigcap_{i=1}^{n} B_{(x_i;y_i)}.$$

An open base for the neighborhood system of $a \in \mathcal{L}_U(V)$ is thus given by the sets

$$B_{(x_1,x_2,\ldots,x_n)}(a) = \{b \in \mathcal{L}_U(V) \mid x_i a = x_i b,\ i = 1,2,\ldots,n\},$$

and for $a = 0$, we have

$$B_{(x_1,x_2,\ldots,x_n)}(0) = \{b \in \mathcal{L}_U(V) \mid x_i b = 0,\ i = 1,2,\ldots,n\},$$

which shows that an open base of the neighborhood system of $0$ is given by the set of annihilators of all finite dimensional subspaces of $V$. Another way of expressing the above neighborhood system can be obtained as follows.

III.6.2 PROPOSITION. *Let* $a \in \mathfrak{L}_U(V)$ *and* $c = \sum_{k=1}^{n} [i_k, \gamma_k, \lambda_k] \in \mathfrak{F}_U(V)$, *where* $n$ *is minimal. Then* $B_{(v_{\lambda_1}, v_{\lambda_k}, \ldots, v_{\lambda_n})}(a) = \{b \in \mathfrak{L}_U(V) \mid ca = cb\}$.

PROOF. First note that $u_{i_1}, u_{i_2}, \ldots, u_{i_n}$ are linearly independent by I.2.5. For any $b \in \mathfrak{L}_U(V)$, we obtain

$$b \in B_{(v_{\lambda_1}, v_{\lambda_2}, \ldots, v_{\lambda_n})}(a) \Leftrightarrow v_{\lambda_i} b = v_{\lambda_i} a \quad \text{for} \quad i = 1, 2, \ldots, n$$

$$\Leftrightarrow \sum_{k=1}^{n} u_{i_k} \gamma_k(v_{\lambda_k}(b - a), u) = 0 \quad \text{for all} \quad u \in U$$

$$\Leftrightarrow \sum_{k=1}^{n} u_{i_k} \gamma_k(v_{\lambda_k}, (v - a)^* u) = 0 \quad \text{for all} \quad u \in U$$

$$\Leftrightarrow c^*(b - a)^* = 0 \Leftrightarrow c(b - a) = 0 \Leftrightarrow cb = ca.$$

In particular, for $a = 0$, we have $B_{(v_{\lambda_1}, v_{\lambda_2}, \ldots, v_{\lambda_n})}(0) = \mathfrak{A}_r(c)$. It follows that the right annihilators of elements of $\mathfrak{F}_U(V)$ form an open base for the neighborhood system of $0$. This gives an intrinsic characterization of the topology $\phi_U(V)$. We introduce some more notation. For $a \in \mathfrak{L}_U(V)$ and $c \in \mathfrak{F}_U(V)$, we let $\mathfrak{n}_c(a) = \{b \in \mathfrak{L}_U(V) \mid ca = cb\}$. Now 6.2 yields

$$\mathfrak{n}_c(a) = B_{(v_{\lambda_1}, v_{\lambda_2}, \ldots, v_{\lambda_n})}(a)$$

if $\{v_{\lambda_1}, v_{\lambda_2}, \ldots, v_{\lambda_n}\}$ is a basis for $Vc$, $\mathfrak{n}_c(0) = \mathfrak{A}_r(c)$.

A ring $\mathfrak{R}$ provided with a topology $\tau$ is a topological ring if the additive group of $\mathfrak{R}$ is a topological group and the multiplication is jointly continuous, to be denoted by $(\mathfrak{R}, \tau)$. We also require that $\tau$ be Hausdorff.

III.6.3 LEMMA. $(\mathfrak{L}_U(V), \phi_U(V))$ *is a topological ring.*

PROOF. Let $a, b \in \mathfrak{L}_U(V)$, $c \in \mathfrak{F}_U(V)$ and consider the basic open set $\mathfrak{n}_c(ab)$. For $d \in \mathfrak{n}_c(a)$ and $e \in \mathfrak{n}_{ca}(b)$, we obtain

$$c(de) = (cd)e = (ca)e = (ca)b = c(ab)$$

which shows that $de \in \mathfrak{n}_c(ab)$. Hence $\mathfrak{n}_c(a)\mathfrak{n}_{ca}(b) \subseteq \mathfrak{n}_c(ab)$ so that the multiplication is jointly continuous. A similar argument shows that $\mathfrak{n}_c(a) - \mathfrak{n}_c(b) \subseteq \mathfrak{n}_c(a - b)$, proving that the additive group of $\mathfrak{L}_U(V)$ is a topological group. Suppose next that $a \neq b$. Then $a - b$ is a nonzero element of $\mathfrak{L}_U(V)$ which by I.3.3 implies the existence of $c \in \mathfrak{F}_U(V)$ such that $c(a - b) \neq 0$. Hence $ca \neq cb$; if $d \in \mathfrak{n}_c(a) \cap \mathfrak{n}_c(b)$, then $ca = cd = cb$, a contradiction. Thus $\mathfrak{n}_c(a) \cap \mathfrak{n}_c(b) = \emptyset$ and the topology $\phi_U(V)$ is Hausdorff.

III.6.4 LEMMA. *In* $\mathfrak{L}_U(V)$, *we have*

i) $\mathfrak{A}_\ell(\mathfrak{I}_T(V)) = \mathfrak{I}_U(T^\perp)$ *if* $T$ *is a subspace of* $U$,

ii) $\mathfrak{A}_r(\mathfrak{I}_U(S)) = \mathfrak{I}_{S^\perp}(V)$ *if* $S$ *is a subspace of* $V$.

PROOF. We will prove i); the proof of ii) is left as an exercise. We may suppose that $U$ is a t-subspace of $V^*$. Item i) follows from the following sequence of equivalent statements

$$a \in \mathfrak{A}_\ell(\mathfrak{I}_T(V)) \Leftrightarrow a\mathfrak{I}_T(V) = 0 \Leftrightarrow ab = 0 \quad \text{for all} \quad b \in \mathfrak{I}_T(V)$$

$$\Leftrightarrow (bU^* \subseteq T \quad \text{implies} \quad ab = 0) \Leftrightarrow Va \subseteq T^\perp.$$

Only the last equivalence needs a proof. Suppose $bU^* \subseteq T$ implies $ab = 0$. Let $u_i \in T$ and $v_\lambda \in V$ be such that $(v_\lambda, u_i) = \gamma \neq 0$ and let $b = [i, \gamma^{-1}, \lambda]$. Then $b \in \mathfrak{L}_U(V)$, $b^*U \subseteq T$ and $b^*u_i = u_i$. Hence for any $v \in V$ using the hypothesis, we obtain

$$(va, u_i) = (va, b^*u_i) = (v, a^*b^*u_i) = (v, (ab)^*u_i) = (v, 0) = 0$$

and since $u_i \in T$ is arbitrary, we have $(va, t) = 0$ for all $t \in T$ so that $Va \subseteq T^\perp$. Conversely, suppose that $Va \subseteq T^\perp$. Then for any $b \in \mathfrak{L}_U(V)$ for which $b^*U \subseteq T$, we immediately obtain for any $u \in U$, $v \in V$, $(v(ab), u) = (va, b^*u) = 0$ since $b^*u \in T$ and $Va \subseteq T^\perp$. Thus $v(ab) = 0$ for all $v \in V$ and hence $ab = 0$.

III.6.5 COROLLARY. *For any* $c \in \mathfrak{F}_U(V)$ *and* $\{v_1, v_2, \ldots, v_n\}$ *a basis of* $Vc$, *we have in* $\mathfrak{L}_U(V)$,

$$\mathfrak{I}_{(Vc)^\perp}(V) = \mathfrak{A}_r(\mathfrak{I}_U(Vc)) = B_{(v_1, v_2, \ldots, v_n)}(0) = \mathfrak{n}_c(0) = \mathfrak{A}_r(c).$$

Note that in view of 2.7, the family of all open subspaces of $V$ constitutes a base for the neighborhood system of 0; an analogous situation occurs in $\mathfrak{L}_U(V)$, viz.

III.6.6 COROLLARY. *The family* $\{\mathfrak{I}_T(V) \mid T$ *open subspace of* $U\}$ *of right ideals of* $\mathfrak{L}_U(V)$ *constitutes an open base for the neighborhood system of* 0 *for the topology* $\phi_U(V)$.

PROOF. Interchanging the roles of $U$ and $V$ is 2.7, we infer that a subspace $T$ of $U$ is open if and only if $T = S^\perp$ for some finite dimensional subspace $S$ of $V$. For each such $S$ there exists $c \in \mathfrak{F}_U(V)$ such that $Vc = S$ as follows from I.2.3-2.5. Hence by 6.5, the family in the present corollary coincides with the family $\mathfrak{A}_r(c)$ where $c$ varies over $\mathfrak{F}_U(V)$, which by the above discussion has the required properties.

III.6.7 EXERCISES.

i) Find necessary and sufficient conditions in order that $\phi_U(V)$ be discrete.

ii) For a dual pair $(U,V)$ and finite dimensional subspaces $T$ of $U$ and $S$ of $V$, what can be said about the topology $\phi_U(V)$ relativized to $\mathfrak{I}_T(V)$ and $\mathfrak{I}_U(S)$, respectively?

iii) Let $(U,V)$ satisfy all the conditions of a dual pair except that the bilinear form may be degenerate. What are necessary and sufficient conditions on the bilinear form in order that $\phi_U(V)$ be Hausdorff.

iv) Let $(U,V)$ be a dual pair. Show that for any subspace $T$ of $U$, we have $T^{\perp} = \bigcap_{a \in \mathfrak{I}_T(V)} N_a$.

v) Show that $(\mathfrak{L}_U(V), \phi_U(V))$ has no proper closed ideals.

For the origin of the results of this section as well as for further discussion, see Jacobson [5],([6], Chapter IX, Section 6),([7], Chapter IV, Section 18).

## III.7 A TOPOLOGY FOR $\mathfrak{F}_U(V)$

The topology in question is the topology $\phi_U(V)$ relativized to $\mathfrak{F}_U(V)$ and henceforth will not be mentioned explicitly. We study here a few topological properties of one-sided ideals of $\mathfrak{F}_U(V)$. The closure of a subset $A$ of $\mathfrak{F}_U(V)$ or $U$ will be denoted by $\bar{A}$. The next two propositions are due to Dieudonné [2], see also Jacobson [5].

III.7.1 <u>PROPOSITION</u>. <u>Every left ideal of $\mathfrak{F}_U(V)$ is closed</u>.

<u>PROOF</u>. Let $a \in \overline{\mathfrak{I}_U(S)}$ and $c \in \mathfrak{F}_U(V)$. Then $h_c(a) \cap \mathfrak{I}_U(S) \neq \emptyset$ and hence there exists $b \in \mathfrak{I}_U(S)$ such that $cb = ca$. Hence $Vb \subseteq S$ so that $Vca = Vcb \subseteq Vb \subseteq S$. Consequently $Vca \subseteq S$ for all $c \in \mathfrak{F}_U(V)$. Let $\gamma v_\lambda \in V$; then for some $u_i \in U$, we have $(v_\lambda, u_i) = \sigma \neq 0$ so that

$$(\gamma v_\lambda)a = \gamma v_\lambda [i, \sigma^{-1}, \lambda] a \in S$$

and thus $a \in \mathfrak{I}_U(S)$. Hence $\overline{\mathfrak{I}_U(S)} \subseteq \mathfrak{I}_U(S)$.

III.7.2 <u>PROPOSITION</u>. <u>For every right ideal</u> $R$ <u>of</u> $\mathfrak{F} = \mathfrak{F}_U(V)$, <u>we have</u> $\bar{R} = \mathfrak{A}_{r,\mathfrak{F}} \mathfrak{A}_{\ell,\mathfrak{F}}(R)$.

<u>PROOF</u>. First note that $\mathfrak{A}_{r,\mathfrak{F}} \mathfrak{A}_{\ell,\mathfrak{F}}(\mathfrak{I}_T(V)) = \mathfrak{I}_{T^{\perp\perp}}(V) = \mathfrak{I}_{\bar{T}}(V)$ by 6.4 and 2.2. Let $a \in \mathfrak{I}_{\bar{T}}(V)$ and $x_1, x_2, \ldots, x_n$ be linearly independent vectors in $V$. Then $a^*U \subseteq \bar{T}$ and letting $S = [x_1, x_2, \ldots, x_n]$, for any $u \in U$, we have $(a^*u + S^{\perp}) \cap T \neq \emptyset$. Further, we can write $a = \sum_{k=1}^{m} [i_k, \gamma_k, \lambda_k]$ with $m$ minimal. It follows that

$$\left(\sum_{k=1}^{m} u_{i_k} \gamma_k (v_{\lambda_k}, u) + S^{\perp}\right) \cap T \neq \emptyset \qquad (u \in U).$$

By I.2.3, there exist $u_1', u_2', \ldots, u_m' \in U$ such that $(v_{\lambda_i}, u_j') = \delta_{ij}$ for $1 \leq i, j \leq m$

since $v_{\lambda_1}, v_{\lambda_2}, \dots, v_{\lambda_m}$ are linearly independent by I.2.5. Computing $a^*u_k'$, we infer the existence of $g_k \in S^\perp$ such that $t_k = u_{i_k}\gamma_k + g_k \in T$. Letting $t_k = u_{j_k}\tau_k$ and $b = \sum_{k=1}^{m} [j_k, \tau_k, \lambda_k]$, we obtain

$$b^*u = \sum_{k=1}^{m} u_{j_k}\tau_k(v_{\lambda_k}, u) = \sum_{k=1}^{m} t_k(v_{\lambda_k}, u) \in T$$

and

$$x_p b = \sum_{k=1}^{m} (x_p, u_{j_k})\tau_k v_{\lambda_k} = \sum_{k=1}^{m} (x_p, t_k) v_{\lambda_k}$$

$$= \sum_{k=1}^{m} (x_p, u_{i_k})\gamma_k v_{\lambda_k} = x_p a$$

for $1 \le p \le n$ so that

$$b \in B_{(x_1, x_2, \dots, x_n)}(a) \cap \mathfrak{J}_T(V) \tag{1}$$

and thus $a \in \overline{\mathfrak{J}_T(V)}$.

Conversely, let $a \in \overline{\mathfrak{J}_T(V)}$. Let $u \in U$, $x_1, x_2, \dots, x_n$ be linearly independent vectors in $V$ and $S = [x_1, x_2, \dots, x_n]$. There exists $b$ satisfying (1), so $x_k b = x_k a$ for $1 \le k \le n$ and $b^*U \subseteq T$. Let $t = b^*u$; then $t = a^*u + (b - a)^*u$, where

$$(x_k, (b - a)^*u) = (x_k(b - a), u) = (0, u) = 0.$$

Thus $t \in (a^*u + S^\perp) \cap T$ which proves that $a^*u \in \overline{T}$ and hence $a^*U \subseteq \overline{T}$, that is $a \in \mathfrak{J}_{\overline{T}}(V)$.

III.7.3 <u>COROLLARY</u>. *For any subspace* $T$ *of* $U$, *we have*

$$\overline{\mathfrak{J}_T(V)} = \mathfrak{J}_{\overline{T}}(V) = \mathfrak{J}_{T^{\perp\perp}}(V) = \mathfrak{A}_{r,\mathfrak{F}}\mathfrak{A}_{\ell,\mathfrak{F}}(\mathfrak{J}_T(V)).$$

It is clear that the set $\mathfrak{A}_{r,\mathfrak{R}}(A)$, where $A$ is a nonempty subset of a ring $\mathfrak{R}$, is always a right ideal, and that $\mathfrak{A}_{r,\mathfrak{R}}(A) = \mathfrak{A}_{r,\mathfrak{R}}(L)$, where $L$ is the left ideal of $\mathfrak{R}$ generated by $A$.

III.7.4 <u>COROLLARY</u> (cf. 2.5). *The following conditions on a subspace* $T$ *of* $U$ *are equivalent.*

i) $T$ *is closed.*

ii) $\mathfrak{J}_T(V)$ *is closed*

iii) $\mathfrak{J}_T(V)$ *is a right annihilator.*

<u>PROOF</u>. Exercise.

This corollary implies that the lattice isomorphism $\chi$ in I.3.10 maps closed subspaces of $U$ onto closed right ideals of $\mathfrak{F}_U(V)$. A slight variation of this corollary is valid for open subspaces, see 7.10, exercise iv).

III.7.5 <u>COROLLARY</u>. *Every principal right ideal of* $\mathfrak{F}_U(V)$ *is closed.*

PROOF. Exercise.

An important kind of right ideal is provided by the following.

III.7.6 DEFINITION. Let $R$ be a right ideal of a ring $\mathfrak{R}$. An element $a \in \mathfrak{R}$ is a modular left identity of $\mathfrak{R}$ relative to $R$ if $r - ar \in R$ for all $r \in \mathfrak{R}$; in such a case $R$ is a modular right ideal of $\mathfrak{R}$

The next three results are new.

III.7.7 PROPOSITION. A right ideal $\mathfrak{I}_T(V)$ of $\mathfrak{F}_U(V)$ is modular if and only if $T$ contains an open subspace.

PROOF. Necessity. By hypothesis $r - ar \in \mathfrak{I}_T(V)$ for some $a \in \mathfrak{F}_U(V)$ and all $r \in \mathfrak{F}_U(V)$. Then $a = \sum_{k=1}^{n} [i_k,\gamma_k,\lambda_k]$, where $n$ is minimal. Let $u \in (Va)^{\perp}$ and $u = u_t\sigma$. For every $v \in V$, we have $(va,u) = 0$ and thus

$$0 = (va,u_t) = (\sum_{k=1}^{n} (v,u_{i_k})\gamma_k v_{\lambda_k},u_t) = \sum_{k=1}^{n} (v,u_{i_k})\gamma_k (v_{\lambda_k},u_t).$$

There exist $v_1,v_2,\ldots,v_n \in V$ such that $(v_j,u_{i_k}) = \delta_{jk}$ for $1 \leq j,k \leq n$ by I.2.3, and hence the last equation implies that $(v_{\lambda_k},u_t) = 0$ for $k = 1,2,\ldots,n$. Let $\gamma$ and $v_\lambda$ be such that $\gamma(v_\lambda,u_t) = \sigma$. Then

$$\begin{aligned} u = u_t\sigma &= u_t\gamma(v_\lambda,u_t) \\ &= [t,\gamma,\lambda]u_t - (\sum_{k=1}^{n} [i_k,\gamma_k,\lambda_k])[t,\gamma,\lambda]u_t \in T \end{aligned}$$

by hypothesis. Consequently $(Va)^{\perp} \subseteq T$, where $\dim Va < \infty$ and thus $(Va)^{\perp}$ is open.

Sufficiency. Since the subspaces $S^{\perp}$, where $\dim S < \infty$, form an open base for the neighborhood system of $0$, the hypothesis implies that $S^{\perp} \subseteq T$ for some finite dimensional subspace $S$ of $V$. There exists an idempotent $e \in \mathfrak{F}_U(V)$ such that $Ve = S$ (e.g., use I.2.3). Thus $(Ve)^{\perp} \subseteq T$, that is $(ve,u) = 0$ for all $v \in V$ implies $u \in T$. For any $v \in V$, $u \in U$ and $r \in \mathfrak{F}_U(V)$, we obtain

$$\begin{aligned} (ve,(r - er)^*u) &= (ve,r^*u - e^*r^*u) = (ve,r^*u) - (ve,e^*r^*u) \\ &= (ve,r^*u) - (ve,r^*u) = 0 \end{aligned}$$

so that $(r - er)^*U \subseteq T$ and therefore $r - er \in \mathfrak{I}_T(V)$.

III.7.8 COROLLARY. $\mathfrak{F}_U(V)$ has an idempotent modular left identity relative to every modular right ideal.

PROOF. This follows from the last part of the proof of 7.7.

III.7.9 THEOREM (cf. 2.9). The following conditions on a subspace $T$ of $U$ are equivalent.

i) $T$ is a closed hyperplane.

ii) $\mathfrak{I}_T(V)$ <u>is a closed maximal right ideal</u>.

iii) $\mathfrak{I}_T(V)$ <u>is a modular maximal right ideal</u>.

<u>PROOF</u>. i) $\Leftrightarrow$ ii). This follows from 7.4 and I.3.4.

i) $\Rightarrow$ iii). This follows from 7.7 in view of 2.9.

iii) $\Rightarrow$ i). Let $e = \sum_{k=1}^{n} [i_k,\gamma_k,\lambda_k] \in \mathfrak{F}_U(V)$ be such that $r - er \in \mathfrak{I}_T(V)$ for all $r \in \mathfrak{F}_U(V)$. Let $u_i \in U\setminus T$ and $r = [j,\delta,\nu]$. Since $T$ is evidently a hyperplane, it follows that $u_j = t + u_i\tau$ for some $t \in T$ and $\tau \in \Delta$. Hence

$$\begin{aligned}(r - er)^*u &= ([j,\delta,\nu] - \sum_{k=1}^{n} [i_k,\gamma_k(v_{\lambda_k},u_j)\delta,\nu])^*u \\ &= (u_j - \sum_{k=1}^{n} u_{i_k}\gamma_k(v_{\lambda_k},u_j))\delta(v_\nu,u) \\ &= \{t - \sum_{k=1}^{n} u_{i_k}\gamma_k(v_{\lambda_k},t) - [\sum_{k=1}^{n} u_{i_k}\gamma_k(v_{\lambda_k},u_i) - u_i]\tau\}\delta(v_\nu,u) \in T\end{aligned}$$

for all $t \in T$, $\tau \in \Delta$, $u \in U$. Consequently

$$\sum_{k=1}^{n} u_{i_k}\gamma_k(v_{\lambda_k},t) + [\sum_{k=1}^{n} u_{i_k}\gamma_k(v_{\lambda_k},u_i) - u_i]\tau \in T$$

for all $t \in T$, $\tau \in \Delta$. For $t = 0$ and $\tau \neq 0$, we obtain

$$\sum_{k=1}^{n} u_{i_k}\gamma_k(v_{\lambda_k},u_i) - u_i \in T \tag{1}$$

and for $\tau = 0$,

$$\sum_{k=1}^{n} u_{i_k}\gamma_k(v_{\lambda_k},t) \in T \qquad (t \in T). \tag{2}$$

By 1.14, we have $T = g^\perp$ for some $g \in U^*$ relative to the dual pair $(U,U^*)$. Hence (1) yields $g(\sum_{k=1}^{n} u_{i_k}\gamma_k(v_{\lambda_k},u_i) - u_i) = 0$ whence

$$gu_i = \sum_{k=1}^{n} (gu_{i_k})\gamma_k(v_{\lambda_k},u_i), \tag{3}$$

and by (2), we obtain

$$\sum_{k=1}^{n} (gu_{i_k})\gamma_k(v_{\lambda_k},t) = 0. \tag{4}$$

Now suppose that $(v_{\lambda_k},u) = 0$ for $k = 1,2,\ldots,n$. Then $u = t + u_i\tau$ which by (3) and (4) gives

$$\begin{aligned}gu &= g(t + u_i\tau) = gt + (gu_i)\tau = (gu_i)\tau \\ &= \sum_{k=1}^{n} (gu_{i_k})\gamma_k(v_{\lambda_k},u_i\tau) = \sum_{k=1}^{n} (gu_{i_k})\gamma_k(v_{\lambda_k},u - t) \\ &= \sum_{k=1}^{n} (gu_{i_k})\gamma_k(v_{\lambda_k},u) = 0.\end{aligned}$$

It follows that for $S = [f_{v_{\lambda_1}}, f_{v_{\lambda_2}}, \ldots, f_{v_{\lambda_n}}]$, we have $S^{\perp} \subseteq g^{\perp}$ so that

$$g \in [g] = g^{\perp\perp} \subseteq S^{\perp\perp} = S$$

by 2.3. Consequently $g \in \text{nat } V$ and thus $g = f_v$ for some $v \in V$. But then $T = g^{\perp} = v^{\perp}$ and hence $T$ is open since $v^{\perp} = f_v^{-1}0$, and thus $T$ is also closed.

III.7.10 **EXERCISES**.

i) Show that a maximal right ideal of $\mathfrak{F}_U(V)$ is either closed or dense.

ii) Prove that every right ideal of $\mathfrak{F}(V)$ is a right annihilator if and only if $V$ is finite dimensional.

iii) Show that for any right ideal $R$ of $\mathfrak{F}_U(V)$, we have

$$\overline{R} = \{a \in \mathfrak{F}_U(V) \mid \bigcap_{r \in R} N_r \subseteq N_a\}.$$

iv) For a dual pair $(U,V)$, prove that the following conditions on a subspace $T$ of $U$ are equivalent.

a) $T$ is open.

b) $\mathfrak{I}_T(V)$ is open.

c) $\mathfrak{I}_T(V)$ is the right annihilator of some element of $\mathfrak{F}_U(V)$.

v) Show that in $\mathfrak{F}(V)$, provided with the finite topology, every left ideal is a left annihilator.

vi) In any ring $\mathfrak{R}$ (not necessarily with identity), write

$$(1 - a)\mathfrak{R} = \{r - ar \mid r \in \mathfrak{R}\}$$

for any $a \in \mathfrak{R}$. Show that for $e = e^2$, we have $(1 - e)\mathfrak{R} = \mathfrak{A}_r(e)$. Also show that for $e = e^2 \in \mathfrak{F}_U(V)$, we have $(1 - e)\mathfrak{F}_U(V) = \mathfrak{I}_{(1-e)^*U}(V)$ where $(1 - e)^*U = \{u - e^*u \mid u \in U\}$ and that $e$ is a modular left identity for $\mathfrak{F}_U(V)$ relative to $\mathfrak{I}_T(V)$ if and only if $(Ve)^{\perp} \subseteq T$.

vii) For any $a \in \mathfrak{F}_U(V)$ show that $\mathfrak{A}_r(a) = \mathfrak{I}_{(Va)^{\perp}}(V)$.

viii) Let $(U,V)$ be a dual pair and $T$ be a subspace of $U$. Show that $\text{codim } T < \infty$ if and only if $\mathfrak{I}_T(V) = \mathfrak{A}_r(a)$ for some $a \in \mathfrak{F}_U(V)$.

The principal references for this section are Behrens ([1], Chapter II, Section 7), Dieudonné [2], Jacobson [5],([6], Chapter IX, Section 8),([7], Chapter I, Section 3 and Chapter IV, Section 18).

## III.8 ANOTHER TOPOLOGY FOR $\mathfrak{L}_U(V)$

For a set $\mathfrak{F}$ of functions on a set $X$ into a uniform space $(Y,\mathfrak{B})$, for every $\beta \in \mathfrak{B}$, we let

$$W(\beta) = \{(f,g) \in \mathfrak{F} \times \mathfrak{F} \mid (f(x),g(x)) \in \beta \text{ for all } x \in X\}.$$

Then the family $\{W(\beta) \mid \beta \in \mathfrak{B}\}$ is a base for a uniformity $\mathcal{W}$ on $\mathfrak{F}$ called the <u>uniformity of uniform convergence</u>; the corresponding topology is the <u>topology of uniform convergence</u>.

We now apply this construction to the uniformity associated with $\tau_U(V)$ and to the set $\mathfrak{L}_U(V)$ of functions on $V$. Specifically we proceed as follows.

1. An open base for the neighborhood system of $0$ in $V$ is given by the family $\{T^\perp \mid T$ is a subspace of $U$, $\dim T < \infty\}$ so that a set $A \subseteq V$ is open if and only if for every $v \in A$ there exists a finite dimensional subspace $T$ of $U$ such that $v + T^\perp \subseteq A$.

2. The uniformity $\mathcal{U}$ on $V$ induced by $\tau_U(V)$ has for a base the family

$$\{\mathcal{U}(A) \mid A \text{ is a } \tau_U(V)\text{-neighborhood of } 0\}$$

where $\mathcal{U}(A) = \{(x,y) \mid x - y \in A\}$. Consequently $\mathcal{U}$ consists of all binary relations $\beta$ on $V$ for which there exists a neighborhood $A$ of $0$ such that $\mathcal{U}(A) \subseteq \beta$.

3. According to the above construction, the uniformity $\mathcal{W}$ of uniform convergence on $\mathfrak{L}_U(V)$ induced by $\mathcal{U}$ has as a base the family $\{W(\beta) \mid \beta \in \mathcal{U}\}$, where for $\beta = \mathcal{U}(A)$,

$$W(\mathcal{U}(A)) = \{(a,b) \in \mathfrak{L}_U(V) \times \mathfrak{L}_U(V) \mid V(a-b) \subseteq A\}.$$

4. For any neighborhood $A$ of $0$ and $a \in \mathfrak{L}_U(V)$, we have

$$W(\mathcal{U}(A))[a] = \{b \in \mathfrak{L}_U(V) \mid V(a-b) \subseteq A\}$$

and thus the topology of uniform convergence on $\mathfrak{L}_U(V)$ (simply <u>uniform topology</u>) consists of those sets $C$ with the property that for every $a \in C$, there exists a neighborhood $A$ of $0$ such that $W(\mathcal{U}(A))[a] \subseteq C$.

For every finite dimensional subspace $T$ of $U$, let

$$\hat{T} = \{a \in \mathfrak{L}_U(V) \mid Va \subseteq T^\perp\}.$$

III.8.1 <u>PROPOSITION</u>. <u>The family $\{\hat{T} \mid T$ is a subspace of $U$, $\dim T < \infty\}$ forms an open base of the neighborhood system of $0$ for the uniform topology on $\mathfrak{L}_U(V)$.</u>

<u>PROOF</u>. Let $T$ be a finite dimensional subspace of $U$. For any $a \in \hat{T}$, we have $Va \subseteq T^\perp$ so that

$$\begin{aligned} W(\mathcal{U}(T^\perp))[a] &= \{b \in \mathfrak{L}_U(V) \mid V(a-b) \subseteq T^\perp\} \\ &= \{b \in \mathfrak{L}_U(V) \mid -vb \in -va + T^\perp \text{ for all } v \in V\} \\ &= \{b \in \mathfrak{L}_U(V) \mid Vb \subseteq T^\perp\} = \hat{T} \end{aligned}$$

and thus $\hat{T}$ is open. Next let $C$ be an open set in the uniform topology. Then for every $a \in C$, there exists a neighborhood $A$ of $0$ such that $W(\mathcal{U}(A))[a] \subseteq C$. Since $A$ is a neighborhood of $0$, there exists a finite dimensional subspace $T$ of

$U$ such that $T^{\perp} \subseteq A$. Hence

$$a+\hat{T} = \{a+b \mid b \in \mathcal{L}_U(V),\ Vb \subseteq T^{\perp}\} = \{c \in \mathcal{L}_U(V) \mid V(c-a) \subseteq T^{\perp}\}$$
$$= \{c \in \mathcal{L}_U(V) \mid V(a-c) \subseteq T^{\perp}\} = W(\mathcal{U}(T^{\perp}))[a] \subseteq C$$

as required.

III.8.2 <u>COROLLARY</u>. <u>The family of sets</u>

$$P_{(u_1,u_2,\ldots,u_n)}(a) = \{b \in \mathcal{L}_U(V) \mid b^*u_k = a^*u_k \ \underline{\text{for}}\ k = 1,2,\ldots,n\}$$

<u>where</u> $\{u_1,u_2,\ldots,u_n\}$ <u>ranges over all linearly independent finite subsets of</u> $U$, <u>constitutes an open base for the neighborhood system of</u> $a \in \mathcal{L}_U(V)$ <u>in the uniform topology</u>.

<u>PROOF</u>. It suffices to show that $a+\hat{T} = P_{(u_1,u_2,\ldots,u_n)}(a)$, where $\{u_1,u_2,\ldots,u_n\}$ is a basis of $T$. We have

$$b \in a+\hat{T} \Leftrightarrow b = a+c,\ vc \in T^{\perp} \text{ for some } c \in \mathcal{L}_U(V) \text{ and all } v \in V$$
$$\Leftrightarrow (v(b-a),u_k) = 0 \text{ for all } v \in V \text{ and } k = 1,2,\ldots,n$$
$$\Leftrightarrow (v,(b^*-a^*)u_k) = 0 \text{ for all } v \in V \text{ and } k = 1,2,\ldots,n$$
$$\Leftrightarrow b^*u_k = a^*u_k \text{ for } k = 1,2,\ldots,n \Leftrightarrow b \in P_{(u_1,u_2,\ldots,u_n)}(a).$$

For $a \in \mathcal{L}_U(V)$ and $c \in \mathfrak{F}_U(V)$, let $\mathfrak{m}_c(a) = \{b \in \mathcal{L}_U(V) \mid bc = ac\}$.

III.8.3 <u>PROPOSITION</u>. <u>For</u> $a \in \mathcal{L}_U(V)$ <u>and</u> $c = \sum_{k=1}^{n} [i_k,\gamma_k,\lambda_k] \in \mathfrak{F}_U(V)$, <u>where</u> $n$ <u>is minimal, we have</u> $\mathfrak{m}_c(a) = P_{(u_{i_1},u_{i_2},\ldots,u_{i_n})}(a)$.

<u>PROOF</u>. Note that $v_{\lambda_1},v_{\lambda_2},\ldots,v_{\lambda_n}$ are linearly independent by I.2.5. For any $b \in \mathcal{L}_U(V)$, we obtain

$$b \in P_{(u_{i_1},u_{i_2},\ldots,u_{i_n})}(a) \Leftrightarrow b^*u_k = a^*u_k \text{ for } k = 1,2,\ldots,n$$
$$\Leftrightarrow \sum_{k=1}^{n} (v,(b^*-a^*)u_k)\gamma_k v_{\lambda_k} = 0 \text{ for all } v \in V$$
$$\Leftrightarrow \sum_{k=1}^{n} (v(b-a),u_k)\gamma_k v_{\lambda_k} = 0 \text{ for all } v \in V$$
$$\Leftrightarrow v(b-a)c = 0 \text{ for all } v \in V \Leftrightarrow (b-a)c = 0$$
$$\Leftrightarrow bc = ac \Leftrightarrow b \in \mathfrak{m}_c(a).$$

In the special case $a = 0$, we have

$$P_{(u_{i_1},u_{i_2},\ldots,u_{i_n})}(0) = \mathfrak{m}_c(0) = \mathfrak{A}_\ell(c).$$

Consequently the uniform topology on $\mathcal{L}_U(V)$ has an open base for the neighborhood system of $0$ consisting of left annihilators of elements of $\mathfrak{F}_U(V)$. Since the corresponding statement for the finite topology for $\mathfrak{L}_U(V)$ is valid for right annihilators of elements of $\mathfrak{F}_U(V)$, it follows that by the left-right duality, we immediately obtain all the statements for the uniform topology from the corresponding ones for the finite topology. This also shows that the uniform topology is intrinsic and can be defined simply as a topology on a primitive ring with a nonzero socle having an open base for the neighborhood system of 0 consisting of left annihilators of elements of the socle. It is interesting that these two topologies on $\mathfrak{L}_U(V)$ have been defined in quite different ways, but, in the end, they turned out to be symmetric relative to the left-right duality. In view of these remarks, it is not surprising that the following connection between the two is valid.

III.8.4 PROPOSITION. The mapping $a \to a^*$ is a uniform isomorphism and hence a homeomorphism of $\mathfrak{L}_U(V)$ with the finite topology onto $\mathfrak{L}_V(U)$ with the uniform topology induced by the V-topology on U.

PROOF. We know that the mapping $\varphi: a \to a^*$ is an isomorphism of $\mathfrak{L}_U(V)$ onto $\mathfrak{L}_V(U)$, see I.1.2. Further, for $a \in \mathfrak{L}_U(V)$ and $c \in \mathfrak{F}_U(V)$, we have

$$\begin{aligned} \mathfrak{n}_c(a)\varphi &= \{b \in \mathfrak{L}_U(V) \mid cb = ca\}\varphi \\ &= \{b^* \in \mathfrak{L}_V(U) \mid c^*b^* = c^*a^*\} = \mathfrak{m}_{c^*}(a^*) \end{aligned}$$

which proves that $\varphi$ is also a homeomorphism. Furthermore, we have

$$\begin{aligned} \mathcal{U}(\mathfrak{U}_r(c))\varphi &= \{(x,y) \mid x - y \in \mathfrak{U}_r(c)\}\varphi = \{(x\varphi, y\varphi) \mid x - y \in \mathfrak{U}_r(c)\} \\ &= \{(x\varphi, y\varphi) \mid x\varphi - y\varphi \in \mathfrak{U}_r(c\varphi)\} = \mathcal{U}(\mathfrak{U}_r(c\varphi)) \end{aligned}$$

which shows that $\varphi$ is also a uniform isomorphism.

Some of the results in this section can be found in Jacobson ([7], Chapter IV, Section 18), others are new.

## III.9 COMPLETE PRIMITIVE RINGS

If we consider a primitive ring $\mathfrak{R}$ with a nonzero socle as a ring of linear transformations, then $\mathfrak{F}_U(V) \subseteq \mathfrak{R} \subseteq \mathfrak{L}_U(V)$ for some t-subspace U of $V^*$. Hence the two topologies on $\mathfrak{L}_U(V)$ induce two topologies on $\mathfrak{R}$, and since they are necessarily uniform topologies for some uniformities, we may ask the following questions: (1) What are the completions of the uniform spaces induced by these topologies? (2) What are necessary and sufficient conditions for completeness of these uniform spaces? We consider first the finite topology on $\mathfrak{L}_U(V)$ and using the answer to the first question, reply to the second.

III.9.1 LEMMA. If $\mathfrak{L}(V)$ is given the finite topology, then a subset S of $\mathfrak{L}(V)$ is dense if and only if S is n-fold transitive for every positive integer n.

PROOF. This follows easily from the description of a base of the finite topology in Section 6.

Note that 9.1 says that the "density" defined in I.2.9 coincides with the density in the finite topology of $\mathfrak{L}(V)$.

III.9.2 PROPOSITION. $\mathfrak{L}(V)$ is complete in the finite topology.

PROOF. The reasoning here is quite analogous to that in 4.1; indeed, one considers $\mathfrak{L}(V)$ as a subset of $V^V$, where $V$ has the discrete topology, instead of $V^*$ as a subset of $\Delta^V$, where $\Delta$ has the discrete topology. The details of the proof are left as an exercise.

The next result is due to Dieudonné [2].

III.9.3 COROLLARY. Let $\mathfrak{R}$ be a ring such that $\mathfrak{F}_U(V) \subseteq \mathfrak{R} \subseteq \mathfrak{L}_U(V)$ for some t-subspace $U$ of $V^*$. Then

i) (identity map, $(\mathfrak{L}(V),\mathfrak{f})$) is the completion of $\mathfrak{R}$ in the finite topology.

ii) $(a \to a^*, (\mathfrak{L}(U),\mathfrak{f}))$ is the completion of $\mathfrak{R}$ in the uniform topology.

PROOF. i) By 9.1, $\mathfrak{R}$ is dense in $(\mathfrak{L}(V),\mathfrak{f})$ which is complete by 9.2.

ii) The mapping $\varphi: a \to a^*$ $(a \in \mathfrak{L}_U(V))$ is an algebraic and uniform isomorphism of $\mathfrak{L}_U(V)$ with the uniform topology onto $\mathfrak{L}_V(U)$ with the finite topology by the dual of 8.4. The dual of 9.2 shows that $\mathfrak{L}(U)$ is complete in the finite topology. On the other hand, we have $\mathfrak{F}_V(U) \subseteq \mathfrak{R}\varphi \subseteq \mathfrak{L}_V(U)$, so by 9.1, $\mathfrak{R}\varphi$ is dense in $\mathfrak{L}(U)$. Hence $(\varphi,(\mathfrak{L}(U),\mathfrak{f}))$, must be a completion of $\mathfrak{R}$ in the uniform topology. Both $\mathfrak{R}$ and $(\mathfrak{L}(U),\mathfrak{f})$ are Hausdorff, and thus this completion is unique up to a uniform isomorphism by Kelley ([1], p. 197).

We have seen in Sections 6 and 8 that both topologies on $\mathfrak{R}$ as in 9.3 are intrinsic. In view of this, it makes sense to introduce the following two topologies directly on an abstract ring.

III.9.4 DEFINITION. Let $\mathfrak{R}$ be a primitive ring with a nonzero socle $\mathfrak{S}$. The topology on $\mathfrak{R}$ having as an open base for the neighborhood system of 0 the family $\{\mathfrak{U}_r(c) \mid c \in \mathfrak{S}\}$ is the right socle topology for $\mathfrak{R}$; dually, the topology on $\mathfrak{R}$ having as an open base for the neighborhood system of 0 the family $\{\mathfrak{U}_\ell(c) \mid c \in \mathfrak{S}\}$ is the left socle topology for $\mathfrak{R}$.

In light of the results in Sections 6 and 8, the canonical isomorphism $\mathfrak{R} \to \mathfrak{L}(V)$ is a uniform isomorphism, and thus also a homeomorphism of $\mathfrak{R}$, with the right socle topology, onto a subring of $\mathfrak{L}_U(V)$ containing $\mathfrak{F}_U(V)$, with the finite topology, as well as of $\mathfrak{R}$ with the left socle topology onto the same subring with the uniform topology. The above corollary thus yields

III.9.5 COROLLARY. Let $\mathfrak{R}$ be a primitive ring with a nonzero socle, and let $\psi:\mathfrak{R} \to \mathfrak{L}_U(V)$ be the canonical isomorphism. Then

i) $(\psi,(\mathfrak{L}(V),\mathfrak{j}))$ is the completion of $\mathfrak{R}$ in the right socle topology,

ii) $(\psi(a \to a^*),(\mathfrak{L}(U,\mathfrak{j})))$ is the completion of $\mathfrak{R}$ in the left socle topology.

III.9.6 COROLLARY. A primitive ring $\mathfrak{R}$ with a nonzero socle is complete in the right (respectively left) socle topology if and only if $\mathfrak{R}$ is isomorphic to $\mathfrak{L}(V)$ for some left (respectively right) vector space V. Every algebraic isomorphism of $\mathfrak{R}$ onto $\mathfrak{L}(V)$ is uniform and hence a homeomorphism.

PROOF. The first statement follows immediately from the first (respectively second) part of 9.5. Note that the bracketed statement can also be proved directly by mapping $\mathfrak{R}$ into $\mathfrak{L}_V(U)$, for we know that in view of the hypothesis, $\mathfrak{R}$ is also a left primitive ring with minimal left ideals. Any isomorphism $\psi$ of $\mathfrak{R}$ onto $\mathfrak{L}(V)$ carries $\mathfrak{A}_r(c)$ (respectively $\mathfrak{A}_\ell(c)$) onto $\mathfrak{A}_r(c\psi)$ (respectively $\mathfrak{A}_\ell(c\psi)$) for any element c in the socle of $\mathfrak{R}$ and thus carries the uniform neighborhoods in $\mathfrak{R}$ onto the uniform neighborhoods in $\mathfrak{L}(V)$. Hence $\psi$ must be a uniform isomorphism and thus also a homeomorphism.

We will now derive a topology for a semiprime ring which is an essential extension of its nonzero socle. The procedure is somewhat similar to that for prime rings with a nonzero socle whose topology was derived from its representation as linear transformations on a vector space. We will now go via a complete direct sum.

III.9.7 LEMMA. Let $\{\mathfrak{R}_\alpha\}_{\alpha \in A}$ be a family of prime rings with a nonzero socle each endowed with the right socle topology. Then the ring $\mathfrak{R} = \prod_{\alpha \in A} \mathfrak{R}_\alpha$ endowed with the product topology has the family $\{\mathfrak{A}_r(c) \mid c \in \mathfrak{S}\}$ as an open base for the neighborhood system of 0, where $\mathfrak{S}$ is the socle of $\mathfrak{R}$.

PROOF. Elements of $\mathfrak{R}$ will be written in the form $a = (a_\alpha)$. Let $c \in \mathfrak{S}$ and $a \in \mathfrak{R}$. Then there exist $\alpha_1,\alpha_2,\ldots,\alpha_n \in A$ such that $c_{\alpha_i} \in \mathfrak{S}_{\alpha_i}$ for $i = 1,2,\ldots,n$ and $c_\beta = 0$ otherwise. For every open set $P_\beta$ in $\mathfrak{R}_\beta$, let $\overline{P}_\beta = \{b \in \mathfrak{R} \mid b_\beta \in P_\beta\}$. By definition the family of sets $\{\overline{P}_\beta \mid \beta \in A,\ P_\beta \text{ open in } \mathfrak{R}_\beta\}$ forms a subbase for the product topology in $\mathfrak{R}$. With the notation just introduced, we assert

$$a + \mathfrak{A}_r(c) = \bigcap_{i=1}^{n} \overline{a_{\alpha_i} + \mathfrak{A}_r(c_{\alpha_i})}. \tag{1}$$

For let $b \in a + \mathfrak{A}_r(c)$. Then $b = a + d$ where $cd = 0$. Further, $b_{\alpha_i} = a_{\alpha_i} + d_{\alpha_i}$ with $c_{\alpha_i} d_{\alpha_i} = 0$ and thus $b_{\alpha_i} \in a_{\alpha_i} + \mathfrak{A}_r(c_{\alpha_i})$. Hence $a + \mathfrak{A}_r(c) \subseteq \overline{a_{\alpha_i} + \mathfrak{A}_r(c_{\alpha_i})}$

and thus also $a+\mathfrak{A}_r(c) \subseteq \bigcap_{i=1}^{n} \overline{a_{\alpha_i}+\mathfrak{A}_r(c_{\alpha_i})}$. Conversely, let $b \in \bigcap_{i=1}^{n} \overline{a_{\alpha_i}+\mathfrak{A}_r(c_{\alpha_i})}$. Then $b_{\alpha_i} = a_{\alpha_i} + d_{\alpha_i}$, where $c_{\alpha_i} d_{\alpha_i} = 0$ for $i = 1,2,\ldots,n$; let $d_\beta = b_\beta - a_\beta$ if $\beta \neq \alpha_i$ for $i = 1,2,\ldots,n$. We obtain $b = a+d$ with $cd = (c_\alpha d_\alpha) = 0$. Consequently $\bigcap_{i=1}^{n} \overline{a_{\alpha_i}+\mathfrak{A}_r(c_{\alpha_i})} \subseteq a+\mathfrak{A}_r(c)$ which establishes (1).

Now if $a \in \overline{P}_\alpha$, where $P_\alpha$ is an open set in $\mathfrak{R}_\alpha$, then $a_\alpha \in P_\alpha$ and there exists $c_\alpha \in \mathfrak{S}_\alpha$ such that $a_\alpha + \mathfrak{A}_r(c_\alpha) \subseteq P_\alpha$. Hence for $c \in \mathfrak{R}$ for which $c_\beta = 0$ if $\beta \neq \alpha$, we obtain by (1), $a+\mathfrak{A}_r(c) = \overline{a_\alpha + \mathfrak{A}_r(c_\alpha)} \subseteq \overline{P}_\alpha$.

Furthermore, (1) shows that each $a+\mathfrak{A}_r(c)$ is open in the product topology. Consequently the family of sets $\{a+\mathfrak{A}_r(c) \mid a \in \mathfrak{R},\ c \in \mathfrak{S}\}$ forms a base for the product topology of $\mathfrak{R}$ and thus the family of sets $\{\mathfrak{A}_r(c) \mid c \in \mathfrak{S}\}$ forms an open base for the neighborhood system of $0$.

Now let $\mathfrak{R}$ be a semiprime ring which is an essential extension of its nonzero socle $\mathfrak{S}$. By II.5.9, there exist prime rings $\mathfrak{R}_\alpha$, $a \in A$, with nonzero socles $\mathfrak{S}_\alpha$ and an isomorphism $\varphi$ mapping $\mathfrak{R}$ into a complete direct sum $\prod_{\alpha \in A} \mathfrak{R}_\alpha$ such that $\mathfrak{R}\varphi$ contains the socle of $\prod_{\alpha \in A} \mathfrak{R}_\alpha$. Giving each $\mathfrak{R}_\alpha$ the right socle topology and $\prod_{\alpha \in A} \mathfrak{R}_\alpha$ the product topology, $\mathfrak{R}$ can be given a topology for which $\varphi$ is a homeomorphism of $\mathfrak{R}$ onto $\mathfrak{R}\varphi$. In particular, $\mathfrak{R}\varphi$ has the relativized product topology and thus has an open base for the neighborhood system of $0$ consisting of right annihilators of the elements of the socle by 9.7. This property carries over to $\mathfrak{R}$ which shows that the topology is intrinsic, i.e., does not depend upon the particular representation $\varphi$. It is clear that we can proceed directly as follows (cf. 9.4).

III.9.8 <u>DEFINITION</u>. Let $\mathfrak{R}$ be a semiprime ring essential extension of its nonzero socle $\mathfrak{S}$. The topology on $\mathfrak{R}$ having as an open base for the neighborhood system of $0$ the family $\{\mathfrak{A}_r(c) \mid c \in \mathfrak{S}\}$ is the <u>right socle topology</u> for $\mathfrak{R}$. The <u>left socle topology</u> is defined dually.

The following result is new.

III.9.9 <u>THEOREM</u>. <u>Let $\mathfrak{R}$ be a semiprime ring which is an essential extension of its nonzero socle $\mathfrak{S}$. Let $\varphi$ be an isomorphism of $\mathfrak{R}$ onto a subring of $\prod_{\alpha \in A} \mathfrak{R}_\alpha$ containing the socle of $\prod_{\alpha \in A} \mathfrak{R}_\alpha$, where each $\mathfrak{R}_\alpha$ is a prime ring with a nonzero socle. For each $\alpha \in A$, let $\psi_\alpha$ be an isomorphism of $\mathfrak{R}_\alpha$ onto a subring $\mathfrak{R}'_\alpha$ of $\mathcal{L}_{U_\alpha}(V_\alpha)$ containing $\mathfrak{F}_{U_\alpha}(V_\alpha)$ for some dual pair $(U_\alpha, V_\alpha)$. Define $\psi$ on $\prod_{\alpha \in A} \mathfrak{R}_\alpha$ by $(a_\alpha)\psi = (a_\alpha \psi_\alpha)$. Let $\tau$ be the right socle topology on $\mathfrak{R}$, give each $\mathcal{L}(V_\alpha)$ the</u>

finite topology and let $\phi$ be the product topology on $\prod_{\alpha\in A} \mathfrak{L}(V_\alpha)$. Then $(\varphi\psi, (\prod_{\alpha\in A} \mathfrak{L}(V_\alpha), \phi))$ is the completion of $(\mathfrak{R}, \tau)$.

PROOF. First note that the existence of $\varphi$ is guaranteed by II.5.9 and that of $\psi$ by II.2.8 and I.2.13. Giving $\mathfrak{R}_\alpha$ the right socle topology, $\psi_\alpha$ becomes a uniform isomorphism onto a dense subspace of $\mathfrak{L}(V_\alpha)$ by 9.5. Further, giving $\prod_{\alpha\in A} \mathfrak{R}_\alpha$ the product topology makes $\varphi$ a uniform isomorphism in view of 9.7 and the discussion following it. By Kelley ([1], p. 183), $\psi$ is a uniform isomorphism of $\prod_{\alpha\in A} \mathfrak{R}_\alpha$ onto a subring of $\prod_{\alpha\in A} \mathfrak{L}(V_\alpha)$. Hence $\varphi\psi$ is a uniform isomorphism of $\mathfrak{R}$ onto a subring of $\prod_{\alpha\in A} \mathfrak{L}(V_\alpha)$. In view of 9.2, each $\mathfrak{L}(V_\alpha)$ is complete so that by Kelley ([1], p. 194), $\prod_{\alpha\in A} \mathfrak{L}(V_\alpha)$ is complete. It is easy to see that $\varphi\psi$ maps the socle of $\mathfrak{R}$ onto the socle of $\prod_{\alpha\in A} \mathfrak{L}(V_\alpha)$. By 9.1 and I.2.13, we know that the socle $\mathfrak{F}_{U_\alpha}(V_\alpha)$ of $\mathfrak{R}_\alpha\psi_\alpha$ is dense in $\mathfrak{L}(V_\alpha)$. Hence if $P_i$ is an open subset of $\mathfrak{L}(V_{\alpha_i})$, then there exists $c_i \in P_i \cap \mathfrak{F}_{U_{\alpha_i}}(V_{\alpha_i})$ for $i = 1,2,\ldots,n$. Since the socle of $\prod_{\alpha\in A} \mathfrak{L}(V_\alpha)$ is a direct sum of the socles of all $\mathfrak{L}(V_\alpha)$, as it easily follows from II.5.8, we have that $c = (c_\alpha)$, where $c_{\alpha_i} = c_i$ for $i = 1,2,\ldots,n$ and $c_\beta = 0$ otherwise, is an element of the socle of $\prod_{\alpha\in A} \mathfrak{L}(V_\alpha)$. With the notation of the proof of 9.7, we have that $c$ is also an element of $\bigcap_{i=1}^{n} \overline{P_i}$. This shows that the socle of $\prod_{\alpha\in A} \mathfrak{L}(V_\alpha)$ is dense in $\prod_{\alpha\in A} \mathfrak{L}(V_\alpha)$ which completes the proof of the theorem.

III.9.10 COROLLARY. A ring $\mathfrak{R}$ as in 9.9 is complete in the right socle topology if and only if $\mathfrak{R}$ is isomorphic to a complete direct sum of the rings of all linear transformations on left vector spaces, say $\prod_{\alpha\in A} \mathfrak{L}(V_\alpha)$. Furthermore, if both $\mathfrak{R}$ and $\prod_{\alpha\in A} \mathfrak{L}(V_\alpha)$ are provided with the topologies as in 9.9, then every algebraic isomorphism of $\mathfrak{R}$ onto $\prod_{\alpha\in A} \mathfrak{L}(V_\alpha)$ is a uniform isomorphism and hence a homeormorphism.

PROOF. Exercise.

III.9.11 EXERCISES.

i) Let $\mathfrak{R}$ be a ring satisfying the hypothesis in 9.9. Show directly that the family $\{\mathfrak{U}_r(c) \mid c \in \mathfrak{S}\}$ satisfies the requirements on a family of sets to be a base for the neighborhood system of 0 for a topology on $\mathfrak{R}$, i.e., that for any $a,b \in \mathfrak{S}$, there exists $c \in \mathfrak{S}$ such that $\mathfrak{U}_r(c) \subseteq \mathfrak{U}_r(a) \cap \mathfrak{U}_r(b)$.

ii) What are necessary and sufficient conditions on a ring $\mathfrak{R}$ satisfying the hypothesis in 9.9 in order that $\mathfrak{R}$ be complete both in the left and the right socle topologies?

iii) Let $\mathfrak{R}$ be a semiprime ring with a nonzero socle $\mathfrak{S}$ and let $\mathfrak{T} = \mathfrak{A}_{r,\mathfrak{R}}(\mathfrak{S})$. Show that the family $\{\mathfrak{A}_r(F) \mid F \subseteq \mathfrak{S}+\mathfrak{T}$ and $F$ is finite$\}$ satisfies the requirements for a base of the neighborhood system of 0. Let $\tau$ denote this topology. Show that $(\mathfrak{R},\tau)$ is a Hausdorff topological ring. If $\mathfrak{R}$ is also an essential extension of $\mathfrak{S}$, how does $\tau$ compare with the right socle topology of $\mathfrak{R}$?

A characterization of $\prod_{\alpha\in A} \mathfrak{L}(V_\alpha)$ as a semisimple linearly compact ring can be found in Leptin ([2], Part I). For more information on complete rings, see Dieudonné [2],[3] and Eckstein [1].

# APPENDIX

## ON LINEARLY COMPACT PRIMITIVE AND SEMISIMPLE RINGS

by

Richard Wiegandt

### 0. INTRODUCTION

In ring theory the celebrated Wedderburn-Artin Structure Theorem is of central significance. It states that a ring $\mathfrak{R}$ with d.c.c. on right ideals is semisimple (in this case semisimple means semiprime) if and only if $\mathfrak{R}$ is a finite direct sum of rings of linear transformations on finite dimensional vector spaces over division rings. It was Leptin [2] who succeeded in eliminating both finiteness conditions from this characterization; he proved that linearly compact semisimple rings are just complete direct sums of rings of linear transformations on vector spaces over division rings. The price he had to pay was topological, and a topologically minded mathematician may object that linear compactness is actually a kind of finiteness conditions. Nevertheless, Leptin's result is in a certain sense the best possible generalization of the Wedderburn-Artin Structure Theorem. The study of topological rings started, in fact, earlier (see the cited papers of Kaplansky, Dieudonné and Zelinsky). In particular, Wedderburn-Artin-like theorems were obtained by Kaplansky [3] and Zelinsky [2], but these theorems are consequences of Leptin's result.

It is surprizing that in the flood of mathematical publications only some dozen papers dealing with linearly compact rings have appeared in the last fifteen years since the publication of Leptin's papers. What is even more surprizing, Leptin's already classic results are not available in any textbook, monograph or even in lecture notes. This strange situation may be due to several reasons, one of them is definitely the fact that the study of linearly compact rings requires certain familiarity both with algebra and topology.

It is the purpose of this appendix to discuss Leptin's results concerning linearly compact semisimple rings and several related topics. The preceding lecture notes by Professor M. Petrich, particularly Part III, provide an excellent and natural background for this aim. Taking into consideration the material presented in the preceding lecture notes, the discussion of linearly compact semisimple rings can be approached easily via linearly compact primitive rings. This method differs from that of Leptin, so there are only a few coincidences with the original proofs. Moreover, our approach seems to be more algebraic and less topological; however, the use of inverse limits was inevitable.

Of course, only fragments of the theory of linearly compact rings can be explained in this short appendix. For further results and recent developments, we give references at the end of this note.

Finally, I express my sincere gratitude to Professor M. Petrich for the opportunity of attaching this appendix to his lecture notes and for his critical remarks concerning this appendix.

## 1. MORE ABOUT PRIMITIVE RINGS

In order to describe the structure of linearly compact primitive rings, we need more extensive knowledge of primitive rings. So far only primitive rings with non-zero socle were considered, in what follows we shall not impose this requirement.

Irreducible and faithful $\mathfrak{R}$-modules as well as primitive rings have already been defined in II.1.20 and II.1.21. Concerning irreducible and faithful $\mathfrak{R}$-modules we have the following simple

1.1 LEMMA. If $M$ is an irreducible faithful $\mathfrak{R}$-module and $0 \neq x \in M$, then $x\mathfrak{R} = M$.

PROOF. Since $x\mathfrak{R} \subseteq M$, the irreducibility of $M$ implies that either $x\mathfrak{R} = 0$ or $x\mathfrak{R} = M$. Suppose $x\mathfrak{R} = 0$ and consider the submodule

$$\langle x \rangle = \{kx \mid k = 0, \pm 1, \pm 2, \dots\}.$$

The irreducibility of $M$ implies $\langle x \rangle = M$, further since $x\mathfrak{R} = 0$, we obtain $0 = \langle x \rangle \mathfrak{R} = M\mathfrak{R}$, contradicting the assumption that $M$ is faithful. Hence $x\mathfrak{R} = M$, as asserted.

1.2 LEMMA. Every primitive ring $\mathfrak{R}$ is isomorphic to a subring of the ring $\mathfrak{L}(\Delta, V)$ of all linear transformations on a vector space $V$ over a division ring $\Delta$. Moreover, if $M$ is a faithful irreducible $\mathfrak{R}$-module, then $M^+$ is a vector space over a suitable division ring $\Delta$.

PROOF. We have to construct the division ring $\Delta$ as well as the vector space $V$.

Since $\mathfrak{R}$ is primitive, it has a faithful irreducible $\mathfrak{R}$-module $M$. The set $\mathfrak{L}(M^+)$ of all endomorphisms of the additive group $M^+$ is obviously a ring. For each element $r \in \mathfrak{R}$, the mapping

$$r^*: x \to xr \qquad (x \in M)$$

is clearly an endomorphism of $M^+$. Consider the subset

$$\mathfrak{R}^* = \{r^* \in \mathfrak{L}(M^+) \mid r \in \mathfrak{R}\}$$

of $\mathfrak{L}(M^+)$. The mapping $\varphi: r \to r^*$ is a homomorphism from $\mathfrak{R}$ onto $\mathfrak{R}^*$, for

$$x(r^* + s^*) = xr^* + xs^* = xr + xs = x(r+s)$$

and

$$x(r^*s^*) = (xr^*)s^* = (xr^*)s = xrs$$

hold for every $x \in M^+$ and $r,s \in \mathfrak{R}$. If $r^* = s^*$, then $xr^* = xs^*$ is valid for every $x \in M$, i.e. $x(r-s) = 0$. Since $M$ is faithful, we have $r = s$. Thus $\varphi$ is an isomorphism.

Next, consider the set

$$\Delta = \{a \in \mathfrak{L}(M^+) \mid ar^* = r^*a \text{ for every } r^* \in \mathfrak{R}\},$$

the so-called <u>centralizer of</u> $\mathfrak{R}$ <u>in</u> $\mathfrak{L}(M^+)$ ($\Delta$ is, in fact, the centralizer of $\mathfrak{R}^*$ in $\mathfrak{L}(M^+)$). We claim that $\Delta$ is a division ring. $\Delta$ is obviously a ring with identity. Let $0 \neq d \in \Delta$. Then there is an $x \in M$ such that $xd \neq 0$ and consequently each $y \in M$ has the form

$$y = xdr^* = xr^*d = (xr)d$$

for some $r \in \mathfrak{R}$ which shows that $d$ maps $M^+$ onto itself. Suppose $zd = 0$ for some $z \in M$. For every element $r^* \in \mathfrak{R}$, we have

$$0 = zdr^* = zr^*d = (zr)d.$$

Hence $d$ maps $z\mathfrak{R}$ onto $0$ and by 1.1 it follows that $z = 0$. Thus $d$ is an isomorphism of $M^+$ onto $M^+$ and consequently $d$ has an inverse in $\mathfrak{L}(M^+)$ which must be, of course, in $\Delta$.

These considerations show that $M^+$ is a vector space over the division ring $\Delta$ and by the isomorphism $\varphi: \mathfrak{R} \to \mathfrak{R}^*$, $\mathfrak{R}$ can be embedded in $\mathfrak{L}(\Delta, M^+)$.

1.3 <u>LEMMA</u>. <u>Let</u> $\mathfrak{R}$ <u>be a primitive ring and suppose that</u> $\mathfrak{R}$ <u>is a subring of</u> $\mathfrak{L}(\Delta, V)$ <u>such that</u> $V$ <u>is a faithful irreducible</u> $\mathfrak{R}$<u>-module</u>. <u>If</u> $W$ <u>is a finite dimensional subspace of</u> $V$ <u>and</u> $\mathfrak{A}_{\mathfrak{R}}(W)$ <u>is the annihilator of</u> $W$ <u>in</u> $\mathfrak{R}$, <u>then</u> $x\mathfrak{A}_{\mathfrak{R}}(W) = V$ <u>for every element</u> $x \in V \backslash W$.

<u>PROOF</u>. The proof is by induction on the dimension $n$ of $W$. If $n = 0$, then $\mathfrak{A}_{\mathfrak{R}}(0) = \mathfrak{R}$ and so for every $x \neq 0$ we have $x\mathfrak{R} = V$, for otherwise $V$ would not be a faithful irreducible $\mathfrak{R}$-module.

Let $n \geq 1$ and assume the validity of the statement for $n-1$. Then $V$ splits into a direct sum $W = U + \{x_n\}$ where $U$ is an $(n-1)$-dimensional vector space. Take an arbitrary element $x \in V \backslash W$. We shall exhibit an element $a \in \mathfrak{A}_{\mathfrak{R}}(W)$ such that $xa \neq 0$ which is evidently equivalent to $\mathfrak{A}_{\mathfrak{R}}(W) \neq \mathfrak{A}_{\mathfrak{R}}(V)$. This will imply that $x\mathfrak{A}_{\mathfrak{R}}(W)$ is a nonzero submodule of $V$ and so it must equal $V$. Assume that such an element $a \in \mathfrak{A}_{\mathfrak{R}}(W)$ does not exist, i.e. $\mathfrak{A}_{\mathfrak{R}}(W) = \mathfrak{A}_{\mathfrak{R}}(V)$. By hypothesis

$x_n\mathfrak{A}_{\mathfrak{R}}(U) = V$, so for every element $v \in V$, there is an $a_v \in \mathfrak{A}_{\mathfrak{R}}(U)$ such that $x_n a_v = v$. If $b_v$ is another element with $x_n b_v = v$, then $x_n(a_v - b_v) = 0$ and thus $a_v - b_v \in \mathfrak{A}_{\mathfrak{R}}(U) \cap \mathfrak{A}_{\mathfrak{R}}(x_n) = \mathfrak{A}_{\mathfrak{R}}(W)$. For any $x \in V\backslash W$, define a mapping $\psi_x: V \to V$ by

$$v\psi_x = xa_v \qquad (v \in V).$$

The element $a_v$ is not necessarily unique, nevertheless the mapping $\psi_x$ is single-valued. Indeed, $a_v - b_v \in \mathfrak{A}_{\mathfrak{R}}(W) = \mathfrak{A}(V)$ implies $xa_v - xb_v = x(a_v - b_v) = 0$, and hence $xa_v = xb_v$ holds for every $x \in V$. Consequently $\psi_x$ is single-valued. We show next that $\psi_x$ is a homomorphism of the $\mathfrak{R}$-module $V$ onto itself. Since $a_{v+w}$ is an element with the property

$$x_n a_{v+w} = v + w = x_n a_v + x_n a_w = x_n(a_v + a_w),$$

we have

$$(v+w)\psi_x = xa_{v+w} = xa_v + xa_w = v\psi_x + w\psi_x.$$

Since $x_n a_{vr} = vr = x_n a_v r$, we obtain

$$(vr)\psi_x = xa_{vr} = xa_v r = (v\psi_x)r.$$

Hence $\psi_x$ is indeed a module homomorphism for each $x \in V\backslash W$. Moreover, $(v\psi_x)r = (vr)\psi_x$ shows that $\psi_x$ belongs to the centralizer $\Delta$. Since for every $a \in \mathfrak{A}_{\mathfrak{R}}(U)$, there is a $v \in V$ such that $x_n a = v$, we can write $a = a_v$. Consequently

$$xa = xa_v = v\psi_x = (x_n a)\psi_x = (x_n\psi_n)a$$

which implies $(x - x_n\psi_x)a = 0$ for every $a \in \mathfrak{A}_{\mathfrak{R}}(U)$. By the induction hypothesis, $y\mathfrak{A}_{\mathfrak{R}}(U) = 0$ implies $y \in U$, and we obtain $x - x_n\psi_x \in U \subseteq W$. Since $x_n\Delta$ is the 1-dimensional subspace $\{x_n\}$ generated by $x_n$, $\psi_x \in \Delta$ implies $x_n\psi_x \in \{x_n\}$. Hence $x \in U + \{x_n\} \in W$, contradicting the hypothesis that $x \in V\backslash W$.

Applying 1.2 and 1.3 we can easily derive Jacobson's Density Theorem for primitive rings.

1.4 <u>THEOREM</u>. <u>Every primitive ring $\mathfrak{R}$ is isomorphic to a dense subring of the ring $\mathfrak{L}(\Delta,V)$ of all linear transformations on a vector space $V$ over a division ring $\Delta$.</u>

<u>PROOF</u>. By 1.2 the ring $\mathfrak{R}$ can be embedded into a ring $\mathfrak{L}(\Delta,V)$ such that $V$ is an irreducible faithful $\mathfrak{R}$-module and $\Delta$ is the centralizer of $\mathfrak{R}$ in $\mathfrak{L}(V)$. Take $n$ linearly independent vectors $x_1,\ldots,x_n \in V$ and arbitrary vectors $y_1,\ldots,y_n \in V$. If $n = 1$, then by $x_1\mathfrak{R} = V$, there exists an $r \in \mathfrak{R}$ such that $x_1 r = y_1$ and hence $\mathfrak{R}$ is transitive. Assume $\mathfrak{R}$ is (n-1)-fold transitive. Then there is an element $r \in \mathfrak{R}$ such that $x_i r = y_i$ for $i = 1,\ldots,n-1$, but $x_n r$ may differ from $y_n$. Consider the annihilator $\mathfrak{A} = \mathfrak{A}_{\mathfrak{R}}(x_1,\ldots,x_n)$. Now, 1.3 is applicable and we obtain $x_n\mathfrak{A} = V$. Hence there exists an element $s \in \mathfrak{A}$ such that $x_n s = y_n$. The linear transformation $r + s \in \mathfrak{L}(V)$ is now the desired one, namely

$x_i(r+s) = y_i$ holds for each $i = 1,\dots,n$. Thus $\mathfrak{R}$ is an n-fold transitive subring of $\mathcal{L}(V)$ for every $n \geq 1$ which means that $\mathfrak{R}$ is dense in $\mathcal{L}(V)$

It is also true that every dense subring of $\mathcal{L}(\Delta,V)$ is primitive, but we shall not use this statement. The proof of 1.4 along with 1.3 is actually a version of the usual one (cf. Divinsky [1] and Kertész [1]). For a discussion of the density theorem, consult Jacobson ([7], Chapter II).

## 2. INVERSE LIMITS AND LINEARLY COMPACT MODULES

In this section we continue the preparations for the study of linearly compact primitive and semisimple rings. We shall prove some elementary statements concerning inverse limits of linearly compact modules, further we shall prove that the complete direct sum of linearly compact modules endowed with the product topology, is linearly compact.

Only Hausdorff topologies will be considered.

Let $A$ be a directed set of indices.

2.1 DEFINITION. A system $\Omega = \{M_\alpha, \pi^\alpha_\beta \mid \alpha > \beta \in A\}$ of topological $\mathfrak{R}$-modules $M_\alpha$ and continuous homomorphisms $\pi^\alpha_\beta : M_\alpha \to M_\beta$ is said to be an inverse system if $\pi^\alpha_\beta \pi^\beta_\gamma = \pi^\alpha_\gamma$ holds for every $\alpha > \beta > \gamma$ and $\pi^\alpha_\alpha$ is the identity mapping on $M_\alpha$ for every $\alpha \in A$. The inverse limit $M = \varprojlim \Omega$ of the inverse system $\Omega$ is the topological submodule $M$ of the complete direct sum $\prod_{\alpha \in A} M_\alpha$ consisting all vectors $(\dots,x_\beta,\dots,x_\alpha,\dots) \in \prod_{\alpha\in A} M_\alpha$ such that $x_\alpha \pi^\alpha_\beta = x_\beta$ for every $\alpha > \beta \in A$ and the topology of $M$ is the product topology relativized to $M$.

2.2 PROPOSITION. The inverse limit $M = \varprojlim \Omega$ is a closed submodule of $\prod_{\alpha \in A} M_\alpha$.

PROOF. We shall show that the set $\prod_{\alpha\in A} M_\alpha \setminus M$ is open. Take an arbitrary element $(\dots,x_\alpha,\dots) \in \prod_{\alpha\in A} M_\alpha \setminus M$. There exist indices $\alpha > \beta$ such that $x_\alpha \pi^\alpha_\beta \neq x_\beta$. Let $K = (\dots,K_\beta,\dots,K_\alpha,\dots)$ be an open subset of $\prod_{\alpha\in A} M_\alpha$ such that $K_\beta$ and $K_\alpha$ are open sets with $x_\beta \in K_\beta$, $x_\alpha \in K_\alpha$ and $K_\alpha \pi^\alpha_\beta \cap K_\beta = \emptyset$ and all the other components $K_\gamma$ equal $M_\gamma$. Observe that the topological modules under consideration are Hausdorff and hence also regular spaces so that such $K_\beta$ does exist. It is clear that $K$ does not contain elements of $M$. Hence $M$ is closed.

Let $M$ be a linearly topologized $\mathfrak{R}$-module and $\underline{K} = \{K_\alpha\}_{\alpha\in A}$ be an open base for the neighborhood system of 0 consisting of submodules. Obviously the $\mathfrak{R}$-modules $M_\alpha = M/K_\alpha$ form an inverse system $\Omega$ together with the homomorphisms

$$\pi_\beta^\alpha : M/K_\alpha \to {}^{M/K_\alpha}/_{K_\beta/K_\alpha} \to M/K_\beta \quad \text{for all} \quad \alpha > \beta \in A.$$

2.3 PROPOSITION (Zelinsky [1]). *M can be embedded into* $\varprojlim \Omega$. *Moreover, if M is complete, then M is isomorphic and homeomorphic to* $\varprojlim \Omega$.

PROOF. Taking the mapping $\varphi$, where $x\varphi = (\ldots, x_\alpha, \ldots)$ with $x \in M$, $x_\alpha \in M_\alpha$, we have a homomorphism from $M$ into $\varprojlim \Omega$. Further, $x\varphi = 0$ implies $x_\alpha = 0$ for each $\alpha \in A$, i.e. $x \in \bigcap_{\alpha \in A} K_\alpha$. Since the topology of $M$ is Hausdorff, we must have $x = 0$ and $\varphi$ is an isomorphism.

Suppose $M$ is complete and consider an arbitrary element

$$y = (\ldots, y_\alpha, \ldots) \in \varprojlim \Omega.$$

Choose elements $x_\alpha \in M$ such that $x_\alpha + K_\alpha = y_\alpha$ for each $\alpha \in A$. Then $\{x_\alpha \mid \alpha \in A\}$ forms a Cauchy net which converges to some element $x_o \in M$ since $M$ is complete. For this element $x_o$ we have $x_o + K_\alpha = y_\alpha$ for each $\alpha \in A$. Hence $\varphi$ maps $M$ onto $\varprojlim \Omega$.

Clearly $\varphi$ is continuous. Take an open submodule $K_\alpha \in \underline{K}$. Since in each $M_\alpha$ the topology is discrete and $K_\beta \subseteq K_\alpha$ if $\beta > \alpha$, the image $K_\alpha\varphi$ of $K_\alpha$ is open in $\varprojlim \Omega$. Thus $\varphi$ is a homeomorphism.

Let $A$ be a set and $F$ be the set of all finite subsets of $A$. Then $F$ is directed by inclusion. Consider the finite direct sums $\prod_{\alpha \in f} M_\alpha$ $(f \in F)$ of topological $\mathfrak{R}$-modules endowed with the product topology. If $g \subseteq f \in F$, then $\prod_{\alpha \in f} M_\alpha$ can be mapped naturally onto $\prod_{\alpha \in g} M_\alpha$ by a projection $\pi_g^f$. Now $\Omega = \{\prod_{\alpha \in f} M_\alpha, \pi_g^f \mid g \subseteq f \in F\}$ forms an inverse system of finite direct sums.

2.4 PROPOSITION. $\varprojlim \Omega$ *is isomorphic and homeomorphic to the complete direct sum* $\prod_{\alpha \in A} M_\alpha$ *endowed with the product topology.*

PROOF. Define a mapping $\varphi : \prod_{\alpha \in A} M_\alpha \to \varprojlim \Omega$ by

$$(\ldots, x_\alpha, \ldots)\varphi = (\ldots, x_f, \ldots)$$

where $x_f = x_{\alpha_1} + \ldots + x_{\alpha_n}$ for the indices $\alpha_1, \ldots, \alpha_n \in f$. Obviously $\varphi$ maps $\prod_{\alpha \in A} M_\alpha$ isomorphically onto $\varprojlim \Omega$ and is continuous. Taking into account that the product topology is the weakest one such that projecting onto the components we get back the topology of the components, it is obvious that $\varphi$ is a homeomorphism.

We continue this section with some elementary properties of linearly compact modules. The following assertion, which will be useful later, provides the opportunity for another definition of linearly compact modules.

2.6 PROPOSITION (Leptin [1]). An $\mathfrak{R}$-module M is linearly compact if and only if it is linearly topologized and satisfies the following condition:

(*) every family of open subvarieties which has the finite intersection property, has nonempty intersection.

PROOF. Since every open submodule in any linear topology is closed, linearly compactness implies the validity of condition (*).

Conversely, suppose that (*) is fulfilled. Take an arbitrary family $\underline{F} = \{a_\alpha + Z_\alpha\}_{\alpha \in A}$ of closed subvarieties having the finite intersection property. Form the family $\underline{G} = \{a_\alpha + Z_\alpha + K_\lambda \mid \alpha \in A, \lambda \in \Lambda\}$, where $\{K_\lambda\}_{\lambda \in \Lambda}$ is an open base for the neighborhood system of 0 consisting of submodules of M. Ovbiously $\underline{G}$ has the finite intersection property and it consists of open subvarieties $a_\alpha + Z_\alpha + K_\lambda$. (As it is easy to verify, a submodule is always open whenever it contains an open submodule of M.) Hence by condition (*), we have

$$0 \neq \bigcap_{\alpha,\lambda} \underline{G} = \bigcap_{\alpha,\lambda} (a_\alpha + Z_\alpha + K_\lambda) = \bigcap_\alpha (a_\alpha + Z_\alpha) = \bigcap_\alpha \underline{F}$$

and M is linearly compact.

2.7 PROPOSITION. If N is a submodule of a linearly topologized $\mathfrak{R}$-module M and N is linearly compact in the relativized topology, then N is closed in M.

PROOF. Let $\underline{K} = \{K_\alpha\}_{\alpha \in A}$ be an open base for the neighborhood system of 0 in M such that each $K_\alpha$ is a submodule of M. Take an element $a \in \overline{N}$. Now for each $\alpha$ the intersection $(a + K_\alpha) \cap N$ is nonempty. Take also an element $a_\alpha \in (a + K_\alpha) \cap N$. Clearly $a_\alpha + (K_\alpha \cap N) \subseteq (a + K_\alpha) \cap N$. For any element $b_\alpha \in (a + K_\alpha) \cap N$, we have $b_\alpha - a_\alpha \in K_\alpha \cap N$ and so $b_\alpha \in a_\alpha + (K_\alpha \cap N)$. Hence also $(a + K_\alpha) \cap N \subseteq a_\alpha + (K_\alpha \cap N)$. Moreover,

$$\underline{F} = \{a_\alpha + (K_\alpha \cap N) \mid \alpha \in A\} = \{(a + K_\alpha) \cap N \mid \alpha \in A\}$$

is a family of open subvarieties of N with the finite intersection property. Since N is linearly compact, the intersection $\cap \underline{F}$ has an element in N. Hence

$$\bigcap_{\alpha \in A} (a_\alpha + (K_\alpha \cap N)) = \bigcap_{\alpha \in A} ((a + K_\alpha) \cap N) \subseteq \bigcap_{\alpha \in A} (a + K_\alpha) = a$$

implies that $a \in N$ and thus $N = \overline{N}$.

2.8 PROPOSITION. Let $\varphi$ be a continuous homomorphism of a linearly compact $\mathfrak{R}$-module M into a linearly topologized $\mathfrak{R}$-module N. Then $M\varphi$ is linearly compact in the relativized topology.

PROOF. The relativized topology of $M\varphi$ is clearly linear. If $\underline{K} = \{a_\alpha + K_\alpha\}_{\alpha \in A}$ is any family of closed subvarieties of $M\varphi$ with the finite intersection property, then $\underline{K}\varphi^{-1} = \{(a_\alpha + K_\alpha)\varphi^{-1}\}_{\alpha \in A}$ is such a family in M. Since M is linearly compact, $\underline{K}\varphi^{-1}$ has a nonempty intersection. Consequently also $\underline{K}$ has a nonempty intersection.

2.9 PROPOSITION. If $A$ and $B$ are linearly compact submodules of a linearly topologized $\mathfrak{R}$-module $M$, then $A + B$ is closed in $M$.

PROOF. By 2.7, $A$ and $B$ are closed submodules of $\mathfrak{R}$ and so the natural homomorphism $\varphi: M \to M/A$ is continuous. In view of 2.8, $B\varphi$ is linearly compact in $M/A$ and thus again by 2.7, $B\varphi$ is closed. Hence $B\varphi\varphi^{-1} = A + B$ is closed in $M$.

2.10 PROPOSITION. The complete direct sum $\prod_{\alpha \in A} M_\alpha$ of linearly compact $\mathfrak{R}$-modules $M_\alpha$ is linearly compact in the product topology.

PROOF. In view of 2.6, it is sufficient to consider only families of open subvarieties with finite intersection property. Let $\underline{F} = \{a_\lambda + K_\lambda\}_{\lambda \in \Lambda}$ be such a family. Each $K_\lambda$ has the form $K_\lambda = \prod_{\alpha \in A} K_{\alpha\lambda}$ where only a finite number of $K_{\alpha\lambda}$ differ from $M_\alpha$. Project $a_\lambda + K_\lambda$ onto the $\alpha$-th component for each $\lambda: a_\lambda + K_\lambda \to a_{\alpha\lambda} + K_{\alpha\lambda}$. Thus we have a family $\underline{F}_\alpha = \{a_{\alpha\lambda} + K_{\alpha\lambda}\}_{\lambda \in \Lambda}$ of open subvarieties of $M_\alpha$. Since $M_\alpha$ is linearly compact, $\bigcap_\lambda \underline{F}_\alpha$ is nonempty for each $\alpha$. Let $a_\alpha \in \bigcap_\lambda \underline{F}_\alpha$ and take the element $a_o = (\ldots, a_\alpha, \ldots) \in \prod_{\alpha \in A} M_\alpha$. Then $a_o$ is obviously contained in each $a_\lambda + \prod_{\alpha \in A} K_{\alpha\lambda}$ since only a finite number of $K_{\alpha\lambda}$'s differ from $M_\alpha$ and $\underline{F}$ has the finite intersection property. Hence $a_o \in \bigcap \underline{F}$ proving that $\prod_{\alpha \in A} M_\alpha$ is linearly compact.

## 3. LINEARLY COMPACT PRIMITIVE RINGS

We now turn to structure theorems for linearly compact primitive rings.

3.1 PROPOSITION. If $\mathfrak{R}$ is a primitive ring endowed with a linearly compact topology $\tau$, then there exists an irreducible faithful $\mathfrak{R}$-module $M'$ which is a continuous image of $\mathfrak{R}$. Moreover $M' \cong \mathfrak{R}/K$, where $K$ is an open maximal right ideal of $\mathfrak{R}$.

PROOF. Since $\mathfrak{R}$ is primitive, there is an irreducible faithful $\mathfrak{R}$-module $M$. For any element $0 \neq m \in M$, we have $m\mathfrak{R} = M$ and hence there exists an element $e \in \mathfrak{R}$ such that $me = m$. Consider the set

$$J = \{y = x - ex \mid x \in \mathfrak{R}\}.$$

Then $J$ is obviously a right ideal of $\mathfrak{R}$. The mapping $x\varphi = x - ex$ maps $\mathfrak{R}$ onto $J$ homomorphically and continuously, and $\varphi$ relativized to $J$ is the identical mapping. As it is easy to verify, the topology induced by $\varphi$ on $J$ is just the relativized topology. Hence $J$ is linearly compact and, by 2.7, closed in $\mathfrak{R}$.

Take an open right ideal $K$ of $\mathfrak{R}$ such that $e \notin K$ and $J \subseteq K$ and $K$ is maximal excluding $e$. By Zorn's lemma such right ideal $K$ does exist (the topology of $\mathfrak{R}$ is Hausdorff, even regular!). This right ideal $K$ is a maximal right ideal of $\mathfrak{R}$, for if $K'$ is a right ideal properly containing $K$, then $e \in K'$ implying that

$$r = (r - er) + er \in L + K' = K'$$

for every $r \in \mathfrak{R}$ and $K' = \mathfrak{R}$.

The factor module $\mathfrak{R}/K$ is clearly a faithful irreducible $\mathfrak{R}$-module and $\mathfrak{R}/K$ is a continuous image of $\mathfrak{R}$.

Consider the finite topology $\phi(M)$ of $\mathfrak{R}$, i.e. the topology with the open base for the neighborhood system of $0$ consisting of the right ideals

$$J(m) = \{x \in \mathfrak{R} \mid mx = 0\}$$

for every $m \in M = \mathfrak{R}/K$. We can take $m = r + K$, so $x \in J(m)$ means $rx \in K$. Since $K$ is open, so is $J(m)$. Moreover, $\bigcap_{m \in M} J(m) = 0$ since $M$ is faithful. We have proved

3.2 PROPOSITION. *The finite topology $\phi(M)$ of $\mathfrak{R}$ is Hausdorff and weaker than the linearly compact topology $\tau$.*

3.3 PROPOSITION. *If $\mathfrak{R}$ is a linearly compact primitive ring, then there is a continuous isomorphism $\varphi$ of $\mathfrak{R}$ onto $\mathfrak{L}(V)$ for some vector space $V$ over a division ring $\Delta$. $\mathfrak{L}(V)$ is endowed with the finite topology.*

PROOF. Applying 1.4 and 1.2, we obtain that $\mathfrak{R}$ is a dense subring of $\mathfrak{L}(V)$ in the finite topology $\phi(V)$, where $V \cong \mathfrak{R}/K$ is a vector space over the centralizer $\Delta$ of $\mathfrak{R}$ in $\mathfrak{L}(V)$. Since $\mathfrak{R}$ is linearly compact also in this weaker topology, by 2.7, $\mathfrak{R}$ is closed in $\mathfrak{L}(V)$. Consequently $\mathfrak{R} = \bar{\mathfrak{R}} = \mathfrak{L}(V)$.

REMARK. *If the topology of the linearly compact primitive ring $\mathfrak{R}$ is a weakest one, then the mapping $\varphi: \mathfrak{R} \to \mathfrak{L}(V)$ is also a homeomorphism.*

3.4 PROPOSITION. *Let $V$ be a vector space over a division ring $\Delta$. Then $\mathfrak{L}(V)$ is a complete direct sum $\prod_{\alpha \in A} J_\alpha$ of minimal idempotent $\mathfrak{L}(V)$-isomorphic right ideals $J_\alpha$. The product topology $\tau$ of $\prod_{\alpha \in A} J_\alpha$ induces the finite topology $\phi(V)$ of $\mathfrak{L}(V)$, and $\phi(V)$ is a weakest linearly compact (Hausdorff) topology of $\mathfrak{L}(V)$.*

PROOF. Take a basis $\{x_\alpha \mid \alpha \in A\}$ of $V$ and consider the right annihilators

$$\mathfrak{A}(\alpha_1, \ldots, \alpha_n) = \{a \in \mathfrak{L}(V) \mid x_{\alpha_i} a = 0,\ i = 1, \ldots, n\}$$

and

$$J(\alpha_1, \ldots, \alpha_n) = \{a \in \mathfrak{L}(V) \mid x_\beta a = 0,\ \beta \in A \setminus \{\alpha_1, \ldots, \alpha_n\}\}$$

for every finite subset $\{\alpha_1, \ldots, \alpha_n\}$ of $A$. It is easy to verify that $J(\alpha_1, \ldots, \alpha_n)$ is an $\mathfrak{R}$-module and a direct sum $J(\alpha_1, \ldots, \alpha_n) = \sum_{i=1}^{n} J_{\alpha_i}$ of $\mathfrak{R}$-modules $J_{\alpha_i} = J(\alpha_i)$. Further, each $J_{\alpha_i}$ is an idempotent minimal right ideal of $\mathfrak{L}(V)$. By II.1.5, each $J_{\alpha_i}$ is generated by an idempotent $e_{\alpha_i}$, i.e. $J_{\alpha_i} = e_{\alpha_i}\mathfrak{L}(V)$. If the image of $x_{\alpha_i}$ under $e_{\alpha_i} \in \mathfrak{L}(V)$ is given by

$$x_{\alpha_i}e_{\alpha_i} = \sum_{j=1}^{\ell} \rho_j x_{\gamma_j} \qquad (\rho_j \in \Delta),$$

then $e_{\alpha_i}^2 = e_{\alpha_i}$ implies

$$0 \neq x_{\alpha_i}e_{\alpha_i} = x_{\alpha_i}e_{\alpha_i}^2 = \sum_{j=1}^{\ell} \rho_j x_{\gamma_j} e_{\alpha_i}.$$

Hence $\gamma_j = \alpha_i$ for one of the $\gamma_j$'s involving $x_{\alpha_i}e_{\alpha_i} = \rho_i x_{\alpha_i}$ $(\rho_i \neq 0)$ for $i = 1,\ldots,n$. Consider a linear transformation $a \in \mathfrak{L}(V)$ such that $x_{\alpha_k} a = \frac{1}{\rho_k \rho_i} x_{\alpha_i}$. Now we have $x_{\alpha_k} e_{\alpha_k} a e_{\alpha_i} = x_{\alpha_i} \neq 0$ which shows that $e_{\alpha_k} \mathfrak{L}(V) e_{\alpha_i} \neq 0$. The mapping $\varphi: e_{\alpha_k} r \to e_{\alpha_k} a e_{\alpha_i} r$ is an $\mathfrak{L}(V)$-homomorphism between $J_{\alpha_k}$ and $J_{\alpha_i}$, and $\varphi$ is not the zero mapping. Since $J_{\alpha_k}$ and $J_{\alpha_i}$ are irreducible $\mathfrak{L}(V)$-modules (as they are minimal right ideals of $\mathfrak{L}(V)$), $\varphi$ is an isomorphism. Hence all the $J_\alpha$'s are isomorphic $\mathfrak{L}(V)$-modules.

We prove next that $\mathfrak{L}(V)$ is a direct sum of $J(\alpha_1,\ldots,\alpha_n)$ and $\mathfrak{A}(\alpha_1,\ldots,\alpha_n)$. Clearly $J(\alpha_1,\ldots,\alpha_n) \cap \mathfrak{A}(\alpha_1,\ldots,\alpha_n) = 0$. Take an arbitrary linear transformation $a \in \mathfrak{L}(V)$ and choose a linear transformation $b \in \mathfrak{L}(V)$ such that

$$x_\beta b = \begin{cases} x_{\alpha_i} & \text{if } \beta \in \{\alpha_1,\ldots,\alpha_n\} \\ 0 & \text{otherwise} \end{cases}.$$

We now have $b \in J(\alpha_1,\ldots,\alpha_n)$ and $a - b \in \mathfrak{A}(\alpha_1,\ldots,\alpha_n)$ implying

$$a = b + (a - b) \in J(\alpha_1,\ldots,\alpha_n) + \mathfrak{A}(\alpha_1,\ldots,\alpha_n).$$

Thus $\mathfrak{L}(V)$ is the required direct sum, and in addition

$$J(\alpha_1,\ldots,\alpha_n) \cong \mathfrak{L}(V)/\mathfrak{A}(\alpha_1,\ldots,\alpha_n).$$

Consider the inverse system

$$\Omega = \{\prod_{\alpha_i \in f} J_{\alpha_i}, \pi_g^f \mid f,g \text{ are finite subsets of } A\},$$

where the direct sums $\prod_{\alpha_i \in f} J_{\alpha_i}$ are endowed with the product topology (which is of course discrete). Choosing

$$\underline{A} = \{\mathfrak{A}(\alpha_1,\ldots,\alpha_n) \mid \text{for every finite subset } \{\alpha_1,\ldots,\alpha_n\} \in A\}$$

for an open base for neighborhood system of $0$ in $\mathfrak{L}(V)$, we obtain the finite topology $\phi(V)$ of $\mathfrak{L}(V)$. By III.9.2, $\mathfrak{L}(V)$ is complete in $\phi(V)$. Hence 2.3 and 2.4 are applicable and we obtain

$$\mathfrak{L}(V) \cong \varprojlim \Omega \cong \prod_{\alpha \in A} J_\alpha.$$

Now 3.2 implies that $\phi(V)$ is a weakest topology of $\mathfrak{L}(V)$ and by 2.10, $\mathfrak{L}(V)$ is linearly compact in $\phi(V)$.

Combining 3.3 and 3.4, we arrive at

3.5 THEOREM. *For a ring* $\mathfrak{R}$, *the following statements are equivalent.*

i) $\mathfrak{R}$ *is a primitive ring endowed with a weakest linearly compact (Hausdorff) topology.*

ii) $\mathfrak{R}$ *is isomorphic and homeomorphic to the ring* $\mathfrak{L}(\Delta,V)$ *of all linear transformations on a vector space* $V$ *over a division ring* $\Delta$, *where* $\mathfrak{L}(\Delta,V)$ *is endowed with the finite topology* $\phi(V)$.

iii) $\mathfrak{R}$ *is isomorphic and homeomorphic to a complete direct sum* $\prod_{\alpha\in A} J_\alpha$ *of* $\mathfrak{R}$-*isomorphic idempotent minimal right ideals* $J_\alpha$, *where* $\prod_{\alpha\in A} J_\alpha$ *is endowed with the product topology and each* $J_\alpha$ *is discrete.*

It is now clear that a linearly compact primitive ring always has a nonzero socle.

A topological ring $\mathfrak{R}$ will be called *topologically simple*, if 0 and $\mathfrak{R}$ are its only closed ideals and $\mathfrak{R}^2 \neq 0$, cf. I.3.5.

3.6 PROPOSITION. *A linearly compact primitive ring is topologically simple.*

PROOF. In view of 3.5, it suffices to show that any ring $\mathfrak{L}(V)$, endowed with the finite topology $\phi(V)$, is topologically simple. Ideals of $\mathfrak{L}(V)$ were found in I.4.3. It follows from I.2.10 that every nonzero ideal of $\mathfrak{L}(V)$ is n-fold transitive for $n = 1,2,\ldots$, and so by III.9.1, is dense in $\mathfrak{L}(V)$ in the finite topology.

## 4. LINEARLY COMPACT SEMISIMPLE RINGS

Before dealing with linearly compact semisimple rings, we will prove some preliminary propositions. An ideal $P$ of a ring $\mathfrak{R}$ is called a *primitive ideal*, if $\mathfrak{R}/P$ is a primitive ring. Hence $\mathfrak{R}$ is primitive if and only if 0 is a primitive ideal of $\mathfrak{R}$.

4.1 PROPOSITION. *If* $P$ *is a primitive ideal of a linearly compact ring* $\mathfrak{R}$, *then* $P$ *is a maximal closed ideal of* $\mathfrak{R}$.

PROOF. We will prove that $P$ is closed in $\mathfrak{R}$. Since $\mathfrak{R}/P$ is primitive, there exists an irreducible faithful $\mathfrak{R}/P$-module $M$. Then $M$ can be considered as an $\mathfrak{R}$-module such that $P$ is just the annihilator of $M$ in $\mathfrak{R}$. As in the proof of 3.1, one can show that $M$ is a linearly topologized $\mathfrak{R}$-module in the discrete topology and that $M \cong \mathfrak{R}/K$ holds with a suitable open maximal right ideal $K$ of $\mathfrak{R}$. Let $a \in \mathfrak{R}\backslash P$. Taking into account that $M$ is a faithful $\mathfrak{R}/P$-module, we deduce that there is an element $m \in M$ with $ma \neq 0$ and a neighborhood $U$ of $a$ with $0 \notin mU$. Hence $U \cap P = \emptyset$ which shows that $P$ is closed in $\mathfrak{R}$. Finally, 3.6 implies that $P$ is a maximal closed ideal of $\mathfrak{R}$.

A set $\mathcal{P}$ consisting of primitive ideals $P_\alpha$, $\alpha \in A$, of a linearly compact ring $\mathfrak{R}$ is said to be an idependent system if for every finite subset $\{P_{\alpha_1},\dots,P_{\alpha_n}\}$ of $\mathcal{P}$, we have $\mathfrak{R}/\bigcap_{i=1}^{n} P_{\alpha_i} \cong \prod_{i=1}^{n} \mathfrak{R}/P_{\alpha_i}$. Using Zorn's lemma, it follows that every linearly compact ring $\mathfrak{R}$ has a maximal independent system $\mathcal{P}_0$ of primitive ideals.

4.2 PROPOSITION. Let $\mathcal{P}_1 = \{P_{1\alpha}\}_{\alpha\in A}$ and $\mathcal{P}_2 = \{P_{2\beta}\}_{\beta\in B}$ be two maximal independent systems of primitive ideals of a linearly compact ring $\mathfrak{R}$ endowed with a weakest linearly compact topology. Then $\bigcap_{\alpha\in A} P_{1\alpha} = \bigcap_{\beta\in \beta} P_{2\beta}$.

PROOF. Suppose that the statement is not ture. Without loss of generality, we may assume that $\bigcap_{\alpha\in A} P_{1\alpha} \not\subseteq \bigcap_{\beta\in B} P_{2\beta}$. There is a maximal closed ideal $D_2 = P_{2\beta_o} \in \mathcal{P}_2$ such that $\bigcap_{\alpha\in A} P_{1\alpha} \not\subseteq D_2$. Hence for any finite subset $\{\alpha_1,\dots,\alpha_n\}$ of $A$, we have $D_1 = \bigcap_{i=1}^{n} P_{1\alpha_i} \not\subseteq D_2$. Since $D_2$ is a maximal closed ideal, by 2.9 we have

$$D_2 \neq D_1 + D_2 = \overline{D_1 + D_2} = \mathfrak{R}.$$

Thus $\mathfrak{R}/D_1 \cap D_2$ splits in a direct sum of $D_1/D_1 \cap D_2 \cong \mathfrak{R}/D_2$ and $D_2/D_1 \cap D_2 \cong \mathfrak{R}/D_1$. Since the topology of $\mathfrak{R}$ is a weakest one and the isomorphism

$$\varphi:\mathfrak{R}/D_1 \cap D_2 \to \mathfrak{R}/D_1 \oplus \mathfrak{R}/D_2$$

is continuous, $\varphi$ is a homeomorphism. Consequently $\{\mathcal{P}_1, D_2\}$ is an independent system of primitive ideals of $\mathfrak{R}$, contradicting the maximality of $\mathcal{P}_1$.

Since for any primitive ideal $P$ of $\mathfrak{R}$ there is a maximal independent system $\mathcal{P}$ containing $P$, 4.2 immediately implies

4.3 PROPOSITION. Let $\mathcal{P} = \{P_\alpha \mid \alpha \in A\}$ be a maximal independent system of primitive ideals of a linearly compact ring $\mathfrak{R}$. Then $\bigcap_{\alpha\in A} P_\alpha$ equals the intersection of all primitive ideals of $\mathfrak{R}$.

4.4 DEFINITION. A ring $\mathfrak{R}$ is semisimple if the intersection $\mathcal{J}(\mathfrak{R})$ of all primitive ideals of $\mathfrak{R}$ equals $0$.

In a terminology consistent with II.5.4, a semisimple ring should be called semiprimitive. In view of II.5.4, it is clear that a ring $\mathfrak{R}$ is semisimple if and only if $\mathfrak{R}$ is a subdirect sum of primitive rings. The intersection $\mathcal{J}(\mathfrak{R})$ of all primitive ideals of $\mathfrak{R}$ is referred as to the Jacobson radical of the ring $\mathfrak{R}$. For more information on this subject, consult Jacobson ([7], Chapter I).

Linearly compact semisimple rings have several interesting characterizations. Some of them are presented in the following result. Characterizations ii) and iii) are due to Leptin [2], whereas iv) can be found in Wiegandt [1].

4.5 THEOREM. The following conditions on a ring $\mathfrak{R}$ are equivalent.

i) $\mathfrak{R}$ <u>is a linearly compact semisimple ring endowed with a weakest linearly compact (Hausdorff) topology</u>.

ii) $\mathfrak{R} \cong \prod_{\alpha\in\Delta} \mathfrak{L}(\Delta_\alpha, V_\alpha)$, <u>algebraically and topologically, where the topology of the right hand side is the product topology of the finite topologies</u> $\phi(V_\alpha)$ <u>of</u> $\mathfrak{L}(V_\alpha)$, $\alpha \in A$.

iii) $\mathfrak{R} \cong \prod_{\gamma\in C} J_\gamma$, <u>where each</u> $J_\gamma$ <u>is an idempotent minimal right ideal, the topology of the right hand side is the product topology and each</u> $J_\gamma$ <u>is discrete</u>.

iv) $\mathfrak{R}$ <u>is a linearly compact regular ring endowed with a weakest linearly compact (Hausdorff) topology (regularity is meant in the algebraic sense of</u> I.3.5).

<u>PROOF</u>. i) $\Rightarrow$ ii). Choose a maximal independent system $\rho = \{P_\alpha \mid \alpha \in A\}$ of primitive ideals of $\mathfrak{R}$. By 4.3, we have $\bigcap_{\alpha\in A} P_\alpha = 0$ and for any finite subset $\{\alpha_1,\ldots,\alpha_n\}$ of $A$, we have

$$\mathfrak{R}/\bigcap_{i=1}^{n} P_{\alpha_i} \cong \prod_{i=1}^{n} \mathfrak{R}/P_{\alpha_i}.$$

These finite direct sums obviously form an inverse system $\Omega$. Taking into account III.5.9, and 2.3 and 2.4 of this appendix, we deduce

$$\mathfrak{R} \cong \varprojlim \Omega \cong \prod_{\alpha\in A} \mathfrak{R}/P_\alpha.$$

In view of 3.5, $\mathfrak{R}/P_\alpha \cong \mathfrak{L}(\Delta_\alpha, V_\alpha)$ holds for each $\alpha \in A$. Since the topology of $\mathfrak{R}$ is weakest, the isomorphism $\mathfrak{R} \to \prod_{\alpha\in A} \mathfrak{R}/P_\alpha$ is also a homeomorphism.

ii) $\Rightarrow$ iii). This follows directly from 3.5.

iii) $\Rightarrow$ iv). By II.1.14, for any $\gamma \in C$, there is an idempotent $e_\gamma \in J_\gamma$ such that $J_\gamma = e_\gamma\mathfrak{R}$ and $\Delta_\gamma = e_\gamma\mathfrak{R}e_\gamma$ is a division ring with the identity $e_\gamma$. Hence for every $a \in J_\gamma$, there is an $r \in \mathfrak{R}$ such that $a = e_\gamma r$ and $e_\gamma r e_\gamma$ has an inverse $(e_\gamma r e_\gamma)^{-1} = e_\gamma s e_\gamma$ in $\Delta_\gamma$. Consequently, for $a \in J_\gamma$ and $e_\gamma s \in J_\gamma$, we have

$$a(e_\gamma s)a = (e_\gamma r)(e_\gamma s)(e_\gamma r) = (e_\gamma r e_\gamma)(e_\gamma s e_\gamma)r = e_\gamma r = a.$$

Thus for each $\gamma \in C$, $J_\gamma$ is a regular ring. Thus $\prod_{\gamma\in C} J_\gamma$ are regular rings and hence $\mathfrak{R}$ is also. Note that the discrete topology in each $J_\gamma$ is a weakest one, so the product topology yields again a weakest one.

iv) $\Rightarrow$ i). Consider an arbitrary element $z \in \mathcal{J}(\mathfrak{R})$. We shall establish the existence of an element $r \in \mathfrak{R}$ such that $z = r - zr$. To this end, consider the right ideal

$$J = \{r - zr \mid r \in \mathfrak{R}\}.$$

According to II.7.6, $J$ is a modular right ideal of $\mathfrak{R}$. Suppose $z \notin J$. Then Zorn's lemma insures the existence of a modular right ideal $M$ which is maximal with respect to the exclusion of $z$. In fact, $M$ is a maximal right ideal, for any right ideal $N$ properly containing $M$ contains $z$ and thus

$$r = (r - zr) + zr \in M + N \subseteq N$$

holds for every $r \in \mathfrak{R}$. Obviously $M$ is modular. Consider next the two-sided ideal

$$P = \{y \in \mathfrak{R} \mid \mathfrak{R}y \subseteq M\}.$$

Since $\mathfrak{R}$ is regular, there is an $x \in \mathfrak{R}$ with $yxy = y$. Hence for every $y \in P$, it follows that $y = yxy \in M$ implying $P \subseteq M$. Moreover, $\mathfrak{R}/M$ is a faithful $\mathfrak{R}/P$-module. Consequently $P$ is a primitive ideal of $\mathfrak{R}$. By hypothesis, we have

$$z \in \mathcal{J}(\mathfrak{R}) \subseteq P \subseteq M$$

contradicting the assumption $z \notin M$. Consequently $z \in J$ and thus there exists an element $r \in \mathfrak{R}$ such that $z = r - zr$.

Again take an arbitrary element $a \in \mathcal{J}(\mathfrak{R})$. Since $\mathfrak{R}$ is regular, there is an element $x \in \mathfrak{R}$ such that $axa = a$. Then $xa \in \mathcal{J}(\mathfrak{R})$. By the above, there is an element $r \in \mathfrak{R}$ such that $xa = r - xar$, and we obtain

$$a = axa = ar - axar = ar - ar = 0.$$

Thus $\mathcal{J}(\mathfrak{R}) = 0$, i.e. $\mathfrak{R}$ is semisimple.

4.6 COROLLARY. *A semisimple ring $\mathfrak{R}$ endowed with a weakest linearly compact topology is primitive if and only if $\mathfrak{R}$ is topologically simple.*

PROOF. Straightforward by 3.6 and 4.5.

4.7 COROLLARY (Leptin [2]). *A linearly compact ring $\mathfrak{R}$ is topologically simple if and only if $\mathfrak{R} \cong \mathfrak{L}(\Delta, V)$.*

PROOF. Trivial by 4.5 and 4.6.

4.8 COROLLARY. *A ring $\mathfrak{R}$ is linearly compact, semisimple and its ideals form an open base for the neighborhood system of 0 if and only if $\mathfrak{R} \cong \prod_{\alpha \in A} \mathfrak{L}(\Delta_\alpha, V_\alpha)$ and $\dim V_\alpha$ is finite for all $\alpha \in A$.*

PROOF. Necessity. By 4.5, we have $\mathfrak{R} \cong \prod_{\alpha \in A} \mathfrak{L}(V_\alpha)$. Consider an open ideal $U$ of $\mathfrak{R}$. Without loss of generality, we may assume that

$$\mathfrak{R}/U \cong \prod_{\text{finite}} \mathfrak{L}(V_\alpha).$$

Since $\prod_{\text{finite}} \mathfrak{L}(V_\alpha)$ is discrete, each component $\mathfrak{L}(V_\alpha)$ is discrete in the finite topology. Hence 3.5 implies that each $\mathfrak{L}(V_\alpha)$ splits into a finite product $\prod_{i=1}^{n_\alpha} J_i$ of minimal idempotent $\mathfrak{R}$-isomorphic right ideals $J_i$. In this case the dimension of each $V_\alpha$ must be finite, for $\dim V_\alpha = n_\alpha$. Sufficiency is obvious.

4.9 COROLLARY. *A ring $\mathfrak{R}$ is semisimple and linearly compact in the discrete topology if and only if* $\mathfrak{R} \cong \prod_{\text{finite}} \mathcal{L}(V_\alpha)$ *and* $\dim V_\alpha$ *is finite for all* $\alpha$.

PROOF. For necessity, 4.8 is applicable, and the number of the components $\mathcal{L}(V_\alpha)$ must be finite since $\mathfrak{R}$ is discrete. Sufficiency is again trivial.

4.10 COROLLARY. *A ring $\mathfrak{R}$ is primitive and linearly compact in the discrete topology if and only if* $\mathfrak{R} \cong \mathcal{L}(\Delta, V)$ *where* $\dim V$ *is finite.*

PROOF. Trivial in view of 3.5.

Finally, 4.6 and 3.5 immediately imply

4.11 COROLLARY. *A linearly compact ring $\mathfrak{R}$ is simple if and only if* $\mathfrak{R} \cong \mathcal{L}(\Delta, V)$ *where* $\dim V$ *is finite.*

Even though primitive rings are always prime and, by II.1.25, for rings with nonzero socle also the converse is valid, for linearly compact rings in 3.5 and 4.5 the terms "primitive" and "semisimple" can no longer be replaced by the terms "prime" and "semiprime", respectively. For let $\mathfrak{P}$ denote the ring of all p-adic integers. If $\underline{P} = \{p^i\mathfrak{P} \mid i = 1,2,\ldots\}$ is considered as an open base for a neighborhood system 0 in $\mathfrak{P}$, then $\underline{P}$ induces a linearly compact topology on $\mathfrak{P}$ (cf. Leptin [1]). Further $\mathfrak{P}_1 = p\mathfrak{P}$, as a closed submodule of $\mathfrak{P}$, is linearly compact. (Prove that a closed submodule of a linearly compact $\mathfrak{R}$-module is always linearly compact in the relativized topology.) Since the $\mathfrak{P}$-submodules of $\mathfrak{P}_1$ are exactly the $\mathfrak{P}_1$-submodules, $\mathfrak{P}_1$ is linearly compact as a $\mathfrak{P}_1$-module. As it is easy to see, $\mathfrak{P}_1$ has no primitive ideals (cf. Wiegandt [3]). On the other hand, for any two nonzero ideals $J, K$ of $\mathfrak{P}_1$, we have $J = p^n\mathfrak{P}$, $K = p^m\mathfrak{P}$ for some $n,m > 1$, and thus $JK \neq 0$ which shows that 0 is a prime ideal of $\mathfrak{P}_1$. Hence $\mathfrak{P}_1$ is linearly compact, prime, but is not semisimple. In addition, the socle of $\mathfrak{P}$ equals 0.

For further discussion of semisimple linearly compact rings, we refer to Leptin [2], Wiegandt [2] and Eckstein [1].

An obvious generalization of linear compactness is the notion of local linear compactness. Since every semisimple locally linearly compact ring has a nonzero socle (see Wiegandt [4]), the study of their structure can be easily handled, which is not the case in the theory of locally compact rings (cf. Skornjakov [1],[2]).

Linearly compact nonsemisimple rings were investigated by Leptin [2], Wiegandt [3] and Arnautov [1], and particular attention was paid to radical rings (where $\mathfrak{R} = \mathcal{J}(\mathfrak{R})$). Further results on linear compactness were obtained by Wolfson [1], Fischer and Gross [1], Fuchs [1], and Warner [1]. For recent developments we refer to Andrunakievič, Arnautov and Ursu [1], Warner [2], Widiger [1] and Wiegandt [5].

## CURRENT ACTIVITY

From the material of these Lectures, we may easily extract the following categories:

1. Objects: pairs of dual vector spaces,
   Morphisms: semilinear isomorphisms with a surjective adjoint;
2. Objects: weakly topologized vector spaces,
   Morphisms: iseomorphisms;
3. Objects: maximal primitive rings with a nonzero socle,
   Morphisms: ring isomorphisms;
4. Objects: complete lattices satisfying the double covering condition (U-closed subspaces of V),
   Morphisms: lattice isomorphisms;
5. Objects: multiplicative semigroups of objects in 3. above (characterized abstractly),
   Morphisms: semigroup isomorphisms.

Further categories may be obtained by taking certain subobjects in 3., 4., and 5. Under some mild restrictions on objects of these categories, most of the resulting categories are either equivalent or isomorphic. A comprehensive study of the relationship among these and many other categories forms the subject of an unpublished manuscript by the author entitled "Categories of vector and projective spaces, semigroups, rings and lattices". These Lectures thus represent a step toward a better understanding of the relationship among these different branches of modern algebra anu projective geometry.

The recent work on semigroups of linear transformations includes mainly investigations of homomorphisms and antiisomorphisms of various semigroups of linear transformations, see Fajans [1],[2], Jodeit and Lam [1], Mihalev and Šatalova [1], Pierce [1]. Multiplicative semigroups of certain rings have been studied by Petrich [4], Satyanarayana [1], a comprehensive survey of this subject can be found in Peinado [1]. Semigroup methods have been successfully used in ring theory by Eckstein [2]. Further classes of rings with unique addition have been recently found by Martindale [1] and Stephenson [1]. Linear compactness for rings has been investigated by Andrunakievič, Arnautov and Ursu [1], Arnautov [1], Warner [1],[2], Widiger [1], Wiegandt [5].

# BIBLIOGRAPHY


Andrunakievič, V.A., Arnautov, V.I. and Ursu, M.I.

[1] Wedderburn decomposition of hierarchically linearly compact rings, Doklady Akad. Nauk SSSR 211(1973),15-18 (in Russian).

Arnautov, V.I.

[1] Radicals in rings with neighborhoods of zero, Mat. Issled. 3(1969),3-17 (in Russian).

Baer, R.

[1] Linear algebra and projective geometry, Academic Press, New York, 1952.

Behrens, E.A.

[1] Ring theory, Academic Press, New York, 1972.

Bourbaki, N.

[1] Eléments de Mathématiques, Livre V, Chap. IV, Hermann, Paris, 1955.

Clifford, A.H. and Preston, G.B.

[1] The algebraic theory of semigroups, Math Surveys No. 7, Amer. Math. Soc., Vol. I (1961), Vol. II (1967), Providence.

Dieudonné, J.

[1] La dualité dans les espaces vectoriels topologiques, Ann. Ecole Norm. Sup. 59(1942),107-139.

[2] Sur le socle d'un anneau et les anneaux simples infinis, Bull. Soc. Math. France 70(1942),46-75.

[3] Les idéaux minimaux dans les anneaux associatifs, Proc. Int. Cong. Math. Cambridge, Mass. 2(1950),44-48.

[4] On the automorphisms of classical groups, Mem. Amer. Math. Soc. No. 2, 1950.

[5] Linearly compact spaces and double vector spaces over sfields, Amer. J. Math. 73(1951),13-19.

Divinsky, N.J.

[1] Rings and radicals, Univ. Toronto Press, 1965.

Eckstein, F.

[1] Complete semisimple rings with ideal neighborhoods of zero, Archiv Math. 18 (1967),586-590.

[2] Semigroup methods in ring theory, J. Algebra 12(1969),177-190.

Eidelheit, M.

[1] On isomorphisms of rings of linear operators, Studia Math. 9(1940),97-105.

Erdos, J.A.

[1] On products of idempotent matrices, Glasgow Math. J. 8(1967),118-122.

Everett, C.J.

[1] An extension theory for rings, Amer. J. Math. 64(1942),363-370.

Faith, C.

[1] Rings with minimum condition on principal ideals, Archiv Math. 10(1959), 327-330.

Faith, C. and Utumi, Y.

[1] On a new proof of Litoff's theorem, Acta Math. Acad. Sci. Hung. 14(1963), 369-371.

Fajans, V.G.

[1] Isomorphisms of semigroups of invertible linear transformations leaving a conus invariant, Izv. Vysš. Učebn. Zaved. Mat. 12(91)(1969),93-98 (in Russian).

[2] Isomorphisms of semigroups of affine transformations, Sibir. Mat. Ž. 11(1970), 193-198 (in Russian); English transl. by Consultants Bureau 11(1970),154-158.

Fischer, H.R. and Gross, H.

[1] Quadratic forms and linear topologies, I, Math. Ann. 157(1964),296-325.

Fuchs, L.

[1] Note on linearly compact abelian groups, J. Austral. Math. Soc. 9(1969), 433-440.

Fuchs, L. and Szele, T.

[1] Contribution to the theory of semisimple rings, Acta Math. Acad. Sci. Hung. 3(1952),233-239.

Gewirtzman, L.

[1] Anti-isomorphisms of the endomorphism rings of a class of free modules, Math. Ann. 159(1965),278-284.

[2] Anti-isomorphisms of the endomorphism rings of torsion-free modules, Math. Zeitschr. 98(1967),391-400.

Gluskin, L.M.

[1] The associative system of square matrices, Doklady Akad. Nauk SSSR 97(1954), 17-20 (in Russian).

[2] Automorphisms of multiplicative semigroups of matrix algebras, Usp. Mat. Nauk 11(1956),199-206 (in Russian).

[3] On semigroups of matrices, Izv. Akad. Nauk SSSR 22(1958),439-448 (in Russian).

[4] Semigroups and rings of endomorphisms of linear spaces, Izv. Akad. Nauk SSSR 23(1959),841-870 (in Russian); Amer. Math. Soc. Translations 45(1965),105-137.

[5] Ideals of semigroups of transformations, Mat. Sbornik 47(1959),111-130 (in Russian).

[6] Semigroups and rings of endomorphisms of linear spaces, II., Izv. Akad. Nauk SSSR 25(1961),809-814 (in Russian); Amer. Math. Soc. Translations 45(1965), 139-145.

[7] On endomorphisms of modules, Algebra i Mat. Logika, Kiev (1966),3-20 (in Russian).

Goldman, O.

[1] A characterization of semisimple rings with the descending chain condition, Bull. Amer. Math. Soc. 52(1946),1021-1027.

Grillet, P.A. and Petrich, M.

[1] Ideal extensions of semigroups, Pacific J. Math. 26(1968),493-508.

Halezov, E.A.

[1] Automorphisms of semigroups of matrices, Doklady Akad. Nauk SSSR 96(1954), 245-248 (in Russian).

[2] Isomorphisms of semigroups of matrices, Uč. Zap. Ivanov. Ped. Inst. 5(1954), 42-56 (in Russian).

Hall, M.

[1] The theory of groups, MacMillan, New York, 1959.

Hotzel, E.

[1] On simple rings with minimal one-sided ideals (unpublished). Abstract: Notices Amer. Math. Soc. 17(1970),238.

Jacobson, N.

[1] Normal semi-linear transformations, Amer. J. Math. 61(1939),45-58.

[2] The theory of rings, Math. Surveys No. II, Amer. Math. Soc., New York, 1943.

[3] Structure theory of simple rings without finiteness assumptions, Trans. Amer. Math. Soc. 57(1945),228-245.

[4] The radical and semi-simplicity for arbitrary rings, Amer. J. Math. 67(1945), 300-320.

[5] On the theory of primitive rings, Ann. Math. 48(1947),8-21.

[6] Lectures in abstract algebra, Vol. II, Linear algebra, Van Nostrand, New York, 1953.

[7] Structure of rings, Amer. Math. Soc. Coll. Publ. Vol. 37, Providence, 1956.

Jaffard, P.

[1] Détermination de certains anneaux, C. R. Acad. Sci., Paris 229(1949),805-806.

Jans, P.

[1] Rings and homology, Holt, Rinehart and Winston, New York, 1964.

Jodeit, M. and Lam, T.Y.

[1] Multiplicative maps of matrix semigroups, Archiv Math. 20(1969),10-16.

Johnson, B.E.

[1] An introduction to the theory of centralizers, Proc. London Math. Soc. 14 (1964),299-320.

Johnson, R.E.

[1] Rings with unique addition, Proc. Amer. Math. Soc. 9(1958),57-61.

Johnson, R.E. and Kiokemeister, F.

[1] The endomorphisms of the total operator domain of an infinite module, Trans. Amer. Math. Soc. 62(1947),404-430.

Kaplansky, I.

[1] Topological rings, Amer. J. Math. 69(1947),153-183.

[2] Locally compact rings, Amer. J. Math., Part I, 70(1948),477-459; Part II, 73(1951),20-24; Part III, 74(1952),929-935.

[3] Ring isomorphisms of Banach algebras, Canad. J. Math. 6(1954),374-381.

[4] Fields and rings, Univ. of Chicago Press, Chicago and London, 1969.

Kelley, J.L.

[1] General topology, Van Nostrand, New York, 1955.

Kertesz, A.

[1] Vorlesungen über artinsche Ringe, Akadémiai Kiadó, Budapest, 1968.

Köthe, G.

[1] Topological vector spaces I, Springer, New York, 1969.

Lambek, J.

[1] Lectures on rings and modules, Blaisdell, Waltham, 1966.

Lefschetz, S.

[1] Algebraic topology, Amer. Math. Soc. Coll. Publ. Vol. 27, New York, 1942.

Leptin, H.

[1] Über eine Klasse linear kompakter abelscher Gruppen, Abh. Math. Sem. Hamburg, Part I, 19(1954),23-40; Part II, 19(1954),221-243.

[2] Linear kompakte Moduln und Ringe, Math. Zeitschr., Part I, 62(1955),241-267; Part II, 66(1957),289-327.

Mackey, G.W.

[1] Isomorphisms of normed linear spaces, Ann. Math. 43(1942),244-266.

[2] On infinite-dimensional linear spaces, Trans. Amer. Math. Soc. 37(1945), 155-207.

MacLane, S.

[1] Extensions and obstructions for rings, Illinois J. Math. 2(1958),316-345.

Martindale, W.S.

[1] When are multiplicative mappings additive? Proc. Amer. Math. Soc. 21(1969), 695-698.

McCoy, N.H.

[1] The theory of rings, McMillan, New York, 1970 (7th printing).

Mihalev, A.V.

[1] Isomorphisms of endomorphism semigroups of modules, Algebra i Logika, Part I, 5, no. 5 (1966),59-67; Part II, 6, no. 2 (1967),35-47 (in Russian).

Mihalev, A.V. and Šatalova, M.A.

[1] Automorphisms and antiautomorphisms of the semigroup of invertible matrices with nonnegative elements, Mat. Sbornik 81(1970),600-609 (in Russian); Transl. Math. USSR-Sbornik 10(1970),547-555.

Morita, K.

[1] Category-isomorphisms and endomorphism rings of modules, Trans. Amer. Math. Soc. 103(1962),451-469.

Müller, U. and Petrich, M.

[1] Erweiterungen eines Ringes durch eine direkte Summe zyklischer Ringe, J. Reine Angew. Math. 248(1971),47-74.

[2] Translationshülle and wesentlichen Erweiterungen eines zyklishen Ringes, J. Reine Angew. Math 249(1971),34-52.

Neumann, J. von

[1] On regular rings, Proc. Nat. Acad. Sci. 22(1936),707-713.

Ornstein, D.

[1] Dual vector spaces, Ann. Math. 69(1959),520-534.

Peinado, R.E.

[1] On semigroups admitting ring structure, Semigroup Forum 1(1970),189-208.

Petrich, M.

[1] The translational hull of a completely 0-simple semigroup, Glasgow Math. J. 9(1968),1-11.

[2] The semigroup of endomorphisms of a linear manifold, Duke Math. J. 36(1969), 145-152.

[3] The translational hull in semigroups and rings, Semigroup Forum 1(1970), 283-360.

[4] Dense extensions of completely 0-simple semigroups, J. Reine Angew. Math., Part I, 258(1973),103-125; Part II, 259(1973),109-131.

[5] Introduction to semigroups, Merrill, Columbus, 1973.

Pierce, S.

[1] Multiplicative maps of matrix semigroups over Dedekind rings, Archiv Math. 24 (1973),25-29.

Pless, V.

[1] The continuous transformation ring of biorthogonal bases spaces, Duke Math. J. 25(1958),365-371.

Plotkin, B.I.

[1] Automorphism groups of algebraic systems, Nauka, Moscow, 1966 (in Russian).

Rédei, L.

[1] Algebra, Vol. 1, Pergamon, Oxford and New York, 1967.

Rédei, L. and Steinfeld, O.

[1] Über Ringe mit gemeinsamer multiplikativer Halbgruppe, Comment. Math. Helv. 26(1952),146-151.

Ribenboim, P.

[1] Rings and modules, Interscience, New York, 1969.

Rickart, C.E.

[1] One-to-one mappings of rings and lattices, Bull. Amer. Math. Soc. 54(1948), 758-764.

[2] Isomorphic groups of linear transformations, Amer. J. Math., Part I, 72(1950), 451-464; Part II, 73(1951),697-716.

[3] Isomorphisms of infinite-dimensional analogues of the classical groups, Bull. Amer. Math. Soc. 57(1951),435-448.

Rjabuhin, Ju. M.

[1] Rings with unique multiplication, Izv. Akad. Nauk Mold. SSR, No. 1 (1963), 77-78 (in Russian).

Rosenberg, A.

[1] Subrings of simple rings with minimal ideals, Trans. Amer. Math. Soc. 73 (1952),115-138.

Rosenberg, A. and Zelinsky, D.

[1] Galois theory of continuous transformation rings, Trans. Amer. Math. Soc. 79 (1955),429-452.

Satyanarayana, M.

[1] On semigroups admitting ring structure, Semigroup Forum, Part I, 3(1971), 43-50; Part II, 6(1973),189-197.

Skornjakov, L.A.

[1] Einfache lokal bikompakte Ringe, Math. Zeitschr. 87(1965),241-251.

[2] Locally bicompact biregular rings, Mat. Sbornik 69(1966),663 (in Russian).

Steinfeld, O.

[1] On Litoff's theorem, Ann. Univ. Sci. Budapest 13(1970),101-102.

Stephenson, W.

[1] Unique addition rings, Canad. J. Math. 21(1969),1455-1461.

Szász, F.

[1] Über Ringe mit Minimalbedingung fur Hauptrechtsideale. I, Publ. Math. Debrecen 7(1960),54-64.

[2] Über Ringe mit Minimalbedingung fur Hauptrechtsideale, Acta Math. Acad. Sci. Hung., Part II, 12(1961),417-431; Part III, 14(1963),447-461.

Szász, G.

[1] Introduction to lattice theory, Academic Press, New York and London, 1963.

Warner, S.

[1] Linearly compact noetherian rings, Math. Ann. 178(1968),53-61.

[2] Linearly compact rings and modules, Math. Ann. 197(1972),29-43.

Widiger, A.

[1] Die Struktur einer Klasse linear kompakter Ringe, Coll. Math. Soc. János Bolyai, 6, Rings, Modules and Radicals, Keszthely (Hungary) (1971), 501-505, North Holland, 1973.

Wiegandt, R.

[1] Über linear kompakte reguläre Ringe, Bull. Acad. Pol. Sci. 13(1965),445-446.

[2] Über halbeinfache linear kompakte Ringe, Studia Sci. Math. Hung. 1(1966), 31-38.

[3] Über transfinit nilpotente Ringe, Acta Math. Acad. Sci. Hung. 17(1966), 101-114.

[4] Über lokal linear kompakte Ringe, Acta. Sci. Math. Szeged 28(1967),255-260.

[5] Radicals coinciding with the Jacobson radical on linearly compact rings, Beiträge zur Algebra und Geometrie, Halle 1(1971),195-199.

Wolfson, K.G.

[1] An ideal theoretic characterization of the ring of all linear transformations, Amer. J. Math. 75(1953),358-386.

[2] Some remarks on $\nu$-transitive rings and linear compactness, Proc. Amer. Math. Soc. 5(1954),617-619.

[3] Annihilator rings, J. London Math. Soc. 31(1956),94-104.

[4] A class of primitive rings, Duke Math. J. 22(1955),157-163.

[5] Isomorphisms of the endomorphism ring of a free module over a principal left ideal domain, Michigan Math. J. 9(1962),69-75.

[6] Isomorphisms of the endomorphism rings of torsion-free modules, Proc. Amer. Math. Soc. 13(1962),712-714.

Zelinsky, D.

[1] Rings with ideal nuclei, Duke Math. J. 18(1951),431-442.

[2] Linearly compact modules and rings, Amer. J. Math. 75(1953),79-90.

[3] Every linear transformation is a sum of nonsingular ones, Proc. Amer. Math. Soc. 5(1954),627-630.

## LIST OF SYMBOLS


| | |
|---|---|
| $(\Delta,V)$ | 1 |
| $V$, $(V,\Delta)$, $\Delta^+$, $V^* = (V^*,\Delta)$ | 2 |
| $\mathfrak{S}(V) = \mathfrak{S}(\Delta,V)$, $\mathfrak{L}(V) = \mathfrak{L}(\Delta,V)$, $\mathfrak{I}(V)$, $\mathfrak{I}'(V)$, $\dim V$, $A \oplus B$, rank $a$, $N_a$ | 3 |
| $(U,V) = (U,\Delta,V) = (U,\Delta,V;\mathfrak{B})$, $f_u$, nat $U$ | 4 |
| $\mathfrak{F}(\Delta,V)$, $\mathfrak{L}_U(V)$, $\mathfrak{F}_U(\Delta,V)$, $\mathfrak{S}_U(\Delta,V)$, $\mathfrak{F}_{n,U}(\Delta,V)$, $a^*$ | 6 |
| $A\backslash B$, $\iota_X$, $\mathfrak{M}\mathfrak{R}$, $S^-$, $\mathfrak{M}\Delta^-$, $[i,\gamma,\lambda]$ | 7 |
| $\mathfrak{C}(A)$, $i_B(A)$ | 11 |
| $\mathfrak{I}_{U'}(V')$ | 16 |
| $\mathfrak{R}^2$, $\mathfrak{U}_\ell(\mathfrak{R})$, $\mathfrak{U}_r(\mathfrak{R})$ | 19 |
| $\mathfrak{R}_n$ | 22 |
| $E_S$, $\mathfrak{P}(U,V)$ | 23 |
| $G_e$ | 25 |
| $S/I$, $J(a)$, $I(a)$, $\mathfrak{c}^+$, $\mathfrak{S}_{\mathfrak{a}}(\Delta,V)$ | 27 |
| $\mathfrak{L}_{\mathfrak{a}}(V)$ | 28 |
| $\mathfrak{m}^o(I,G,M;P)$, $L_n(\Delta)$, $Q_{n,U}(\Delta,V)$ | 29 |
| $\mathfrak{K}_{V'}(V)$, $(\omega,a)$, $(\Delta,V) \cong (\Delta',V')$ | 34 |
| $(b,\omega)$ | 35 |
| $(U,\Delta,V) \cong (U',\Delta',V')$, $\zeta_{(\omega,a)}$ | 38 |
| $\mathfrak{S}_{\mathfrak{a},U}(\Delta,V)$ | 43 |
| $\epsilon_g$, $m_\gamma$ | 44 |
| $C(S)$ | 45 |
| $\mathfrak{R}^+$ | 46 |
| $\Sigma$, $\Lambda$, $M$, $I$ | 47 |
| $\mathfrak{L}$, $\mathfrak{a}(S)$, $\mathfrak{J}(S)$, $HK$ | 48 |
| $\Theta$ | 49 |
| $V^+$ | 54 |
| $\Omega(S)$, $\lambda_a$, $\rho_a$, $\pi_a$, $\Pi(S)$, $\tau = \tau(K:S)$, $\lambda^k$, $\rho^k$ | 55 |
| $S^2$ | 56 |
| $\mathfrak{U}(S)$ | 60 |
| $\delta_{(\lambda,\rho)}$, $\circ$, $\varphi_{(a,a')}$, $\mathcal{Q}(\mathfrak{R})$, $\mathfrak{G}(\mathfrak{R})$ | 62 |
| $AB$, $aB$, $Ab$ | 65 |
| $A^n$, $\mathfrak{a}_{\ell,\mathfrak{R}}(B)$, $\mathfrak{U}_{r,\mathfrak{R}}(B)$ | 66 |
| $AB$, $A^n$, $aB$ | 67 |
| $\sum_{i \in I} A_i$ | 68 |

| | |
|---|---|
| $AB$ | 70 |
| $\mathfrak{U}_{\mathfrak{R}}(I)$ | 74 |
| $\lambda \circ \rho$, $\mathfrak{L}$, $\mathfrak{R}$, $\mathfrak{D}$ | 75 |
| $\mathfrak{M}(I,\Delta,\Lambda;P)$ | 76 |
| $\Delta_{\varkappa}$ | 80 |
| $\mathfrak{R}^{\mathfrak{R}}$, $\mathfrak{R}_{\mathfrak{R}}$ | 81 |
| $\bigoplus_{i\in I} A_i$ | 91 |
| $\sum_{\alpha\in A} \oplus S_\alpha$ | 92 |
| $(a)$ | 95 |
| $\prod_{\alpha\in A} \mathfrak{R}_\alpha$, $\pi_\alpha$ | 96 |
| $\prod_{\alpha\in A} S_\alpha$ | 97 |
| $(a)_r$ | 102 |
| $\prod_{\alpha\in A} X_\alpha$, $X^A$, $(X,\tau)$ | 112 |
| $S_{(\beta;y)}$, $B_{(\alpha_1,\ldots,\alpha_n;x_1,\ldots,x_n)}$, $\tau_U(V) = \tau_U(\Delta,V)$<br>$\prod_{\alpha\in A} X_\alpha$, $B_{(v_1,\ldots,v_n;\delta_1,\ldots,\delta_n)}$, $C_{(v_1,\ldots,v_n;\delta_1,\ldots,\delta_n)}$, | 113 |
| $S^\perp$, $T^\perp$, $[A]$, $[w_1,w_2,\ldots,w_n]$, $v^\perp$, $u^\perp$ | 114 |
| $(\Delta,V,\tau)$ | 115 |
| codim $S$ | 119 |
| $a^*$ | 127 |
| $\lambda[A]$, $\lambda[x]$, $(X,\mathcal{U})$ | 128 |
| $(S,\leq)$, $\{S_n \mid n \in D\}$, $(f,(X^*,\mathcal{U}^*))$ | 129 |
| $\phi$ | 130 |
| $\phi_U(V)$, $B_{(x_1,\ldots,x_n;y_1,\ldots,y_n)}$, $B_{(x_1,\ldots,x_n)}(a)$ | 136 |
| $\mathfrak{n}_c(a)$, $(\mathfrak{R},\tau)$ | 137 |
| $\overline{A}$ | 139 |
| $W(\beta)$, $\mathcal{U}(A)$, $\mathcal{H}$, $\hat{T}$ | 144 |
| $P_{(u_1,\ldots,u_n)}(a)$, $\mathfrak{m}_c(a)$ | 145 |
| $\langle x\rangle$, $\mathfrak{L}(M^+)$, $M^+$, $r^*$, $\mathfrak{R}^*$ | 153 |
| $\mathfrak{U}_{\mathfrak{R}}(W)$ | 154 |
| $\pi^\alpha_\beta$, $\varprojlim \Omega$ | 156 |
| $\mathfrak{U}(\alpha_1,\ldots,\alpha_n)$, $J(\alpha_1,\ldots,\alpha_n)$, $J_{\alpha_i}$ | 160 |
| $\phi(V)$ | 161 |
| $\mathcal{J}(\mathfrak{R})$ | 163 |

# INDEX

adjoint of a linear transformation, 5
adjoint of a semilinear transformation, 35
affine transformation, 54
algebra, 46
annihilator, 60
atomic ring, 69

basis, 3
bilinear form, 4
biorthogonal sets, 81
bitranslation, 55
Boolean ring, 46

Cauchy net, 129
center, 45
centralizer, 48
centralizer of a primitive ring, 154
characteristic, 16
circle composition, 62
codimension, 119
compatible (topology-with the group structure),
complement, 3
complete direct sum, 96
complete lattice, 125
complete uniform space, 129
completely 0-simple semigroup, 71
completion of a uniform space, 129
congruence (induced by), 54
conjugate of a linear transformation, 6
conjugate of a semilinear transformation, 127
conjugate space, 2
contains small sets, 133
converges (a net ___), 129

dense extension, 54
dense set of linear transformations, 12
dense uniform subspace, 129
diagonal, 128
dimension, 3

direct product of rings, 96
direct product of semigroups, 97
direct sum, 91
directed set, 129
directs (a binary relation ___), 129
discrete direct sum, 96
division ring, 1
doubly transitive, 12
dual pair, 4
dual space, 2

endomorphism, 3
essential extension, 60
eventually in (net ___), 129
extension, 54
external direct sum, 96

faithful module, 71
finite intersection property, 132
finite topology of $X^A$, 113
finite topology of $\mathcal{L}_U(V)$, 136
finite topology of $V^*$, 113

general linear group, 29
generalized inner automorphism, 62
generates (a set ___), 81
generating set, 3
Green's relations, 75

homogeneous component, 94
homothetic transformation, 54
hyperplane, 115

ideal, 6
idealizer, 11
idempotent ideal, 66
identity relation, 128
in (net ___ a set), 129
independent family of subrings, 91
independent system of primitive ideals, 163
induced by a semilinear isomorphism, 41
induced congruence, 54
inner automorphism, 48
inner bitranslation, 55

inner left (right) translation, 55
inverse limit, 156
inverse system of topological modules, 156
invertible matrix, 53
irreducible module, 71
isomorphic dual pairs, 38
isomorphic vector spaces, 34

Jacobson radical, 163

large ideal, 60
left annihilator, 19,66,91
left ideal, 27
left $\mathfrak{R}$-module, 73
left row-independent matrix, 76
left socle, 69
left socle topology, 147,149
left translation, 55
left vector space, 1
linear form, 2
linear transformation, 2
linear variety, 132
linearly compact module, 132
linearly independent set, 3
linearly topologized module, 131
locally matrix ring, 25
$\Lambda \times I$-matrix, 76

maximal dense extension, 54
maximal subgroup, 25
minimal left (right, two-sided) ideal, 67
modular left identity, 141
modular right ideal, 141
multiplication induced by a scalar, 44

n-fold transitive, 12
natural image, 4
natural isomorphism of V into $V^{**}$, 26
net, 129
nilpotent element, 74
nilpotent ideal, 66
nondegenerate bilinear form, 4
null space, 3

open function, 126
order isomorphism, 23
orthogonal set of idempotents, 15
orthogonal sum, 92

pair of dual vector spaces, 4
permutable translations, 61
prime ideal, 71
prime ring, 71
primitive ideal, 162
primitive idempotent, 81
primitive regular semigroup, 97
primitive ring, 71
principal factor, 27
product of relations, 75
product of topological spaces, 112
product topology, 113
projection homomorphism, 96
proper ideal, 32

quasi-inner automorphism, 62

range, 3
rank, 3
Rees matrix ring, 76
Rees matrix semigroup, 29
Rees quotient, 27
regular element (semigroup, ring), 19
relative uniformity, 129
relativization of a uniformity, 129
$\mathfrak{R}$-homomorphism, 75
right annihilator, 19,66,91
right column-independent matrix, 76
right ideal 27
right $\mathfrak{R}$-module, 70
right socle, 69
right socle topology, 147,149
right translation, 55
right vector space, 2
ring of endomorphisms, 3
ri-semigroup, 85
$\mathfrak{R}$-isomorphic modules, 75
$\mathfrak{R}$-isomorphism, 75

r-maximal ring, 86
r-maximal semigroup, 86
$\mathfrak{R}$-module, 70
row finite matrix, 52
$\mathfrak{R}$-submodule, 70

sandwich matrix, 29
scalar multiplication, 2
scalars, 2
semidirect product, 50
semigroup, 3
semigroup of endomorphisms, 3
semigroup socle, 73
semilinear automorphism, 34
semilinear isomorphism of dual pairs, 38
semilinear isomorphism of vector spaces, 34
semilinear transformation, 34
semiprime ring, 66
semiprime ideal, 71
semiprime semigroup, 67
semisimple ring, 163
simple ring, 19
socle, 68
split extension, 51
subdirect sum (product), 96
subspace, 3
subvariety, 132
sum of subrings, 68

topological group, 114
topological module, 131
topological ring, 137
topological vector space, 115
topologically simple ring, 162
topology of a uniformity, 128
topology of uniform convergence, 144
total subspace, 4
transitive, 12
translation, 54
translational hull, 55
t-subspace, 4

uniform isomorphism, 129
uniform semigroup (ring), 63
uniform space, 128
uniform subspace, 129
uniform topology, 144
uniformity, 128
uniformly continuous function, 129
uniformly equivalent uniform spaces, 129
uniformity of uniform convergence, 144
unique addition, 45
U-topology, 113

vectors, 2
V-topology, 113

weakly reductive semigroup, 54
weakly topologized vector space, 131

zero ring, 19

Vol. 215: P. Antonelli, D. Burghelea and P. J. Kahn, The Concordance-Homotopy Groups of Geometric Automorphism Groups. X, 140 pages. 1971. DM 16,-

Vol. 216: H. Maaß, Siegel's Modular Forms and Dirichlet Series. VII, 328 pages. 1971. DM 20,-

Vol. 217: T. J. Jech, Lectures in Set Theory with Particular Emphasis on the Method of Forcing. V, 137 pages. 1971. DM 16,-

Vol. 218: C. P. Schnorr, Zufälligkeit und Wahrscheinlichkeit. IV, 212 Seiten. 1971. DM 20,-

Vol. 219: N. L. Alling and N. Greenleaf, Foundations of the Theory of Klein Surfaces. IX, 117 pages. 1971. DM 16,-

Vol. 220: W. A. Coppel, Disconjugacy. V, 148 pages. 1971. DM 16,-

Vol. 221: P. Gabriel und F. Ulmer, Lokal präsentierbare Kategorien. V, 200 Seiten. 1971. DM 18,-

Vol. 222: C. Meghea, Compactification des Espaces Harmoniques. III, 108 pages. 1971. DM 16,-

Vol. 223: U. Felgner, Models of ZF-Set Theory. VI, 173 pages. 1971. DM 16,-

Vol. 224: Revètements Etales et Groupe Fondamental. (SGA 1). Dirigé par A. Grothendieck XXII, 447 pages. 1971. DM 30,-

Vol. 225: Théorie des Intersections et Théorème de Riemann-Roch. (SGA 6). Dirigé par P. Berthelot, A. Grothendieck et L. Illusie. XII, 700 pages. 1971. DM 40,-

Vol. 226: Seminar on Potential Theory, II. Edited by H. Bauer. IV, 170 pages. 1971. DM 18,-

Vol. 227: H. L. Montgomery, Topics in Multiplicative Number Theory. IX, 178 pages. 1971. DM 18,-

Vol. 228: Conference on Applications of Numerical Analysis. Edited by J. Ll. Morris. X, 358 pages. 1971. DM 26,-

Vol. 229: J. Väisälä, Lectures on n-Dimensional Quasiconformal Mappings. XIV, 144 pages. 1971. DM 16,-

Vol. 230: L. Waelbroeck, Topological Vector Spaces and Algebras. VII, 158 pages. 1971. DM 16,-

Vol. 231: H. Reiter, $L^1$-Algebras and Segal Algebras. XI, 113 pages. 1971. DM 16,-

Vol. 232: T. H. Ganelius, Tauberian Remainder Theorems. VI, 75 pages. 1971. DM 16,-

Vol. 233: C. P. Tsokos and W. J. Padgett. Random Integral Equations with Applications to stochastic Systems. VII, 174 pages. 1971. DM 18,-

Vol. 234: A. Andreotti and W. Stoll. Analytic and Algebraic Dependence of Meromorphic Functions. III, 390 pages. 1971. DM 26,-

Vol. 235: Global Differentiable Dynamics. Edited by O. Hájek, A. J. Lohwater, and R. McCann. X, 140 pages. 1971. DM 16,-

Vol. 236: M. Barr, P. A. Grillet, and D. H. van Osdol. Exact Categories and Categories of Sheaves. VII, 239 pages. 1971. DM 20,-

Vol. 237: B. Stenström, Rings and Modules of Quotients. VII, 136 pages. 1971. DM 16,-

Vol. 238: Der kanonische Modul eines Cohen-Macaulay-Rings. Herausgegeben von Jürgen Herzog und Ernst Kunz. VI, 103 Seiten. 1971. DM 16,-

Vol. 239: L. Illusie, Complexe Cotangent et Déformations I. XV, 355 pages. 1971. DM 26,-

Vol. 240: A. Kerber, Representations of Permutation Groups I. VII, 192 pages. 1971. DM 18,-

Vol. 241: S. Kaneyuki, Homogeneous Bounded Domains and Siegel Domains. V, 89 pages. 1971. DM 16,-

Vol. 242: R. R. Coifman et G. Weiss, Analyse Harmonique Non-Commutative sur Certains Espaces. V, 160 pages. 1971. DM 16,-

Vol. 243: Japan-United States Seminar on Ordinary Differential and Functional Equations. Edited by M. Urabe. VIII, 332 pages. 1971. DM 26,-

Vol. 244: Séminaire Bourbaki - vol. 1970/71. Exposés 382-399. IV, 356 pages. 1971. DM 26,-

Vol. 245: D. E. Cohen, Groups of Cohomological Dimension One. V, 99 pages. 1972. DM 16,-

Vol. 246: Lectures on Rings and Modules. Tulane University Ring and Operator Theory Year, 1970-1971. Volume I. X, 661 pages. 1972. DM 40,-

Vol. 247: Lectures on Operator Algebras. Tulane University Ring and Operator Theory Year, 1970-1971. Volume II. XI, 786 pages. 1972. DM 40,-

Vol. 248: Lectures on the Applications of Sheaves to Ring Theory. Tulane University Ring and Operator Theory Year, 1970-1971. Volume III. VIII, 315 pages. 1971. DM 26,-

Vol. 249: Symposium on Algebraic Topology. Edited by P. J. Hilton. VII, 111 pages. 1971. DM 16,-

Vol. 250: B. Jónsson, Topics in Universal Algebra. VI, 220 pages. 1972. DM 20,-

Vol. 251: The Theory of Arithmetic Functions. Edited by A. A. Gioia and D. L. Goldsmith VI, 287 pages. 1972. DM 24,-

Vol. 252: D. A. Stone, Stratified Polyhedra. IX, 193 pages. 1972. DM 18,-

Vol. 253: V. Komkov, Optimal Control Theory for the Damping of Vibrations of Simple Elastic Systems. V, 240 pages. 1972. DM 20,-

Vol. 254: C. U. Jensen, Les Foncteurs Dérivés de $\varprojlim$ et leurs Applications en Théorie des Modules. V, 103 pages. 1972. DM 16,-

Vol. 255: Conference in Mathematical Logic - London '70. Edited by W. Hodges. VIII, 351 pages. 1972. DM 26,-

Vol. 256: C. A. Berenstein and M. A. Dostal, Analytically Uniform Spaces and their Applications to Convolution Equations. VII, 130 pages. 1972. DM 16,-

Vol. 257: R. B. Holmes, A Course on Optimization and Best Approximation. VIII, 233 pages. 1972. DM 20,-

Vol. 258: Séminaire de Probabilités VI. Edited by P. A. Meyer. VI, 253 pages. 1972. DM 22,-

Vol. 259: N. Moulis, Structures de Fredholm sur les Variétés Hilbertiennes. V, 123 pages. 1972. DM 16,-

Vol. 260: R. Godement and H. Jacquet, Zeta Functions of Simple Algebras. IX, 188 pages. 1972. DM 18,-

Vol. 261: A. Guichardet, Symmetric Hilbert Spaces and Related Topics. V, 197 pages. 1972. DM 18,-

Vol. 262: H. G. Zimmer, Computational Problems, Methods, and Results in Algebraic Number Theory. V, 103 pages. 1972. DM 16,-

Vol. 263: T. Parthasarathy, Selection Theorems and their Applications. VII, 101 pages. 1972. DM 16,-

Vol. 264: W. Messing, The Crystals Associated to Barsotti-Tate Groups: With Applications to Abelian Schemes. III, 190 pages. 1972. DM 18,-

Vol. 265: N. Saavedra Rivano, Catégories Tannakiennes. II, 418 pages. 1972. DM 26,-

Vol. 266: Conference on Harmonic Analysis. Edited by D. Gulick and R. L. Lipsman. VI, 323 pages. 1972. DM 24,-

Vol. 267: Numerische Lösung nichtlinearer partieller Differential- und Integro-Differentialgleichungen. Herausgegeben von R. Ansorge und W. Törnig, VI, 339 Seiten. 1972. DM 26,-

Vol. 268: C. G. Simader, On Dirichlet's Boundary Value Problem. IV, 238 pages. 1972. DM 20,-

Vol. 269: Théorie des Topos et Cohomologie Etale des Schémas. (SGA 4). Dirigé par M. Artin, A. Grothendieck et J. L. Verdier. XIX, 525 pages. 1972. DM 50,-

Vol. 270: Théorie des Topos et Cohomologie Etale des Schémas. Tome 2. (SGA 4). Dirigé par M. Artin, A. Grothendieck et J. L. Verdier. V, 418 pages. 1972. DM 50,-

Vol. 271: J. P. May, The Geometry of Iterated Loop Spaces. IX, 175 pages. 1972. DM 18,-

Vol. 272: K. R. Parthasarathy and K. Schmidt, Positive Definite Kernels, Continuous Tensor Products, and Central Limit Theorems of Probability Theory. VI, 107 pages. 1972. DM 16,-

Vol. 273: U. Seip, Kompakt erzeugte Vektorräume und Analysis. IX, 119 Seiten. 1972. DM 16,-

Vol. 274: Toposes, Algebraic Geometry and Logic. Edited by. F. W. Lawvere. VI, 189 pages. 1972. DM 18,-

Vol. 275: Séminaire Pierre Lelong (Analyse) Année 1970-1971. VI, 181 pages. 1972. DM 18,-

Vol. 276: A. Borel, Représentations de Groupes Localement Compacts. V, 98 pages. 1972. DM 16,-

Vol. 277: Séminaire Banach. Edité par C. Houzel. VII, 229 pages. 1972. DM 20,-

Vol. 278: H. Jacquet, Automorphic Forms on GL(2). Part II. XIII, 142 pages. 1972. DM 16,–

Vol. 279: R. Bott, S. Gitler and I. M. James, Lectures on Algebraic and Differential Topology. V, 174 pages. 1972. DM 18,–

Vol. 280: Conference on the Theory of Ordinary and Partial Differential Equations. Edited by W. N. Everitt and B. D. Sleeman. XV, 367 pages. 1972. DM 26,–

Vol. 281: Coherence in Categories. Edited by S. Mac Lane. VII, 235 pages. 1972. DM 20,–

Vol. 282: W. Klingenberg und P. Flaschel, Riemannsche Hilbertmannigfaltigkeiten. Periodische Geodätische. VII, 211 Seiten. 1972. DM 20,–

Vol. 283: L. Illusie, Complexe Cotangent et Déformations II. VII, 304 pages. 1972. DM 24,–

Vol. 284: P. A. Meyer, Martingales and Stochastic Integrals I. VI, 89 pages. 1972. DM 16,–

Vol. 285: P. de la Harpe, Classical Banach-Lie Algebras and Banach-Lie Groups of Operators in Hilbert Space. III, 160 pages. 1972. DM 16,–

Vol. 286: S. Murakami, On Automorphisms of Siegel Domains. V, 95 pages. 1972. DM 16,–

Vol. 287: Hyperfunctions and Pseudo-Differential Equations. Edited by H. Komatsu. VII, 529 pages. 1973. DM 36,–

Vol. 288: Groupes de Monodromie en Géométrie Algébrique. (SGA 7 I). Dirigé par A. Grothendieck. IX, 523 pages. 1972. DM 50,–

Vol. 289: B. Fuglede, Finely Harmonic Functions. III, 188. 1972. DM 18,–

Vol. 290: D. B. Zagier, Equivariant Pontrjagin Classes and Applications to Orbit Spaces. IX, 130 pages. 1972. DM 16,–

Vol. 291: P. Orlik, Seifert Manifolds. VIII, 155 pages. 1972. DM 16,–

Vol. 292: W. D. Wallis, A. P. Street and J. S. Wallis, Combinatorics: Room Squares, Sum-Free Sets, Hadamard Matrices. V, 508 pages. 1972. DM 50,–

Vol. 293: R. A. DeVore, The Approximation of Continuous Functions by Positive Linear Operators. VIII, 289 pages. 1972. DM 24,–

Vol. 294: Stability of Stochastic Dynamical Systems. Edited by R. F. Curtain. IX, 332 pages. 1972. DM 26,–

Vol. 295: C. Dellacherie, Ensembles Analytiques, Capacités, Mesures de Hausdorff. XII, 123 pages. 1972. DM 16,–

Vol. 296: Probability and Information Theory II. Edited by M. Behara, K. Krickeberg and J. Wolfowitz. V, 223 pages. 1973. DM 20,–

Vol. 297: J. Garnett, Analytic Capacity and Measure. IV, 138 pages. 1972. DM 16,–

Vol. 298: Proceedings of the Second Conference on Compact Transformation Groups. Part 1. XIII, 453 pages. 1972. DM 32,–

Vol. 299: Proceedings of the Second Conference on Compact Transformation Groups. Part 2. XIV, 327 pages. 1972. DM 26,–

Vol. 300: P. Eymard, Moyennes Invariantes et Représentations Unitaires. II, 113 pages. 1972. DM 16,–

Vol. 301: F. Pittnauer, Vorlesungen über asymptotische Reihen. VI, 186 Seiten. 1972. DM 18,–

Vol. 302: M. Demazure, Lectures on p-Divisible Groups. V, 98 pages. 1972. DM 16,–

Vol. 303: Graph Theory and Applications. Edited by Y. Alavi, D. R. Lick and A. T. White. IX, 329 pages. 1972. DM 26,–

Vol. 304: A. K. Bousfield and D. M. Kan, Homotopy Limits, Completions and Localizations. V, 348 pages. 1972. DM 26,–

Vol. 305: Théorie des Topos et Cohomologie Etale des Schémas. Tome 3. (SGA 4). Dirigé par M. Artin, A. Grothendieck et J. L. Verdier. VI, 640 pages. 1973. DM 50,–

Vol. 306: H. Luckhardt, Extensional Gödel Functional Interpretation. VI, 161 pages. 1973. DM 18,–

Vol. 307: J. L. Bretagnolle, S. D. Chatterji et P.-A. Meyer, Ecole d'été de Probabilités: Processus Stochastiques. VI, 198 pages. 1973. DM 20,–

Vol. 308: D. Knutson, λ-Rings and the Representation Theory of the Symmetric Group. IV, 203 pages. 1973. DM 20,–

Vol. 309: D. H. Sattinger, Topics in Stability and Bifurcation Theory. VI, 190 pages. 1973. DM 18,–

Vol. 310: B. Iversen, Generic Local Structure of the Morphisms in Commutative Algebra. IV, 108 pages. 1973. DM 16,–

Vol. 311: Conference on Commutative Algebra. Edited by J. W. Brewer and E. A. Rutter. VII, 251 pages. 1973. DM 22,–

Vol. 312: Symposium on Ordinary Differential Equations. Edited by W. A. Harris, Jr. and Y. Sibuya. VIII, 204 pages. 1973. DM 22,–

Vol. 313: K. Jörgens and J. Weidmann, Spectral Properties of Hamiltonian Operators. III, 140 pages. 1973. DM 16,–

Vol. 314: M. Deuring, Lectures on the Theory of Algebraic Functions of One Variable. VI, 151 pages. 1973. DM 16,–

Vol. 315: K. Bichteler, Integration Theory (with Special Attention to Vector Measures). VI, 357 pages. 1973. DM 26,–

Vol. 316: Symposium on Non-Well-Posed Problems and Logarithmic Convexity. Edited by R. J. Knops. V, 176 pages. 1973. DM 18,–

Vol. 317: Séminaire Bourbaki – vol. 1971/72. Exposés 400–417. IV, 361 pages. 1973. DM 26,–

Vol. 318: Recent Advances in Topological Dynamics. Edited by A. Beck, VIII, 285 pages. 1973. DM 24,–

Vol. 319: Conference on Group Theory. Edited by R. W. Gatterdam and K. W. Weston. V, 188 pages. 1973. DM 18,–

Vol. 320: Modular Functions of One Variable I. Edited by W. Kuyk. V, 195 pages. 1973. DM 18,–

Vol. 321: Séminaire de Probabilités VII. Edité par P. A. Meyer. VI, 322 pages. 1973. DM 26,–

Vol. 322: Nonlinear Problems in the Physical Sciences and Biology. Edited by I. Stakgold, D. D. Joseph and D. H. Sattinger. VIII, 357 pages. 1973. DM 26,–

Vol. 323: J. L. Lions, Perturbations Singulières dans les Problèmes aux Limites et en Contrôle Optimal. XII, 645 pages. 1973. DM 42,–

Vol. 324: K. Kreith, Oscillation Theory. VI, 109 pages. 1973. DM 16,–

Vol. 325: Ch.-Ch. Chou, La Transformation de Fourier Complexe et L'Equation de Convolution. IX, 137 pages. 1973. DM 16,–

Vol. 326: A. Robert, Elliptic Curves. VIII, 264 pages. 1973. DM 22,–

Vol. 327: E. Matlis, 1-Dimensional Cohen-Macaulay Rings. XII, 157 pages. 1973. DM 18,–

Vol. 328: J. R. Büchi and D. Siefkes, The Monadic Second Order Theory of All Countable Ordinals. VI, 217 pages. 1973. DM 20,–

Vol. 329: W. Trebels, Multipliers for (C, α)-Bounded Fourier Expansions in Banach Spaces and Approximation Theory. VII, 103 pages. 1973. DM 16,–

Vol. 330: Proceedings of the Second Japan-USSR Symposium on Probability Theory. Edited by G. Maruyama and Yu. V. Prokhorov. VI, 550 pages. 1973. DM 36,–

Vol. 331: Summer School on Topological Vector Spaces. Edited by L. Waelbroeck. VI, 226 pages. 1973. DM 20,–

Vol. 332: Séminaire Pierre Lelong (Analyse) Année 1971-1972. V, 131 pages. 1973. DM 16,–

Vol. 333: Numerische, insbesondere approximationstheoretische Behandlung von Funktionalgleichungen. Herausgegeben von R. Ansorge und W. Törnig. VI, 296 Seiten. 1973. DM 24,–

Vol. 334: F. Schweiger, The Metrical Theory of Jacobi-Perron Algorithm. V, 111 pages. 1973. DM 16,–

Vol. 335: H. Huck, R. Roitzsch, U. Simon, W. Vortisch, R. Walden, B. Wegner und W. Wendland, Beweismethoden der Differentialgeometrie im Großen. IX, 159 Seiten. 1973. DM 18,–

Vol. 336: L'Analyse Harmonique dans le Domaine Complexe. Edité par E. J. Akutowicz. VIII, 169 pages. 1973. DM 18,–

Vol. 337: Cambridge Summer School in Mathematical Logic. Edited by A. R. D. Mathias and H. Rogers. IX, 660 pages. 1973. DM 42,–

Vol: 338: J. Lindenstrauss and L. Tzafriri, Classical Banach Spaces. IX, 243 pages. 1973. DM 22,–

Vol. 339: G. Kempf, F. Knudsen, D. Mumford and B. Saint-Donat, Toroidal Embeddings I. VIII, 209 pages. 1973. DM 20,–

Vol. 340: Groupes de Monodromie en Géométrie Algébrique. (SGA 7 II). Par P. Deligne et N. Katz. X, 438 pages. 1973. DM 40,–

Vol. 341: Algebraic K-Theory I, Higher K-Theories. Edited by H. Bass. XV, 335 pages. 1973. DM 26,–

Vol. 342: Algebraic K-Theory II, "Classical" Algebraic K-Theory, and Connections with Arithmetic. Edited by H. Bass. XV, 527 pages. 1973. DM 36,–